Formulation and Production of Carbonated Soft Drinks

Formulation and Production of Carbonated Soft Drinks

Edited by

ALAN J. MITCHELL
formerly of
Coca-Cola and Schweppes Beverages Ltd.

Blackie
Glasgow and London

Published in the USA by
avi, an imprint of
Van Nostrand Reinhold
New York

Blackie and Son Ltd
Bishopbriggs, Glasgow G64 2NZ
and
7 Leicester Place, London WC2H 7BP

Published in the United States of America by
AVI, an imprint of
Van Nostrand Reinhold
115 Fifth Avenue
New York, New York 10003

Distributed in Canada by
Nelson Canada
1120 Birchmount Road
Scarborough, Ontario M1K 5G4, Canada

16 15 14 13 12 11 10 9 8 7 6 5 4 3 2 1

© 1990 Blackie and Son Ltd
First published 1990

*All rights reserved.
No part of this publication may be reproduced,
stored in a retrieval system, or transmitted,
in any form or by an means –
graphic, electronic or mechanical, including photocopying,
recording, taping – without the written permission of the Publishers*

British Library Cataloguing in Publication Data

Formulation and production of carbonated soft drinks.
1. Soft drinks production
I. Mitchell, Alan J.
663.6

ISBN 0-216-92915-6

Library of Congress Cataloging-in-Publication Data

Formulation and production of carbonated soft drinks / edited by Alan J. Mitchell.
  p.   cm.
Includes bibliographical references (p.).
ISBN 0-442-30287-8
1. Carbonated beverages.   I. Mitchell, Alan J.
TP630.F67 1990
663'.62—dc20   89-29394
                                                                   CIP

Phototypesetting by Thomson Press (India) Limited, New Delhi
Printed in Great Britain by Thomson Litho Limited, East Kilbride, Scotland

Preface

There is comparatively little published material on the technology of the soft drinks industry. The excellent house journals produced by the larger machinery manufacturers and the technical articles which appear in the magazines issued by bottlers' associations in every part of the world are of great benefit and interest to technical staff. However, there is a need for a basic book of reference which includes important aspects of the packaging of carbonated soft drinks; this particular sector of the industry has been chosen, not only because carbonated soft drinks vastly outsell all other soft drinks and mineral waters combined, but also because it is sometimes considered to be the most challenging area of beverage production.

In this book an attempt has been made to review the major stages of carbonated soft drinks production. There are chapters dealing with the composition and combination of the elemental constituents, the various primary packages and the many methods of protecting these packages during the distribution process. This extremely broad spread of activities has been compressed into one convenient volume, but the treatment is by no means exhaustive – each chapter could warrant a complete book in its own right. We have aimed to provide an integrated appraisal of the production of carbonated soft drinks. This book will give the newcomer to the industry an indication of the complexities of the various technologies employed and will provide a basis for experienced technicians who wish to specialise and become more proficient in a particular field. It will also assist soft drinks personnel who are employed outside the production sphere of operations – for example, those in the distribution, sales, marketing and financial departments who require an insight into the technicalities of the processing of soft drinks.

The subject matter concentrates on the production line and the general raw material of manufacture. The expanding segment of dispense technology has been omitted intentionally, since it is considered to be outside the scope of this volume. No specific chapter has been devoted to the legal aspects of the subject as these are so wide-ranging across continents that any cursory treatment would be unsatisfactory. Important restrictions by regulatory bodies on certain substances are mentioned, but it is vital that local enquiries are made to ascertain current controls, if any doubt exists.

Although written for the carbonated soft drink technologist, the book contains much information which will be of interest to other sectors of the beverage industry. In particular, details of primary containers, handling, packing and packaging are also relevant to the wine, spirits, beer and non-carbonated soft drink industries.

My personal thanks are due to the authors who contributed the specialised chapters. They are all well-established experts in their field and give the reader not only the benefit of their experience but also an insight into possible future developments. As in any book of this nature, it must be made clear that the views expressed by the authors are their own and not necessarily the views of the companies which employ them.

I should also express my gratitude to my wife, Rose, and my son-in-law, Duncan, for their support and assistance in those areas of which they are well aware.

A.J.M.

Contributors

P.M. Beesley	British Sugar plc, PO Box 26, Oundle Road, Peterborough PE2 9QU
P.W. Binns	Heuft Ltd., Unit 26, Innage Park, Holly Lane, Atherstone CV9 2HA
R. Hall	Canadean Ltd, Refuge House, 2–4 Henry Street, Bath BA1 1JT
G. Hopkins	Polly Peck International plc, Berkeley Square, London W1X 5DB
R. Hutchinson	20 Henley Gardens, Yateley, Camberley GU17 7LG
D. Kaye	Krones UK Ltd, Phoenix House, Liverpool Street, Salford M5 4LT
R.E. Leighton	Meadow House, Woollensbrook, Hoddesdon EN11 9BN
W. McCarthy	Satec–PWS Ltd, Weston Road, Crewe CW1 1DE
D. McDonald	P.R. Ashurst & Associates, Leighton Cottage, Sandy Rise, Chalfont Heights, Chalfont St Peter SL9 9TR
A.J. Mitchell	White Cottage, Devil's Highway, Crowthorne, Berkshire RG11 6SR
K. O'Donnell	Forum Chemicals Ltd., 41–51 Brighton Road, Redhill RH1 6YS
C. Pettitt	H. Erben Ltd, Hadleigh, Ipswich IP7 6AS
I.S. Roberts	Corporate Packaging Department, Coca-Cola Company, Atlanta, Georgia 30301, USA
B. Taylor	Barnett and Foster International, Denington Estate, Wellingborough NN8 2QJ
M. Turner	Britvic Soft Drinks Ltd, Lichfield Road Industrial Estate, Tamworth B79 7TE

Contents

1 Introduction — 1
A.J. MITCHELL

2 The growth and development of carbonated soft drinks — 5
R. HALL

    2.1 The story so far — 5
    2.2 Future prospects — 13

3 Water treatment — 16
R. HUTCHINSON and W. McCARTHY

    3.1 Introduction — 16
    3.2 Coagulation treatment — 18
        3.2.1 Filtration — 21
        3.2.2 Sterilisation — 22
        3.2.3 Dechlorination — 23
        3.2.4 Plant performance monitoring — 24
        3.2.5 Approximate plant size — 26
    3.3 Ion exchange treatment — 27
        3.3.1 Dealkalisation — 27
        3.3.2 Organic removal — 29
        3.3.3 Nitrate removal — 31
    3.4 Reverse osmosis — 32
    3.5 Alternative sterilisation methods — 34
    3.6 Future developments — 35
    References — 36

4 Carbohydrate sugars — 37
P.M. BEESLEY

    4.1 Introduction — 37
        4.1.1 History — 37
    4.2 Carbohydrate sugars — 37
        4.2.1 Granulated sugar — 38
        4.2.2 Liquid sugar — 39
        4.2.3 Glucose syrup: high fructose syrup — 40
    4.3 Quality — 41
        4.3.1 Trade requirement — 41
        4.3.2 Quality assurance management — 43
        4.3.3 Sugar analysis — 43
    4.4 Transportation and delivery — 44
        4.4.1 Bulk delivery of granulated sugar — 44
        4.4.2 Bulk delivery of liquid carbohydrate sugars — 47
        4.4.3 Security of delivery — 48

4.5	Storage			48
	4.5.1	Granulated sugar in bags		49
	4.5.2	Granulated sugar in bulk		49
	4.5.3	Liquid carbohydrate sugars		50
4.6	On-site dissolving of granulated sugar			52
	4.6.1	Batch dissolving		52
	4.6.2	Continuous dissolving		52
	4.6.3	High capacity dissolving		53
References				54

5 High-intensity sweeteners   56
K. O'DONNELL

5.1	Introduction		56
	5.1.1	Use of intense sweeteners	56
5.2	Current sweeteners		57
	5.2.1	Acesulfame K	57
	5.2.2	Aspartame	59
	5.2.3	Cyclamate	62
	5.2.4	Saccharin	64
	5.2.5	Stevioside/stevia	66
	5.2.6	Thaumatin	69
	5.2.7	Dihydrochalcones	70
5.3	Potential new sweeteners		71
	5.3.1	Alitame	71
	5.3.2	Sucralose	73
5.4	Sweetener approval and regulation		76
5.5	Future use of intense sweeteners		77
References			78

6 Flavourings and emulsions   81
G. HOPKINS

6.1	Flavourings		81
	6.1.1	Legislation	81
	6.1.2	Creation	82
	6.1.3	Production	82
6.2	Emulsions		84
	6.2.1	Manufacture	85
6.3	Application of flavourings and emulsions		87
	6.3.1	Selection	87
	6.3.2	Methods of use	88
6.4	Evaluations		88

7 Acids, colours, preservatives and other additives   90
B. TAYLOR

7.1	Introduction		90
7.2	Acids		92
	7.2.1	Carbonic acid	93
	7.2.2	Citric acid	93
	7.2.3	Tartaric acid	94
	7.2.4	Phosphoric acid	95
	7.2.5	Lactic acid	95
	7.2.6	Acetic acid	96

	7.2.7	Malic acid	96
	7.2.8	Fumaric acid	96
	7.2.9	Ascorbic acid	96
7.3	Colours		97
7.4	Preservatives		100
	7.4.1	Micro-organisms and soft drinks	102
	7.4.2	Sulphur dioxide	103
	7.4.3	Benzoic acid and benzoates	104
	7.4.4	Esters of para-hydroxy-benzoic acid	104
	7.4.5	Sorbic acid and sorbates	105
7.5	Other additives		105
	7.5.1	Emulsifiers	105
	7.5.2	Stabilisers	106
	7.5.3	Saponins	106
	7.5.4	Anti-oxidants	106
7.6	The safety of food additives		107

8 Syrup room operation

M.J. TURNER

			108
8.1	Introduction		108
8.2	Syrup room design		108
	8.2.1	Wall finishes	108
	8.2.2	Floors and drainage	109
	8.2.3	Ceilings and lighting	109
	8.2.4	Heating, ventilating and air conditioning	109
8.3	Syrup room equipment		109
	8.3.1	Storage, mixing tanks and systems	109
	8.3.2	Pipework, fittings and connections	113
	8.3.3	Ingredient flow	114
	8.3.4	Pumps	115
	8.3.5	Measurement of liquid	116
	8.3.6	Filtration of ingredients	117
	8.3.7	Ultraviolet sterilisation	118
	8.3.8	Pasteurisation	119
	8.3.9	Homogenisation	121
8.4	Syrup room materials, storage and handling		123
	8.4.1	Sugar	123
	8.4.2	High-fructose glucose (corn) syrup	123
	8.4.3	Acids	123
	8.4.4	Sweeteners	124
	8.4.5	Preservatives	124
	8.4.6	Flavourings	124
	8.4.7	Colours	125
	8.4.8	Fruit juices and comminuted bases	125
8.5	Syrup room CIP systems and detergents		125
	8.5.1	Design of a CIP unit	126
	8.5.2	Rate of flow in pipelines for CIP	130
	8.5.3	Calculation of Reynolds number	130
	8.5.4	Choice of detergents	131
8.6	Automation and computerisation in syrup rooms		132
	8.6.1	Typical system description	133
	8.6.2	Typical operating sequence for syrup manufacture	135
8.7	Multiple component mixing plant		136
	8.7.1	Construction	136
	8.7.2	Control and operation	137
8.8	Future developments		139

9 Containers and closures — 140
I.S. ROBERTS

- 9.1 Introduction — 140
  - 9.1.1 Basic package types — 140
  - 9.1.2 The function of packaging — 141
  - 9.1.3 Pressurisation — 141
- 9.2 Glass bottles — 142
  - 9.2.1 Raw materials — 142
  - 9.2.2 Glass production — 143
  - 9.2.3 Container formation — 143
  - 9.2.4 Glass bottle strength — 146
  - 9.2.5 Surface coatings — 146
  - 9.2.6 Future developments — 148
- 9.3 Plastic bottles — 148
  - 9.3.1 Raw material — 149
  - 9.3.2 PET bottle production — 150
  - 9.3.3 PET bottle properties — 155
- 9.4 Metal cans — 156
  - 9.4.1 Raw materials — 157
  - 9.4.2 Can production — 158
  - 9.4.3 Double seaming — 160
  - 9.4.4 Can dimensions and production rates — 160
  - 9.4.5 Which metal? — 161
  - 9.4.6 Can lacquering and corrosion — 162
  - 9.4.7 Easy-open ends — 163
  - 9.4.8 Steel can ends — 165
  - 9.4.9 Future can developments — 166
- 9.5 Closure systems — 166
  - 9.5.1 Crowns — 168
  - 9.5.2 Roll-on closures — 168
  - 9.5.3 Plastic closures — 170
  - 9.5.4 Seal formation — 171
  - 9.5.5 Proper application — 171
- 9.6 Future trends — 172

10 Handling empty containers — 174
C. PETTITT

- 10.1 Introduction — 174
- 10.2 Depalletisers — 175
  - 10.2.1 Wooden boxes and plastic crates — 175
  - 10.2.2 Bulk bottle supplies — 176
  - 10.2.3 Cans — 179
  - 10.2.4 Pallet magazines and conveyors — 180
- 10.3 Unscrambling machines — 181
  - 10.3.1 Vertical unscrambler — 181
  - 10.3.2 Rotary unscrambler — 182
- 10.4 Decraters — 182
  - 10.4.1 Gripper heads — 183
  - 10.4.2 Rotary decrater — 184
  - 10.4.3 In-line continuous decrater — 184
  - 10.4.4 In-line lift-up decrater — 184
- 10.5 Crate washers — 185
  - 10.5.1 In-line crate washers — 186
  - 10.5.2 'S' type and twist crate washer — 186
- 10.6 Decapping machines — 187
  - 10.6.1 Decapping heads — 188

CONTENTS xiii

	10.6.2	In-line decappers	188
	10.6.3	Rotary decappers	189
10.7	Bottle washing		189
	10.7.1	Types of bottle washer	190
	10.7.2	Typical treatment sequence	191
	10.7.3	Label removal	193
10.8	Container rinsing		194
	10.8.1	In-line bottle rinsers	194
	10.8.2	Rotary bottle rinsers	195
	10.8.3	Can rinsers	196
10.9	Container conveyors		196
	10.9.1	Slat band chain conveyors	198
	10.9.2	Roller conveyors	198
	10.9.3	Air driven conveyors	199
	10.9.4	Conveyor accumulators	200
	10.9.5	Conveyor combiners	201
	10.9.6	Can conveying	202
	10.9.7	Container flow monitoring and conveyor speed regulation	202

11 Carbonation and filling 203
A.J. MITCHELL

11.1	Introduction		203
11.2	Carbonation		204
	11.2.1	The nature and effects of carbonation	204
	11.2.2	Properties of carbon dioxide	205
	11.2.3	Equilibrium pressure	207
	11.2.4	Measurement of carbonation	208
	11.2.5	Carbonation determination	208
11.3	Carbonators		212
	11.3.1	Designs of carbonators	213
	11.3.2	Air exclusion	219
11.4	Proportioners		220
11.5	Fillers and filling valves		224
	11.5.1	Basic filling valve operation	228
	11.5.2	Filling valve development and the influence of ambient filling	230

12 Container decoration 243
D. KAYE

12.1	Introduction		243
12.2	The aims of the container decoration		244
12.3	Main types of decoration application machinery		246
	12.3.1	Indirect-transfer labelling machines	247
	12.3.2	Direct-transfer labelling machines	255
12.4	Special container decoration applications and systems		257
	12.4.1	Combination labelling machines	258
	12.4.2	Reel-fed labelling	259
	12.4.3	Tamper-evident devices	263
	12.4.4	Metallic neck decoration	265
	12.4.5	Self-adhesive labelling	267
	12.4.6	Container pre-labelling	267
	12.4.7	Can labelling	268
	12.4.8	Container orientation	269
12.5	Factors influencing the efficiency and quality of container decoration		269
	12.5.1	Specification of packaging materials	270
	12.5.2	Adhesives	275
	12.5.3	Plant layout	277

	12.5.4	Machine condition		278
12.6	Future developments			279

13 Container inspection equipment

P.W. BINNS

13.1	Introduction			281
13.2	Empty container inspection			282
	13.2.1	Pallet inspection		282
	13.2.2	Bottles in crates		284
	13.2.3	Empty crate inspection		284
	13.2.4	Bottle inspection – sorting		284
	13.2.5	Empty bottle inspection		287
	13.2.6	Empty can inspection		293
13.3	Filled container inspection			293
	13.3.1	Bottle inspection		293
	13.3.2	Filled can inspection		294
	13.3.3	Labelled bottle inspection		295
	13.3.4	Crate/pack inspection		295
13.4	Filler and closure management			296
13.5	Rejection systems			297
13.6	Future developments			299

14 Secondary and tertiary packaging

R.E. LEIGHTON and A.J. MITCHELL

14.1	Introduction			300
14.2	Returnable bottles and crates			300
14.3	Automatic recraters			302
	14.3.1	The modern recrater		303
14.4	Non-returnable container packaging			306
14.5	Tray erectors, tray loaders and shrink-wrappers			308
	14.5.1	Tray erectors		308
	14.5.2	Tray loaders		309
	14.5.3	Shrink-wrappers		310
14.6	Wrap-around carton machines			312
14.7	Integrated shrink-wrap packaging machines			313
14.8	Multipacks			317
14.9	Palletisers			318
	14.9.1	Low-level sweep palletiser		319
	14.9.2	High-level sweep palletiser		321
	14.9.3	Low-level hook palletiser		321
	14.9.4	Palletiser applications		322
14.10	Over-wrappers			325

15 Effective application of quality control

D. McDONALD

15.1	Introduction			328
15.2	Evolution of QC in the soft drinks industry			329
	15.2.1	Concept of quality		329
	15.2.2	Evolution of soft drinks QC		330
15.3	The small-to-medium-sized business			331
	15.3.1	Contract packing		331
	15.3.2	Setting up a cost-effective system for QC		331
	15.3.3	Product and packaging innovation		332

15.4		National operations with multiple plants	333
	15.4.1	Impact of industry concentration	333
	15.4.2	Organisation of QC at plant level	333
	15.4.3	Centralised organisation for quality	335
	15.4.4	Bottling versus canning QC requirements	335
	15.4.5	Equipment selection for quality	336
	15.4.6	Development of in-line quality monitoring equipment	337
	15.4.7	Potential quality problem areas	337
	15.4.8	Product recall	342
	15.4.9	Water quality and treatment	343
	15.4.10	Statistical QC	345
	15.4.11	Microbiology	345
	15.4.12	Dispensed soft drinks	347
15.5		International QC and QA of soft drink operations	349
	15.5.1	The franchise system	349
	15.5.2	Technical services	350
	15.5.3	The international quality assurance laboratory	351
	15.5.4	Ingredient quality	352
	15.5.5	Packaging quality	353
	15.5.6	Trouble-shooting, the theory in practice	354
15.6		The future	355
	15.6.1	Influence of packaging	356
	15.6.2	New ingredients, formulation and sanitation requirements	356
	15.6.3	Role of the soft drinks associations	357
	15.6.4	The final word	357
References			358

Index 359

1 Introduction

A.J. MITCHELL

The colossal scale of the output of carbonated soft drinks worldwide is almost impossible to imagine. What is certain is that no one really knows the exact total consumption. Various estimates have been made: for example, Houghton[1] suggested a figure of 88 billion litres for the year 1979, rising to 102 billion litres by 1982. It would be wrong and too simplistic to extrapolate these figures to the present time, but we know from Hall's reliable statistics in Chapter 2 that the total consumption in the combined United States and European markets increased by 40% in the period 1978/88 and, applying the resultant annual growth rate of 3.4% to Houghton's latest figure, we obtain a world consumption of almost 130 billion litres for the year 1989. In view of the markets that have opened in the eighties, this is almost certainly a conservative estimate, but is sufficiently accurate to illustrate the extent of the present carbonated soft drink industry. If this total were consumed by the United States alone, then every man, woman and child would drink almost $1\frac{1}{2}$ litres each day of the year – approximately triple the actual intake; applied to the United Kingdom, the equivalent figure would be an awesome $6\frac{1}{2}$ litres per day. The instantaneous flow-rate of 130 billion litres per year is equal to 4122 litres every second which, if pumped, would require a pipeline over 2 metres in diameter. In monetary terms, the world market is worth in excess of £160 000m or over $250 000m.

Having gained an appreciation of the magnitude of the carbonates market, it is equally fascinating to realise that over 40% of the total world output is consumed in Canada, USA and Mexico. Patterns of yearly per capita consumption vary enormously, with the USA an easy leader at 174 litres, and Mexico in a lagging second place at 97 litres – an example of two countries with quite dissimilar economies and yet the top two consumers in the world. The UK, at around 64 litres per person per year, does not even figure in the top ten, presumably due to the popularity of concentrated squashes, which are little known in most other areas of the world.

Assuming that carbonated drinks are readily available, with production centres established and the supply of raw materials organised, the factors which appear to influence per capita consumption are:

- personal wealth (disposal income)
- climatic conditions

- availability of an alternative liquid refreshment (drinking water supply)
- severity of liquor laws (licensing regulations and drink/drive restrictions).

Naturally, the sheer availability of any soft drink, coupled with aggressive and/or subtle promotional advertising, will have the desired effect in most markets. Even in the poorest countries, the acquisition of a soft drink is something of a status symbol as well as a means of enjoyment, however transient.

This huge market for carbonated soft drinks is therefore based on providing not only refreshment but also pleasure to young and old alike. Among the millions of fortunate consumers, only an infinitesimal fraction will have any idea of what comprises their favourite 'fizz', and even less notion of the technologies involved in ensuring that the drink is available to them in absolutely first-class condition. Even the formulation of the world's most popular carbonated soft drink remains a close commercial secret – but is still enjoyed millions of times a day in over 150 countries by trusting and satisfied consumers; such is the confidence and rapport that can be established in the supply of a premium product by an organisation committed to the pursuit of excellence, the raising of standards of community service and involvement in healthy sports.

It is not the product alone, however, which encourages and develops sales. The container and packaging must play their parts in safeguarding the beverage, and they must be appropriate to the intended outlets and afford maximum convenience to the consumer. Packaging technology during the past forty years has risen to these challenges by a series of major breakthroughs, interspersed with periods of refinement and improvement. The milestones are many and varied, including:

- development of palletisation, improving bulk handling in factories and warehouses
- introduction of plastic crates (replacing wooden boxes), encouraging standardisation and allowing automatic decrating, recrating and palletisation
- expansion and development of beverage cans and, later, easy-open aluminium ends of various designs
- introduction of non-returnable bottles, mainly for the supermarket business
- advent of multipacks with 'easy carry' facilities
- introduction of PET (polyethylene terephthalate) bottles, allowing larger capacities to be added to the container range
- application of plastic closures (for glass and PET) with superior sealing properties and tamper-evident facilities
- development of flexible packaging (shrink-wrapped trays) for cans and non-returnable bottles

All these and other developments have been pursued in conjunction with additions to the range of beverages: notably, low-calorie drinks with improved and more-acceptable artificial sweeteners, flavoured mineral waters, high-juice content drinks, fruit-flavoured colas, energy-inducing drinks and many more.

The microchip has effected the greatest improvement in the packaging of soft drinks in the last ten years. Individual machines (particularly those with many sequential operations which function on a 'cascade' arrangement) benefit from microprocessor control in the areas of fault-finding, performance-logging and reliability. Information from the various sectors of a production line may be collated to provide an overall assessment of line performance, obtained on a minute-to-minute basis and either screen-displayed or printed for a permanent record. These techniques may be applied to syrup room operation, raw materials ordering and control and the effective utilisation of mechanical and electrical services. The acquisition of management information in all areas of production is so conveniently accessible that the real danger is an excess of data which cannot be digested. Restraints must be applied to ensure that the right information is offered to the right people.

No one could claim that the soft drinks industry lacks enthusiasm for innovation; progress has been maintained in the years when economic difficulties and energy crises have followed each other in quick succession. Despite all these problems, soft drinks are generally cheaper today (in real terms) than 25 years ago, and this has been achieved by healthy competition, increased productivity, increased standardisation (leading to mechanisation) and a consolidation to faster, highly automated and efficient plants. In most developed countries, the soft drinks industries have undergone a polarisation in which larger organisations have increased in size (by combination or growth) and succeeded by means of large-scale, high-speed production, and the small-to-medium companies, with low overheads, have resisted the temptation to indulge in excessive expansion, remaining competitive in price and service in a prescribed area. Both types of operation can be successful in their own ways, and there is every reason to suppose that this basis for healthy competition will be the pattern for some years to come.

Two important aspects of packaging are energy conservation and ecology. Obviously these are interconnected and, within the soft drinks industry, attitudes vary enormously in different parts of the world. Clearly, the subjects may be touched upon only briefly in this volume and the influence of future developments will be mutually affected. A 'total energy' concept is required, taking into account all energy expended in the life of the package – not just during the treatment it receives in the soft drinks factory but also in the manufacturing and disposal operations. Recycling of containers achieves modest success in those areas where it is taken seriously. The move towards all-PET bottles (including the cap) and two-piece aluminium cans has improved the economic viability of recycling; the future introduction of PET cans with aluminium opening ends will be a complication. A thermally stable PET bottle allowing hot liquor treatment to clean and sterilise it will allow refilling. New legislation may insist on the use of this type of bottle which will radically change bottling lines, secondary packaging and the philosophy of supermarket trading. As in some parts of the world, a deposit system encouraging the return of empty cans and non-returnable bottles to the supermarket may have to be extended. The ecology situation is too crucial to allow the consumers to default on their responsibilities, and enforcement may be the only course.

The developments mentioned in this introduction are all covered in the remainder of the volume. The content is mainly technical, as would be expected from a book dealing with a production process, and will appeal mainly to engineers, chemists and packaging technologists. It is hoped that it will also be interesting to those away from the mainstream of production. As with any treatment of a complex subject, some duplication has been necessary in the various chapters, but this has been minimised.

In every industry there is a conscientious, hard-working core of employees on whom depends much of the success of the business. In soft drinks it is the professional production technologists who carry so much of the responsibility for the quality of the ready-for-the-customer bottle or can. In dedicating this book to members of that soft drinks élite in every part of the world, I can do no better than quote the Chairman and Chief Executive Officer of the Coca-Cola Company, Mr Roberto C. Goizueta, when he addressed the Society of Soft Drink Technologists in Atlanta in 1983. With Mr Goizueta's kind permission I reproduce his comment:

> We don't always get the glamour of marketing people; we don't always get the glory of sales people; we don't always get the clout of financial people. But the simple facts are that you can't market a new product until someone creates a new product; you can't promote a new package until someone designs a new package; you can't use an ingredient until someone finds a use for it and gets it approved; you can't advertise good taste and quality unless someone safeguards flavour and quality; and you can't sell a soft drink at a competitive price (and at a profit) unless someone makes sure it is produced and distributed properly and efficiently.
>
> That 'someone' is you – *the technical professional.*

Reference

1. H.W. Houghton, Preface to *Developments in Soft Drinks Technology* Vol. 3. Elsevier Applied Science Publishers, Barking (1984).

2 The growth and development of carbonated soft drinks

R. HALL

2.1 The story so far

What is so remarkable about carbonated soft drinks? There is no need to consume them – but more and more people do. Although clean tap water is freely available, more and more people are paying for the pleasure of an alternative. Carbonates are gaining 'share of throat' from almost every other beverage. They actually 'outsell' tap water in the United States.

Carbonated soft drinks are a man-made product and a man-made market. They were the first brands to be sold on a genuinely global scale, even before armaments, cars, cigarettes, radios or take-aways. They were among the first Western goods to break into Eastern Europe and China, and yet they are a relative newcomer to the world of beverages, where change is generally a very gradual process. Milk, wine and beer have all been around for well over a thousand years. Tea and coffee were introduced to the wider world some centuries ago. But carbonated soft drinks date back little further than two hundred years.

Their origins are quite varied, though most of them had extraordinary claims made about their medicinal properties in the early days. Artificial carbonation on a commercial scale started with seltzer, soda and other waters in the late 1700s. 'Bewleys Mephitic Julep', one of the earliest, was supposed to be effective against 'putrid fevers, scurvy, dysentery, bilious vomiting, etc.'. By the turn of the century, the Schweppe family had set up operations in Switzerland, Britain and France.

Quinine tonic water was only developed in the 1850s, becoming a means of protecting British forces abroad from malaria. Cola, extracted from the nuts and leaves of the African Cola tree, and Coca, extracted from the leaves of the Bolivian Coca shrub, were not united until 1886. The formula was initially promoted as a brain tonic.

The key to the progress of these products, especially that of Coca-Cola, was built on some guiding principles which hold good for almost any product in any day and age:

1. Creation of demand – by energetic and innovative marketing.

2. Product quality – by careful control of ingredients and manufacture.
3. Wide distribution – through fountains and franchises.

The biggest break came in the Second World War, when Coca-Cola was shipped to American forces, wherever they were posted. The servicemen eventually went home, but the habit stayed.

Thus, in the 1950s, the driving force behind the world's carbonated soft drinks market was dominated by one product – cola, one brand – Coca-Cola, and one image – the American dream. In the United States Pepsi-Cola was also very much in contention. In Europe there were many other flavours, but no such internationally recognised brands – apart from Schweppes. All local bottlers had their local brands and ranges, but were already becoming increasingly dependent on international franchises.

How have things changed today? Despite all the rhetoric, it is worth noting that, outside the United States, carbonated soft drinks are not necessarily always the biggest selling or the fastest growing beverage sector. Before looking at any figures, however, two words of caution are needed. One is that statistics for countries outside the United States, Europe and Japan are not always authoritative and should therefore be treated as estimates. The other is that the internationally accepted use of the word 'carbonates' does not cover all carbonated soft drinks. The categories included in this chapter are:

Cola – regular, diet, caffeine free, cherry
Mixers – tonic, soda, bitter drinks, ginger ale, aperitifs, etc.
Lemon/lemon-lime – lemonade, lemon, cloudy lemon, lemon-lime
Orange – fruit and essence based and
Others – other fruit flavours, cream soda, shandy, other speciality drinks.

Table 2.1 Beverage consumption across the continents: litres per person, 1988

Beverage	United States	Europe	Japan
Milk	79	90	53
Tea	28	50	57
Coffee	95	98	28
Bottled water	24	49	1
Carbonates	174	53	20
Other soft drinks	43	35	42
Beer	89	79	48
Wine, other alcohol	9	43	18
Spirits	6	5	2
Total	547	502	269
Population in millions	246	355	123

Source: Canadean, Beverage Industry

THE GROWTH AND DEVELOPMENT OF CARBONATED SOFT DRINKS

Table 2.2 Beverage consumption across the continents: change in litres per person, 1978–1988

Beverage	United States	Europe	Japan
Milk	−2	−5	+5
Tea	−1	=	−6
Coffee	−8	+12	+13
Bottled water	+16	+22	=
Carbonates	+52	+14	−9
Other soft drinks	−5	+10	+23
Beer	+2	+1	+10
Wine, other alcohol	+2	=	+2
Spirits	−2	−1	−1
Total	+54	+53	+37

Source: Canadean, Beverage Industry

Naturally sparkling and artificially carbonated waters are excluded, even though they represent a sizeable volume, within which flavoured waters are a small but rapidly growing and directly competing part. Some carbonated sports and health drinks may also escape inclusion, but the omissions are thought to be insignificant.

Consumptions per person in the United States, Europe and Japan for all commercial beverages in 1988 are compared in Table 2.1. Carbonates are by far the most popular drinks in the States; Americans consume more than three times as much as the average European and over eight times the average for Japan. In Europe, carbonates are in fourth place, behind coffee, milk and beer, and in Japan they are even further down the scale.

The rate of change in consumption per person during recent years is shown in Table 2.2. From this, it is clear that carbonates have swept all before them in America, but have faced much stiffer competition in Europe and have been left well behind in Japan.

Carbonates have indeed made most of the running throughout North, Central and South America, Africa and the British Commonwealth countries. Among the most affluent nations, bottled water has also been rising rapidly, particularly in Europe, where it looks set to push carbonates into fifth place very soon.

The Japanese experience, therefore, has been rather exceptional. Their leading soft drink is canned coffee. Sports drinks, canned tea and Oolong tea have all recently become very substantial markets in Japan. Some of these have no real equivalent elsewhere and the speed of change in Japanese consumer tastes is notorious.

In order to complete the global picture, Tables 2.3 and 2.4 offer possibly the first published attempt to list the world's top ten carbonates markets and carbonates consumers. Once again, the United States are well ahead. Mexico

Table 2.3 World top ten in carbonates: market size, 1988

Rank	Country	1000 million litres
1	United States	42.7
2	Mexico	8.4
3	China	7.0
4	Brazil	5.1
5	West Germany	4.6
6	United Kingdom	3.5
7	Italy	2.6
8	Japan	2.5
9	Canada	2.4
10	Spain	2.3

Source: Canadean, Salomon Brothers Inc.

Table 2.4 World top ten in carbonates: consumption rates, 1988

Rank	Country	Litres per person
1	United States	174
2	Mexico	97
3	Canada	92
4	Australia	86
5	Norway	76
6	West Germany	76
7	Switzerland	71
8	Venezuela	69
9	Belgium	68
10	Colombia	67

Source: Canadean, Salomon Brothers Inc.

comes second on both counts. Beyond that, the sheer size of China's population makes it the third largest market, but neither India nor Eastern Europe features at all. For the rest, the battle is between wealth and American influence, with most of the remaining places split between Western Europe and South America.

Within the carbonates sector, it is evident from Figure 2.1 that colas hold a pre-eminent position. In 1987 they accounted for nearly 40% of the market in the UK, Europe and Japan and almost 70% in the USA. Fruit flavours tend to be the next most popular, led by orange and lemon.

Two smaller segments are also common to most countries – mixers and lemon-lime. There is also a significant group of other flavours, many of which are linked to fruit although some defy international comparison; for example, the Americans are keen on root beer and the Japanese on fibre drinks.

Why have these patterns been established? In many respects, the answer lies in the three guiding principles identified earlier. Firstly, demand creation has

THE GROWTH AND DEVELOPMENT OF CARBONATED SOFT DRINKS

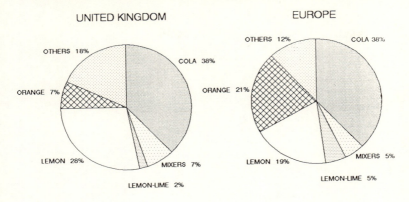

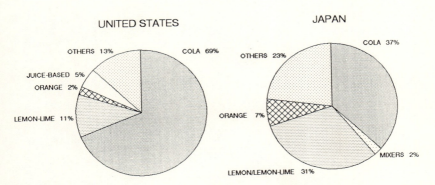

Figure 2.1 Carbonates flavours in selected markets – % shares, 1987. (Source: Canadean; Beverage Industry.)

been maintained and adapted. Today, the major brands present the same powerful marketing images on a world-wide scale. It is difficult to sustain even a national brand without popular music and huge advertising budgets. The intensity of this lifestyle and taste competition has been most acute and most successful in the American 'cola wars', which have excited public interest to an unparalleled extent.

Just as the 'waisted' bottle has remained an essential part of Coca-Cola's image, packaging innovation has also played a vital role in enhancing the appeal of carbonates. In the 1950s, cans were introduced as a convenient and attractive single serve portion. More recently, the availability of PET bottles has made an even greater contribution to the growth of carbonates. PET looks good, is light and does not break. The 2-litre size easily fits into any fridge and is ideal for a family, particularly if sold at a price advantage.

By way of illustration, Figure 2.2 shows how fast cans and PET replaced

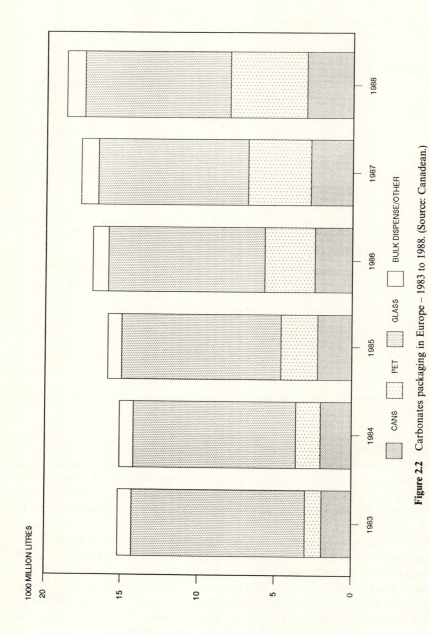

Figure 2.2 Carbonates packaging in Europe – 1983 to 1988. (Source: Canadean.)

glass in Europe between 1983 and 1988, during which sales in cans rose by 55% and the volume in PET rocketed by 450%. In 1988, PET accounted for over 60% of total consumption in Italy, over 50% in Ireland and over 40% in Belgium, France and the UK, and the total market growth would have been very much more pronounced if Denmark, Germany and the Netherlands had not imposed severe restrictions on its use.

Secondly, product quality in its broadest sense has been developed in a number of ways. Massive emphasis has been placed on production technology, leading to extensive rationalisation of bottling plants and vastly reduced unit costs. For example, Coca-Cola's new canning facility at Dunkirk in France will have a substantially greater capacity than the combined current soft drinks canning demand of France, Belgium and the Netherlands.

The leading companies have also undergone considerable diversification, extending the guarantee provided by their main brands into new variants such as cherry or caffeine-free cola and into new flavour categories such as exotic fruits. Schweppes' portfolio now stretches well beyond its traditional tonic and mixers; it ranges from the fruit spectrum to 'Dry' adult drinks as well as 'Bitter' and 'Black' aperitifs.

Ingredients have been hard-pressed to keep up with the pace of change. Many artificial colourings and unnecessary additives have been removed. The quest for natural flavouring has meant more widespread use of real fruit content rather than manufactured essences. Even more significant, though, has been the growth of low-calorie carbonates, boosted by legislation to permit the introduction of a new generation of intense sweeteners, led by aspartame. As may be seen from Figure 2.3, their popularity in the USA is much greater than in Europe, and almost everywhere they continue to gain market share. Since 1980 their share has trebled in Europe from 2.4 to 7.3% in 1988, with UK volume up still further from 4.0 to 14.5%. In the USA they have more than doubled their share, from 12.9 to 26.1%, during the same period, which means that the American consumption of low-calorie drinks is almost as high as the European average for all carbonates.

Thirdly, distribution is now being taken towards its limits. Along with bottling rationalisation there has also been far-reaching rationalisation of distribution networks. At the same time, the task of distribution has become much more complex: a supermarket requires a different pack to a sports centre; air travellers have different requirements to car travellers; people eat more often than previously, but in smaller amounts; and people drink more often than they eat.

The fact that someone wants a drink represents an important consumption opportunity for carbonates. As they seldom miss an opportunity, suppliers of carbonates have designed a host of vending machines, chilled display cabinets and dispensers to cater for every conceivable outlet. In the UK alone, the number of chilled cabinets has risen from under 10 000 to over 60 000 in just five years.

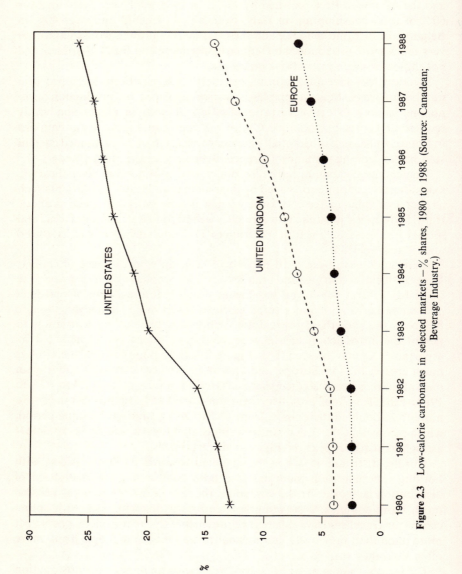

Figure 2.3 Low-calorie carbonates in selected markets — % shares, 1980 to 1988. (Source: Canadean; Beverage Industry.)

2.2 Future prospects

The foregoing is a brief overview of the carbonated soft drinks industry and marketplace of today. But what of its outlook? The following is a selection of key factors that will shape the carbonates world of the 1990s.

Untapped growth. Significant growth in the total market for carbonated soft drinks appears set to continue, all the more so as prosperity increases and East–West relations improve. So far, the USA has convincingly outpaced Europe, despite its consumption rates already being vastly higher. But, in global terms, both may in time be overshadowed by the huge potential of China, Eastern Europe and India as their people gain in wealth and their markets open up to Western influence.

Population structure. In the developed world, population levels are fairly stable, though there is still a gradual increase in the USA and Japan. Elsewhere, populations are rising more quickly, but primarily among those who do not yet have the means to purchase carbonated soft drinks. The main population changes affecting carbonates are therefore related to population structures. Most Western populations are ageing, with declining birth rates and more people living well beyond retirement. In Europe, by the year 2000, there will be 18 million fewer young people aged 10 to 24 than in 1985 and 9 million more over the age of 65. On the other hand, as the cola generation grows older it is taking the habit further up the age spectrum and so the net effect may not prove all that crucial.

Healthy living. Health issues have had an immense impact on eating and drinking habits in the last decade and the current themes may be expected to maintain their importance for some time to come. People want to live better and longer, but are often not prepared to make much effort themselves. To some extent, this puts the onus on manufacturers. Now that artificial preservatives, colouring and flavouring are largely being replaced by natural or nature equivalent alternatives, public demand for purity and safety seems likely to broaden its horizons, especially if the new social disease of consumer terrorism takes root.

Alcohol concern. Growing international concern about the anti-social effects of alcohol is another health-related issue and another with direct implications for carbonated soft drinks. Carbonates will undoubtedly benefit from greater caution about alcohol consumption, but that gain could well be offset by a surge in low- and no-alcohol beers and wines. Certainly, the battle for adult consumers will be long and hard fought.

The environment. Environmental issues already affect the location and construction of new plant as well as energy management and transport, but their impact on packaging is perhaps the most important and most difficult to predict. Two directly conflicting forces are at work: the first is the marketplace, in which retailers and many customers are seeking a variety of types and sizes, which are attractive, light, compact, unbreakable and instantly disposable; the second is the Green lobby and other consumers, who are more anxious about recycling or re-use than anything else. The conflict is most starkly evident in Europe where the whole basis of the European Community is the free passage of goods, but where individual governments are placing restrictions on packaging in completely incongruous ways.

The solution must lie in the industry finding its own environmentally acceptable answers. Otherwise governments will introduce their own arbitrary deposits, taxes, quotas or outright bans.

Lifestyle imagery. It would be hard for carbonated soft drinks to retain their contemporary lifestyle-based advertising imagery if their packaging failed to move with the times. Image building is the cornerstone of carbonates' appeal to the customer. Rather, the spread of global marketing seems set to continue, particularly in places like Eastern Europe and China. The cola wars may well be joined by the 'lemon–lime wars' or just the 'megabrand' wars as other flavours are also subjected to the same competitive pressures.

Market niches. Although global marketing is clearly gaining in importance, a large number of people will still want to find a carbonated soft drink that expresses more individuality. Smaller production runs involve greater expense, but also offer higher added value to manufacturers. They will also help keep smaller and more flexible companies in business. Some market niches are difficult to replicate, such as A.G. Barr's Irn Bru (originally Iron Brew) from Scotland, but there are many other ways of establishing a niche, such as Orangina's distinctive bottle.

Widening availability. The need for 'arm's reach' availability has already been discussed and industry attention will remain focused on this. Beyond the ever-increasing variety of dispensers and packaging, one of the next moves could target on daily consumption occasions that are not normally associated with carbonates, such as breakfast, coffee breaks or bedtime drinks.

Pricing contrasts. Pricing is showing signs of polarising at two extremes. The family-sized pack is becoming more of a price sensitive commodity, often bearing a supermarket's own label. At the other end are the premium niche products, usually in single serve containers. Cans come somewhere in the middle, often discounted, and even more so in multipacks.

Industry concentration. In today's mass market, manufacturers, distributors and retailers are all becoming high-technology low-cost operators, with more and more power concentrated in fewer of them. There are huge economies of scale to be obtained from the latest techniques, and the breaking down of national trade barriers in Europe will only serve to reinforce these pressures. This will leave less room for independent franchise holders for the international brands.

Paradoxically, growing wealth is making consumers more interested in imagery than price, so the independents should be able to find continued success in niche markets as well as regional markets. In Europe, for example, tastes are still very much dominated by regional traditions and this will take many years to change.

Scope for opposites. A recurring feature of these outlook comments has been the recognition of conflicting trends and the polarisation of behaviour, as there will always be opportunities in the antithesis of mainstream developments. Just as the word 'natural' characterises much of the change in the modern marketplace, it should be noted that the low-calorie segment depends heavily on artificial sweeteners. Equally, caffeine has been taken out of colas in one successful variant, but there is also a high-caffeine variant.

So the future for carbonated soft drinks offers us a series of contrasting variations on a theme. In the main, there will be continued growth, with the big becoming bigger and the small becoming fewer but more up-market.

3 Water treatment

R. HUTCHINSON and W. McCARTHY

3.1 Introduction

Water is the principal ingredient in carbonated beverage production, and in some cases it reaches over 90% of the product. It follows from this fact that beverage manufacturers pay particular attention to the quality of the water used in their formulation, to ensure that when the product arrives at the consumer it is consistent in taste and appearance and has not deteriorated during storage. Specifications covering both the physical and chemical properties have been developed by the major manufacturers and have been issued to their licensees to ensure that these criteria are achieved at all times.

The world's water is constantly recycled by the process called the Hydrological Cycle. The process is powered by the sun, which evaporates water from the sea, rivers, lakes and ponds into the atmosphere to form clouds. The climatic conditions allow it to fall back to earth as rain, hail, sleet or snow. When the water is released by the clouds into the atmosphere it is very pure but as it nears the earth's surface it absorbs atmospheric pollutants such as dust, smoke, acid fumes, etc. The precipitation falling on the ground will then also pick up and dissolve inorganic salts, naturally occurring organic matters and bacteria consistent with the geology of the ground on which it falls and through which it permeates. As an example, precipitation falling on chalk will produce a water that is high in dissolved solids and will probably have a high alkalinity and total hardness, while that falling on granite or a similar virtually insoluble stone, will be low in dissolved solids, hardness and alkalinity.

Seasonal variations in water supply can also occur in temperate regions where rainfall differs widely from winter to summer or in cold areas where there is an annual thaw of snow and ice; any water treatment plant must be flexible enough to accommodate these variations and still supply treated water to the specification required.

The public supply in the urbanised areas of the world is acceptable for potable uses, but, although it will meet part of the specifications set by soft drinks manufacturers, it will not necessarily meet their specifications fully without further treatment, and most manufacturers install water-treatment plant in their own factories. Water used from sources other than public supply will certainly require treatment to meet the set specifications.

WATER TREATMENT

A uniform and consistent supply of water will improve the operating efficiency of the factory by allowing a constant manufacturing process to be established and so eliminating the need to stop frequently or to alter the manufacturing conditions to meet any changes in water quality. Typical analyses of municipal water from 'hard' and 'soft' areas are given in Table 3.1.

Some of the impurities found in water can have a significant effect upon the quality of the final carbonated product, as summarised in Table 3.2, and the standards set by carbonated soft drink manufacturers are remarkably similar, as detailed in Table 3.3.[1]

Water authorities are required by statute to distribute a water that is 'pure

Table 3.1 Typical analyses from 'hard' and 'soft' areas

	Hard area	Soft area
Colour (Hazen units)	5	20
Turbidity	0.3	5
Conductivity (μS)	680	250
pH	7.9	7.6
Cations:		
Calcium (Ca^{2+})	88	33
Magnesium (Mg^{2+})	24	9
Sodium (Na^+)	34	24
Potassium (K^+)	7	1
Anions:		
Total alkalinity ($CaCO_3$)	219	92
Chloride (Cl')	49	22
Sulphate (SO_4'')	73	38
Nitrate (NO_3')	31	10
Organic content (OA)	2	10

Table 3.2 Effect of impurities in water on the quality of the product

Impurity	Effect on product	Treatment process
Suspended particles	Foaming at the filler. Loss of carbonation. Visible particles in product	Filtration or coagulation and filtration
Organic matter	Deposits or precipitation in product. Formation of neck rings during storage	Coagulation and chlorination. Organic scavenger using ion exchange
Bacteria	Spoilage by creation of off tastes. Great danger of health risks	Chlorination or ultraviolet radiation
Taste and odour organic compounds	Off-tastes	Chlorination or absorption by activated carbon
High alkalinity	Neutralises acidity. Creates a bland taste	Coagulation. Dealkalisation by ion exchange
Nitrate	Health hazard. Corrosion of cans	Ion exchange by nitrate specific ion exchange resin
Polysaccharides	Precipitation in storage	Reverse osmosis

Table 3.3 Chemical specifications for water to be used in soft drink production (all figures shown are the maximum permitted levels expressed as mg/litre)

Characteristic	Company				
	A	B	C	D	E
Total dissolved solids	500	850	—	500	850
Alkalinity (as $CaCO_3$)	50	50	50	50	50
Chloride (as Cl')	300	250	250	—	250
Sulphate (as SO_4'')	300	250	250	—	250
Iron	Nil	0.1	0.2	Nil	0.3
Aluminium	0.1			Nil	0.2

and wholesome', which, although always suitable for potable needs, does not necessarily mean that the water is clear and colourless. Since the 1960s there has been a marked improvement in the operating efficiency of municipal water works and the aesthetic appearance of public supplies has generally improved, but prior to this time water available to beverage manufacturers often had residual colour and suspended solids and therefore did not meet the manufacturing specifications.

Because of the variations in the quality of raw water, and as river waters may contain a high concentration of suspended solids as well as colour and organics, beverage manufacturers originally issued specifications that almost always demanded coagulation as the principal method of treatment in an attempt to produce a standard water and thereby produce a constant tasting beverage.

From the treatment and removal methods given in Table 3.2, a process sequence emerges that will produce water to the issued specifications that can be used in the manufacture of carbonated beverages.

The actual water-treatment process used will depend upon (a) the source of the raw water, (b) the treatment the water may have had prior to its arrival at the factory and (c) any seasonal variations that may be likely to affect the product. The functions of any water-treatment plant will involve all or at least some of the following:

(1) To remove colour and suspended particles (including colloids).
(2) To remove organic compounds likely to cause or impart off-tastes and odours.
(3) To reduce the alkalinity to a set level.
(4) To remove micro-organisms and bacteria, by sterilisation.
(5) To ensure a consistent product at all times.

3.2 Coagulation

Coagulation is a chemical process where a gelatinous precipitate or floc is formed that will absorb foreign organic matter from the water. In waters that contain suspended solids, the floc will use the solids as a nucleus thereby enlarging the size of the suspended particle, making it suitable for filtering.

The design and rated output of a coagulation tank will vary from one manufacturer to another. Despite variations they will all have four operating zones: mixing, flocculation, blanket concentration and the clear well. In Figure 3.1 the chemicals are shown as being dosed into the clarification tank. In some circumstances the chemicals will be dosed into the water prior to the clarification tank so that the reaction time is increased.

Either ferric sulphate or aluminium sulphate would be used as the coagulant to form the floc. The choice of which coagulant to use will depend upon the characteristics of the water to be treated. In waters that have a total alkalinity below 100 ppm $CaCO_3$ that would generally have a low pH, aluminium sulphate would be used. The side reactions that take place would be used to destroy the low alkalinity concentration present in the water. Waters that have a total alkalinity over 100 ppm $CaCO_3$ would have a higher pH and in this case ferric sulphate, which would be prepared by the oxidation of ferrous sulphate, would be selected as the coagulant. The choice would be based on economics as it would require less chemicals to adjust the pH to around 8.5 than it would to lower the pH to the optimum required for the use of aluminium sulphate.

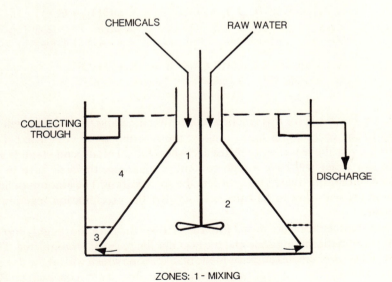

Figure 3.1 Coagulation tank.

The chemical reactions that take place are given below.

(a) *Using aluminium sulphate:*

$$Al_2(SO_4)_3 + H_2O = 2Al(OH)_3 + H_2SO_4$$
Alum — Water — Aluminium hydroxide — Sulphuric acid

$$H_2SO_4 + Ca(HCO_3)_2 = CaSO_4 + 2H_2O + 2CO_2$$
Sulphuric acid — Calcium bicarbonate — Calcium sulphate — Water — Carbon dioxide

The second reaction is the destruction of alkalinity that will be present in the water. Aluminium is only insoluble between pH 5.5 and 7.0. At a pH below 5.5 aluminium will dissolve as an aluminium salt and at a pH over 7.0 aluminium will dissolve as an aluminate. If the characteristics of the water are such that when the above reaction is complete the pH is outside the above limits, then the pH will have to be adjusted to bring it within the recommended range.

(b) *Using ferrous sulphate:*

$FeSO_4$ is oxidised by chlorine to $Fe_2(SO_4)_3$
Ferrous sulphate — Ferric sulphate

$$Fe_2(SO_4)_3 + 3Ca(HCO_3)_2 = 2Fe(OH)_3 + 3Ca(OH)_2 + 6CO_2$$
Ferric sulphate — Calcium bicarbonate — Ferric hydroxide — Calcium hydroxide — Carbon dioxide

$$Ca(OH)_2 + Ca(HCO_3)_2 = 2CaCO_3 + 2H_2O$$
Lime — Calcium alkalinity — Calcium carbonate — Water

The optimum pH for this reaction falls between 7.8 and 8.7 and lime will be added to the coagulation tank to allow the reaction to be completed. The reaction given above is essentially used for the removal of suspended solids but as the raw water will have a high content of alkalinity this will have to be reduced so that the treated water meets the specification. The amount of lime to be added will have to be calculated so that the precipitation reaction is completed.

When the chemicals are dosed into the clarifier they are thoroughly mixed by the rotor in the mixing zone and the flocculation reaction is completed. The action of the rotor creates currents in the blanket zone which will lift the floc off the bottom of the clarifier and prevent the floc from coalescing. The clarification reactions mean that floc is continuously produced during the time that the clarifier is in operation and the volume of floc in the tank will gradually increase. The optimum floc concentration will be about 20% (measured by the one-hour settlement test) and in operation the floc will be controlled between 15 and 25%. The floc concentration will be controlled manually by drawing some of the floc from the bottom of the tank. In order to ensure that the floc concentration does not exceed the recommended levels, regular control tests will have to be carried out and the floc level adjusted accordingly.

Under certain operating conditions (where the alkalinity is low or where the colour content of the raw water is also low) the floc produced could be very light and may be carried over with the clarified water. This will place a high load on the sand filter that follows the clarifier and will mean that the filter will have to be washed more frequently. In these circumstances a small dose of a polyelectrolyte solution is sometimes added to the coagulation tank to improve the reaction. The polyelectrolyte will assist in the formation of the floc by drawing the smaller particles together to make larger particles, which will be trapped by the floc concentration in the bottom of the reaction tank.

3.2.1 Filtration

The clarified water that is drawn from the coagulation tank may contain some fine particles of floc and these will be removed by filtration through a sand filter. Sand filters will be designed to operate at a flow rate of $15\,m^3/h$ per square metre of filter area. This will dictate the size of the filter dependent upon the flow rate through the plant.

A typical sand filter (Figure 3.2) consists of a pressure vessel that contains a bed of filter sand (typically having a particle size between 14 and 25 BS mesh – 0.6 to 1.2 mm) supported on a bed of graded gravel. The water from the clarifier is directed into the top of the filter and as it passes down through the filter sand the matrix formed by the media holds the suspended particles in the water as they are presented to the filter bed. The void space between the media particles would apparently allow particles that are smaller than the rated void

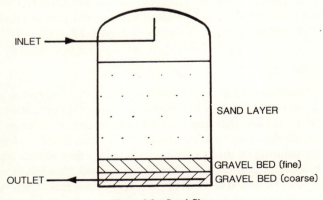

Figure 3.2 Sand filter.

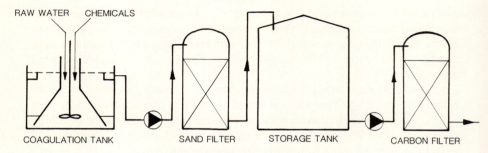

Figure 3.3 Coagulation plant.

size to pass through. In practice, because of a 'bridging' effect across the media particles, small particles are retained and the filtered water is clear.

As the filter media gather more particles the void space between the particles will get progressively smaller and will create a resistance to the liquid flow. In order to maintain the desired flow through the filter media after prolonged use, the inlet pressure would have to be increased. This is not always a practical suggestion and the answer is to clean the filter at regular intervals and remove the restriction to the flow. Filter cleaning is carried out by washing the media using a reverse flow of water and backwashing the floc out of the filter to drain. In some cases where the floc is likely to stick to the media, an air scour is used as the initial stage of the cleaning process. As the air is blown in at the bottom of the filter the air current will scrub the floc from the sand particles. The backwash flow of water will then wash out all the particles from the media and the result will be a very clean filter ready for the next operating cycle. Filters are usually operated up to a pressure differential of 1 bar in excess of the pressure that could be expected from a clean filter. Automatic operating filters will be fitted with a pressure differential switch reading the difference between the inlet and outlet pressures of the filter that will initiate the wash process automatically once the difference across the filter has reached the maximum. A typical make-up of a water-treatment plant using clarification as the primary treatment process is shown in Figure 3.3.

3.2.2 *Sterilisation*

Free chlorine will be used as the sterilising agent for the water prior to its use in the manufacture of the drink. It is usual to use about 10 ppm of free chlorine for sterilisation to ensure that the reaction will reach completion. The free chlorine may be dosed as chlorine gas but it is usually prepared from sodium hypochlorite solution for safety reasons. In the case where ferrous iron salts are used as the coagulation chemical, free chlorine will be used to oxidise the ferrous salt to ferric salt. It is usual in this case to dose more than 10 ppm free chlorine into the coagulation tank so the the ferrous iron is fully oxidised to the ferric state and so that there is the required excess for sterilisation. Where ferric iron is used as the flocculating chemical the residence time in the coagulation tank can be included as part of the contact time necessary with free chlorine for sterilisation. In this case the chlorinated water storage tank can be smaller. Free chlorine will not be used as part of the coagulation reactions where aluminium sulphate is used as the flocculating chemical and will generally be dosed after the sand filter, though in some cases chlorine may be dosed into the clarifier to prevent bacterial growth. In this case the residence time in the clarifier cannot be counted as part of the sterilisation period and storage capacity for the full two-hour reaction time with the chlorine will be necessary.

Approximately 10 ppm of free chlorine will be dosed into the water for sterilisation. The reactions that take place will not only destroy bacteria that may be in the water and render it sterile but will also react with other organic molecules that may remain in the water after the clarification reactions. The residual organics may vary from naturally occurring organic matter to simple organic compounds such as chlorophenols and trihalomethanes that could impart both taste and odour to the final carbonated product. The action of the free chlorine will break down these organic molecules.

3.2.3 *Dechlorination*

Because of the oxidation reactions that are taking place during the sterilisation period some of the chlorine will be consumed but the amount of chlorine dosed will be enough to ensure that there will always be at least a 5 ppm residual at the end of the sterilisation period. The residual free chlorine and the majority of the organic molecules will be removed from the sterilised water prior to use by absorption on activated carbon (Figure 3.4).

The activated carbon will be housed in a pressure filter which is similar in construction and arrangement to the sand filter. The activated carbon will be

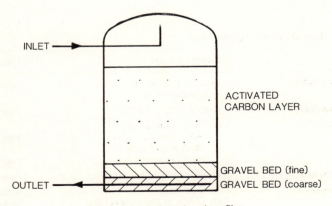

Figure 3.4 Activated carbon filter.

supported in the filter on a bed of graded gravel and the volume of carbon will be calculated to give a contact time of five minutes with the water. This contact time will be long enough to remove all the free chlorine from the water. Activated carbon will also remove other organic molecules that remain in the water after the coagulation reaction or are formed as part of the sterilisation reactions. A five-minute contact time may be enough to remove not only the free chlorine but also a majority of the taste- and odour-forming organic molecules. Some residual organic compounds may, however, remain in the water because the contact time with activated carbon necessary for their total removal will be longer than five minutes. In general, the organic residual is expected to be below the detection threshold with only a short contact time.

3.2.4 *Plant performance monitoring*

In order to ensure correct operation of a typical plant and at the same time to trim the plant for any variation in the raw water supply, the operation of the plant should be monitored regularly (Table 3.4).

Floc concentration. The test will require a sample to be taken from the blanket zone of the clarification tank in a measuring cylinder. The usual volume will be 100 ml. The cylinder will be allowed to stand for 30 min and the floc concentration will be calculated as a percentage. If the concentration is above the recommended figure then floc will have to be discharged. If the floc concentration is below the recommended figure then the clarification process will not be operating at the optimum efficiency, but as the reaction continuously produces floc the concentration will be made up during operation.

Residual alkalinity. This will be checked by simple titration on a known volume of water using a volumetric solution of sulphuric acid.

Table 3.4 Typical monitoring procedure

Test	Sampling point	Frequency
Floc concentration	Blanket zone in the coagulation tank	Daily
Residual alkalinity	Outlet of the sand filter	Two hourly
Chlorine dose	(1) With alum dosing: inlet to storage tank	Two hourly
	(2) With iron dosing: outlet of sand filter	Two hourly
Chlorine residual	Outlet of the carbon filter	Two hourly
Microbiological bacteria	(1) Raw water	Weekly
	(2) Outlet of the carbon filter	Weekly

The total alkalinity (M alk) is determined as the volume of 0.02 N sulphuric acid in millilitres (ml) required to neutralise the sample to pH 3.8. The indicator will be screened methyl orange. When using a 100 ml water sample the M alk in the water will be calculated by multiplying the titration volume by 10, e.g. titration of 0.02 N H_2SO_4 = 5.0 ml

$$M \text{ alk is } 5.0 \times 10 = 50 \text{ ppm}$$

In the cases where ferrous/lime coagulation is used it is necessary to carry out frequent checks to ensure that the correct volume of lime is dosed. This is particularly important where the alkalinity of the raw water is variable. Controls and instrumentation will be able to handle minor variations in the alkalinity of the raw water but will not always be able to cope with wide alkalinity variations of the incoming water. Because of this fact it is important that regular chemical checks are maintained on a plant such as this. Manually operated and controlled plants will not operate without regular chemical checks. The alkalinity measured (P alk) in this case will be directly applicable to the amount of lime dosed into the clarifier.

The alkalinity (P alk) will be determined as the volume of 0.02 N sulphuric acid required to neutralise the sample to pH 8.3. The indicator will be phenolphthalein and the colour change will be from pink to colourless. The volume of 0.02 N acid required to change the colour should be recorded. The P alk represents only a specific part of the alkalinity and the total alkalinity can be established by putting screened methyl orange indicator in the colourless solution and continuing the titration to the 'end point' without zeroing the burette. The M alk is then the total volume of 0.02 N acid used. When using 100 ml of water the calculation of the alkalinity will be obtained by multiplying the titration volume by 10, e.g.

$$P = 2.5 \quad P \text{ alk} = 25$$
$$M = 4.5 \quad M \text{ alk} = 45$$

Correct lime dosing can be calculating by use of the formula 2P − M. The calculation will be:

$$(2 \times 25) - 45 = 5$$

The dosing will be correct if the calculation shows a figure between 0 and 7. If the calculation gives a negative result, then too little lime is being dosed and the dosing rate should be increased; if the calculation gives a result that is greater than 7, then too much lime is being dosed and the dosing rate will need to be reduced.

Free chlorine. This will be checked either by a proprietary automatic chlorine monitoring meter or by a colorimetric test kit. The colorimetric method relies upon a reaction with a reagent and the colour produced is directly proportional to the concentration of free chlorine in the water. The colour produced will be compared against a standard colour disc and the comparison will show the concentration of chlorine in the water. In the case of the final treated water there should be no residual in the water after it has passed through the carbon filter. If any residual is found, then this would be an indication that the activated carbon needs to be changed.

Microbiological testing. Water, both raw and treated as well as raw materials, intermediate samples and the final product will all need to be tested for the presence of bacteria and it is usual to use one of the following methods:

  (i) Direct plating
 (ii) Surface inoculation
(iii) Membrane filtration
 (iv) Broth inoculation
  (v) Microscopic examination.

3.2.5 *Approximate plant size*

The area the plant requires will depend upon the duty at which it will operate. For the purposes of comparison with other processes that will be available, a plant capable of producing $30 \, m^3/h$ of treated water will require the following equipment: clarifier, chemical dosing pumps and storage tanks, pumps, sand filter, two-hour storage tank and an activated carbon filter, all occupying a floor space of $60 \, m^2$. The calculation is based on a single sand and carbon filter, which would mean that when these items require washing there would be no flow to service.

Advantages of coagulation plants

- Low operating costs. Chemicals used are inexpensive.
- Continuous operation possible with single clarifier. Duplicate filters will be necessary to enable continuous operation at all times.
- Produces a water with low suspended solids. Ideal when operating on river waters. This process is sometimes used as pre-treatment to other treatment methods.

Disadvantages of coagulation plants

- Not easily controlled when operating on raw waters with varying alkalinity.
- Coagulation reactions are slow and the tank tends to require a large surface area.
- Not recommended for water supplies that have more than 500 ppm dissolved solids.
- No effect on chloride and sulphate content if raw water levels are above the desired specification.

3.3 Ion exchange

A high percentage of carbonated soft drinks factories still operate clarification plants, and because the process has been in use for many years it has earned itself the description of the 'traditional process'. Over the last decade there has been a noticeable improvement in the quality of water supplied by the water authorities with respect to both suspended solids and residual colour and organics. The traditional process is still used in those water-treatment plants that have been installed in factories that are forced to use river water as their raw water or in factories that will be supplied with variable water of a type that made the soft drinks producers set the standards in the first place. The improvement of municipal supplies in the urban areas of Western Europe, the Americas and Oceania has meant that the principal reason for using coagulation as the primary form of treatment has disappeared and has allowed other processes to be considered without any reduction in specifications.

As ion-exchange processes are easily automated, the numbers of staff required to operate the plants are reduced to a minimum. This has several benefits, particularly concerning the number of personnel required to operate a factory, which is an important point to consider in the design of new factories. Because of the increasing popularity of ion exchange the process can be described as the 'modern process'.

3.3.1 *Dealkalisation*

Municipal waters in the urbanised areas usually meet the specification required by soft drinks manufacturers with respect to the mineral content except for total alkalinity, and such a process that is specific for alkalinity reduction has been in industrial use for many years. The weakly acidic cation exchange resins are based on methacrylic acid with carboxylic active groups. A weakly acid cation exchange resin operating in the hydrogen form is only weakly ionised and will therefore only react with cations associated with bicarbonates or carbonates in solution.

The chemical exchange reaction taking place may be approximated as follows, where 'R' represents the resin structure and 'H' the exchangeable ion:

$$\underset{\text{Regenerated resin}}{2R-H} + \underset{\text{Calcium alkalinity}}{Ca(HCO_3)_2} = \underset{\text{Exhausted resin}}{R_2-Ca} + \underset{\text{Carbon dioxide}}{2CO_2} + \underset{\text{Water}}{2H_2O}$$

The total capacity, and therefore the operating capacity, of weak acid cation resins is very high and the acid requirement for regeneration of the resin is only about 15% over the stoichiometric equivalent. These properties should be compared to strong acid cation exchange resins that have been the mainstay of softening and demineralisation processes for many years where the total capacity is about half that of the dealkalisation resins and the regeneration efficiency about 50%.[2,3]

The quality of the treated water produced by the dealkalisation process will mean that all the alkalinity will have been removed, giving a treated water with a pH about 3.8 and being rich in free carbon dioxide. In actual operating experience it is sometimes found that, owing to variation in manufacture, some resins will salt split and there will be a small reaction with some of the strong anions, which will be converted to the equivalent mineral acid and will cause the pH of the treated water to fall below the neutral point of pH 3.8. This will last for only the initial part of the service run and is not generally considered to be important. As the specifications will allow up to 50 ppm alkalinity in the treated water, and as there is no alkalinity in the treated water leaving the dealkalisation unit, it is possible to blend the treated water with a portion of raw water to raise the alkalinity content to 50 ppm. Any 'acidity' in the dealkalised water will be neutralised by alkalinity in the raw water and blending with raw water will, in fact, reduce the running costs of such a plant while still meeting the specification required.

The resin has a finite operating capacity, and when the capacity has been used up alkalinity will begin to slip through into the treated water. As the alkalinity begins to increase, the pH will also rise and this fact is used as an indicator that the plant requires regeneration. Automatic plants that monitor the pH of the dealkalised water will initiate regeneration of the resin when the pH has risen to 5.5. This point will coincide with approximately 50 ppm of alkalinity in the treated water. On plants where the control is by a throughput meter that will indicate or initiate regeneration, the volume of treated water produced during the service run is set slightly low so that there is no possibility of alkalinity slipping through into the treated water. Strong acids (either hydrochloric or sulphuric) will be used for regeneration of the resins and the reaction that takes place (in which 'R' represents the resin structure) may be approximated as follows:

$$\underset{\text{Exhausted resin}}{R_2-Ca} + \underset{\text{Hydrochloric acid}}{2HCl} = \underset{\text{Regenerated resin}}{R-H_2} + \underset{\text{Calcium chloride}}{CaCl_2}$$

or

$$\underset{\text{Exhausted resin}}{R_2-Ca} + \underset{\text{Sulphuric acid}}{H_2SO_4} = \underset{\text{Regenerated resin}}{R-H_2} + \underset{\text{Calcium sulphate}}{CaSO_4}$$

The use of sulphuric acid as the regenerant acid can cause problems by the production of calcium sulphate in the regeneration effluent. Calcium sulphate is very insoluble and for this reason the injected concentration of sulphuric acid must be limited to 0.8% in order to minimise the risk of calcium sulphate precipitation. Using stronger concentrations will almost ensure that calcium

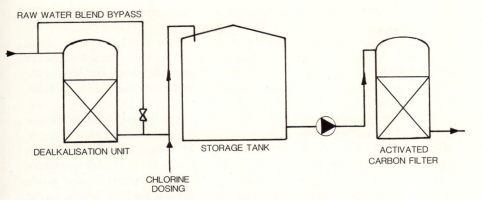

Figure 3.5 Typical ion-exchange treatment plant.

sulphate will be precipitated and if the precipitation occurs in the resin bed it will probably cause the operating capacity of the resin to be reduced and the quality of the treated water produced by the plant to be affected. Owing to the density of calcium sulphate precipitate, and to the fact that it is very insoluble, it is very difficult to clean a resin bed once precipitation has taken place. When hydrochloric acid is used as the regenerant, calcium chloride is produced as the waste product and this, being far more soluble than calcium sulphate, does not cause the same precipitation problems and allows the acid to be injected at a higher concentration thereby shortening the regeneration time.

The dealkalised water will require sterilisation with chlorine, storage for two hours, and then treatment by activated carbon to remove the residual chlorine before use (Figure 3.5). The sterilisation, storage and dechlorination stages will be the same as those installed in the coagulation process. As a matter of comparison with the coagulation process a plant capable of dealkalising $30\,m^3/h$ coupled with the necessary regeneration equipment as well as the chlorination, storage and dechlorination equipment is estimated to require about $30\,m^2$ of floor space. This should be compared with the area of $60\,m^2$ estimated for the coagulation plant. From these figures it will be realised that one of the great advantages of using the dealkalisation process is space saving.

3.3.2 *Organic removal*

Dealkalisation is not the only ion-exchange process that is currently finding a place within the soft drinks industry. The coagulation process not only reduced the alkalinity content of the water but also removed the organic matter as part of the flocculation reactions. The dealkalisation process, however, is specific to the removal of alkalinity from water and all other constituents in the feed water will remain unaffected. Organic matter in the raw water may cause taste and odour problems in the carbonated product, particularly after sterilisation using chlorine. In some areas the naturally occurring organic matter reaches levels where it is likely to give colour to the water, which will necessitate its removal to meet the required specifications. Other waters may also contain significant concentrations of organic matter (colourless to the naked eye) which will depend upon the nature and type of the organic molecule as well as the concentration. In these circumstances organic removal must be considered a necessity.

Removal of organic matter by ion exchange was developed in the 1960s as a method of protecting anion exchange resins in a demineralisation plant in order to maintain the quality of treated water produced. The process uses an organic scavenger resin operating in the chloride form. Natural organic matters are complex organic molecules that contain carboxylic acid active groups, meaning that they will act as weak acid anions, and will be held on the organic scavenger. The organic scavenger is a highly porous resin that will allow the organic molecules easy passage in and out of the structure. Ion-exchange resins with a macroporous or macroreticular structure are particularly suited to this application and offer physical strength coupled with capacity and reversibility of organics during regeneration. As ion-exchange treatment will only remove about 70% of the organic matter in the raw water, it will be a matter of decision by the operator to determine if the residual organic concentration after scavenger treatment is high enough to cause problems with the final product. As a general rule waters having an organic content above 1 ppm as measured by the 4 hours at 27 °C permanganate test will require treatment while those waters that have a concentration at 1 ppm or below will not normally require special treatment.

Regeneration will use 10% sodium chloride solution prepared from concentrated brine, though under certain conditions regeneration will be made with a mixture of 10% sodium chloride and 2% sodium hydroxide, which will give the scavenger a higher operating capacity. The regenerant will displace almost all of the organic matter absorbed in the previous operating cycle but some residue will remain in the resin and will require other regeneration techniques such as hot soaking or alkaline brine regeneration to assist in its removal. Under certain conditions the residue will not be removed by any means and this residue will be classed as irreversible fouling of the resin. The irreversible fouling of the resin will progressively increase and there will come a time when the fouling threshold will reach such a level indicating that the economic life of the resin has been reached, which will mean that the resin must be changed. Alkaline brine regeneration tends to remove more of the absorbed organic matter from the resin and therefore the build up of irreversible fouling will be slower.

Scavenger resins available up to the mid 1980s were based on the conventional polystyrene matrix. Recent developments in resin structure have produced a range of resins based on an acrylic structure.[4] The acrylic matrix allows for effective adsorption of the organic molecule during the service cycle and effective desorption during regeneration. The high resistance of the acrylic matrix to organic fouling, coupled with the inherent kinetic capacity for the organic molecule, makes these new scavenger resins ideally suitable for organic removal.

3.3.3 Nitrate removal

With the increasing use of nitrogen, phosphorus and potassium (NPK) enriched fertilisers for agricultural use, and the changing use of pasture land to arable use, the nitrate concentration in waters has increased. This is particularly so of waters in rural areas where both surface and ground waters are affected. Waters in urbanised areas are also affected and statistics show that the average nitrate content of the River Thames has risen from about 20 ppm NO_3 in 1950 to currently about 40 ppm NO_3. Nitrates in water are best known for causing infantile methaemoglobinaemia (the 'blue baby syndrome') but it is for other reasons that the soft drinks industry removes nitrates.

Some cans are manufactured using tin, and there is a chemical reaction between nitrate and tin that is likely to cause corrosion of the can during storage, with possible spoilage of product and perhaps loss of good will of both the retailer and the customer. Can manufacturers have developed can-lining processes that will protect the can itself and not impart anything into the drink. A number of manufacturers have decided not to risk the possibility of a slight flaw in the can lining and are installing equipment to reduce the nitrate concentration of water.

The removal of nitrate from water has been known in the water-treatment industry for a number of years and has used a strong anion-exchange resin operating in the chloride form. Although the nitrate will be removed from the water, both the nitrate and the sulphate in the water will be exchanged for a chloride ion. In thin waters this will not be a problem, but in waters with a high mineral content the conversion of strong anions to chloride could put the chloride content of the water over the limit set by the drink manufacturer. Because it was likely to upset the anionic balance of the water, and because there was a risk that the mineral content of the water would change, there was a likelihood of a high concentration of nitrate in the treated water because all the nitrate exchanged during the service run would be eluted off the resin once it reached saturation as the resin has a greater affinity for the sulphate ion than it does for the nitrate ion. For this reason the process was not widely adopted in the industry.

As a result of a British government sponsored research project in the 1970s an active resin that will selectively exchange the nitrate ion was developed. This discovery has led to the development of a nitrate specific anion exchange resin that has been commercially available for about five years.[5] The resin is a modified anion exchange resin operating in the chloride form, and although it will exchange the bicarbonate, chloride and sulphate ions at the early part of the service run, it very quickly reaches an equilibrium for these ions and, as a result, the treated water is similar (except for the nitrate content) to the influent water for the remainder of the service run. If this imbalance cannot be tolerated then two similar plants can be operated in parallel, one plant operating 50% of its service cycle in front of the other with, perhaps, a third standby plant to allow continuous operation.

The affinity of the resin for the nitrate ion is such that when the resin has reached its operating capacity the nitrate ion will not be eluted from the resin and at no time will the nitrate concentration in the treated water be higher than that in the influent water unless the operating capacity of the resin is exceeded.

Regeneration of nitrate removal plants uses sodium chloride solution. These plants operate, and are regenerated, in the same way as a softening plant.

Dealkalisation, organic scavenging and nitrate removal are the three ion-exchange processes that have found a place in the production of carbonated soft drinks. Ion exchange has a principal advantage over coagulation and flocculation with regard to the space required for the plant. Not all factories will lend themselves to the installation of an ion-exchange treatment plant, but the advantages of such a system should mean that careful consideration is given to this process at the planning stage of a new production facility.

The principal advantages and disadvantages of ion exchange can be summarised as follows:

Advantages:
- Compact system; requires little floor space.
- Easily adaptable to variable influent water quality.
- Easily automated process; requires only occasional operator attendance.

Disadvantages:
- The feed water must not contain suspended solids or free chlorine. Free chlorine over 0.5 ppm Cl_2 will degrade or break down the resin structure causing quality and/or capacity problems.
- Regeneration effluents will contain concentrated salts. They may also have extremes of pH, which may necessitate treatment prior to discharge.

3.4 Reverse osmosis

If clarification is described as the 'traditional process' and ion exchange is described as the 'modern process' then reverse osmosis, which has not yet found a place in the beverage industry, must be called the *'future process'*.

As industry expands there will be a proportionate increase in the amount of water it uses. The increased demand will lead to a drop in the water table that will cause a rise in the concentration of dissolved solids. The increased demand will also lead to a higher discharge level and a greater re-use of water, which will lead to an increase in the mineral content of public supply waters. In some urbanised areas of Europe this is a fact today, and it will tend to increase over future decades. As the solids contents rise, and if we assume that they rise in proportion to the present mineral content in the water, then water-treatment plants will have to increase in size and chemical consumption in order to keep pace with demand. Statistics show that there is likely to be an increase in consumption of carbonated soft drinks within the same period, in which case water-treatment plants will have to be very large just to keep pace not only with the increased use of water but also with the increase in dissolved solids in the feed water. The potential increase in dissolved solids will give problems similar to those experienced in Middle Eastern countries, where brackish water has had to be treated to produce potable water. Reverse osmosis is a process that will treat water that has a high proportion of dissolved solids and produce a treated water than can be used directly or be further treated to meet the desired specifications set for carbonated beverage production.

Osmosis is a process that has been known for many years. If two solutions of differing densities and salt concentrations are separated by a permeable membrane, the less dense solution will flow through the membrane and will dilute the concentrated solution until an equilibrium is reached. If a pressure is placed on the concentrated solution then the flow will be in the opposite direction. The membrane would be capable of holding back the dissolved ions with the result that the denser solution will become more concentrated. This is reverse osmosis. In the process the less dense water produced will be known as the permeate and the concentrated water will be the reject. By use of this process it is possible to reduce the mineral content in the permeate by about 90%. If the feed water were to have a mineral concentration of 2000 ppm the permeate will have a concentration of 200 ppm, which will usually meet most of the specifications set by the major manufacturers. It is possible by design of the reverse osmosis system to treat a water with higher salt concentrations, and in some cases this will improve the operating efficiency of the equipment. The membranes used in the reverse osmosis systems usually fall into one of three categories: viz. hollow fibre, spiral wound or tubular. Because of the material of construction and, in some cases, the pressure used in operation, fouling is sometimes a problem that will necessitate pre-treatment to ensure constant operation with respect to quality and quantity of permeate produced. The type of fouling will be either inorganic or organic and the type of pre-treatment will depend on the exact type of the fouling compounds. Pre-treatment will vary from simple chemical dosing to complex treatment that may include clarification and filtration. The amount of reject water that will be discharged to drain will depend on the concentration of dissolved solids in the feed water and can be as high as 50% of the feed-water flow. This would mean that to produce $1\,m^3$ of permeate the feed water would have to be $2\,m^3$ and there would be a reject flow of $1\,m^3$. This is an extreme case and, in general, recoveries of 80% plus are not uncommon. There is also a temperature effect upon the operation of a reverse osmosis plant. The optimum operating temperature is 20°C and for every drop of 1°C there would be a drop of about 3% in operating efficiency.

Reverse osmosis plants are very compact and, for comparison, a plant capable of treating 30 m^3/h would require a floor space of about 20 m^2. This would be compared to the floor spaces required for a clarification plant and an ion-exchange plant. With the need to use space as economically as possible, reverse osmosis plants will certainly follow this trend. Reverse osmosis has already found an application in the treatment of water for beverage production. During the times of drought, polysaccharides are known to develop in reservoirs and be fed into the public mains supply. The presence of these compounds in a carbonated beverage will cause precipitation while the product is on the shelf and will render the drink unsaleable. It has been found that these compounds remain unaffected by clarification and while ion-exchange treatment only effects partial removal, reverse osmosis has been proved to be the only treatment process that will effectively remove this type of compound. The occurrence of polysaccharides in raw water is rare in temperate climates, except in cases of drought, but it may be a more significant problem in warmer and drier climes.

Following reverse osmosis treatment, the permeate water will still require sterilisation before it can be used for beverage production. In most cases sterilisation will be carried out by the conventional system of chlorine dosing, storage and dechlorination.

3.5 Alternative sterilisation methods

The sterilisation of water by chlorine is the most widely used method in the beverage production industry. Free chlorine does have some problems attached to its use with respect to the safety requirements necessary during handling and the incubation time and storage facilities necessary to ensure full sterilisation. If the two-hour storage time could be dispensed with, eliminating the need to install a large storage tank, while at the same time still offering a sterilised water, then there would be benefits with respect to the floor space occupied by a water-treatment plant. Other 'food and beverage' industries have experimented, and are now satisfied that ultraviolet sterilisation will produce a water that is acceptable for their process applications. The advantage of ultraviolet sterilisation is that it is immediate, and as it does not require a long incubation period it eliminates the necessity for a large storage tank. The wavelength of the ultraviolet rays can be selected so that not only are bacteria eliminated but naturally occurring organic matter can also be destroyed. The sterilisation reaction is very rapid, meaning that the plant is compact and may be sited close to the point of use, ensuring that the only water that is sterilised is that to be used for beverage production. This could mean that there will be more than one sterilisation point in a factory, but as ultraviolet units are very compact is not considered to be a problem. The ultraviolet sterilisation process requires very little operator attendance and the automatic controls of the system are such that if the lamp intensity drops below an acceptable level (which could mean that the contact time would have to be increased), an alarm condition will be indicated on the control panel local to the unit and perhaps repeated at a central point.

3.6 Future developments

It may appear that there has been a progressive movement from one process to another as water-treatment developments have been made. In some cases this is true, and old technology plants have been replaced with the newest and latest state-of-the-art equipment. The selection of a treatment process will be dependent on a number of factors, such as the source and analysis of the water to be treated, the location of the site and the budget available for new equipment.

All the processes described in this chapter will be used for many years to come. In practice, new developments have been slow to be accepted by the beverage industry; the high capital cost of water-treatment plant coupled with the rugged nature of the equipment means, generally, that there are long periods between investments. It is neither the nature nor the policy of the soft drinks industry to take risks with its principal ingredient, and for this reason it has taken time for process research to be carried out to ensure that any new treatment is suitable for the product.

The recent re-alignment of franchises in Europe, in addition to acquisitions and mergers, has resulted in new investment by the companies involved. This investment usually meant the construction of a new soft drinks factory or at least the refurbishment of an existing facility, bringing it up to the latest specification. New investment has coincided with the acceptance of new processes by the industry, which allows water-treatment plant to occupy less floor space, giving more room to other requirements; the improvement in computer technology means that automatic control of the equipment is simplified, requiring less manpower for operation.

Keeping this in mind, it is perhaps worthwhile speculating on the type of water treatment that will be available and used by the carbonated soft drinks manufacturer in the future. It has been indicated that a clarification plant, although capable of operating continuously, cannot automatically react to variations in input water quality and therefore is difficult to automate and requires operator attendance at frequent intervals. Ion-exchange dealkalisation is adaptable to the variations in the alkalinity content of the incoming water and the process is specific to the reduction of alkalinity which is the prime objective of the treatment plant. However, the ion-exchange unit has only a finite capacity and when it is exhausted the resin will need regeneration with dilute mineral acid. Continuous operation will require standby equipment while regeneration necessitates the storage of large volumes of acid.

Strong acids are not desirable commodities near a beverage plant (on grounds of health and safety) while bulk storage will require an independent facility to separate it from the main production area.

Reverse osmosis is beginning to show itself as worthy of acceptance as a treatment process for beverage production: it has proved to be adaptable in allowing the process to start and stop as required; it will remove particles and organic molecules that cannot be removed by other processes; and as it is a compact and easily automated process, it therefore fulfils all the requirements of the modern beverage facility. The reverse osmosis process is not specific in the removal of alkalinity, but this can be easily remedied by blending the permeate water with untreated water in such proportions that the optimum analysis is achieved. With more attention being paid to the quality of water used in beverage production, an adaptable plant is essential.

No new revolutionary water-treatment process is known to be in the development stage or likely to be available in the next decade. Present-day treatment processes will continue to be used well into the next century, and any developments that take place will be centred around the refinement, optimisation and automation of current processes. Since reverse osmosis is a clean process that does not require strong and potentially dangerous chemicals for its operation, is compact and is easily automated, it is highly probable that future beverage factories will install such a plant as the primary treatment process. The permeate will be blended with pre-treated incoming water to meet the operator's specification. Pre-treatment, such as micron filtration, will ensure that the blend water is of the highest quality. The mixed water will be stored adjacent to the plant but the storage volume will be limited to, approximately, a 15-minute operation as the start-up and attainment of quality water by the reverse osmosis plant will be immediate. The blended water will be fed directly into the production feed line and will be sterilised by ultraviolet radiation and filtered free of bacterial particles prior to its entry into the beverage processing plant.

The time of the adoption of this process scheme is not far away. With constant demand for a purer water and with the quality of the basic raw material being subject to variability due to industrialisation and general changing climatic conditions, soft drinks manufacturers will accept reverse osmosis as the system that meets their needs in every respect.

References

1. H.W. Houghton and D. McDonald in *Developments in Soft Drinks Technology*, Vol. 1, L.F. Green (ed.) Elsevier Applied Science Publishers, Barking (1978).
2. Dow Chemical Co., Technical data sheets for Dowex HCR-S(E) and Dowex CCR-3 (1989).
3. Rohm and Haas Ltd, Technical data sheets for Amberlite IR 120 and Amberlite IRC 84 (now Amberlite 76) (1989).
4. Rohm and Hass Ltd, Technical data sheet for Amberlite IRA 458 (1989).
5. Rohm and Hass Ltd, *Nitrate Removal: The performance of Imac HP441 and Imac HP555* (1987).

4 Carbohydrate sugars

P.M. BEESLEY

4.1 Introduction

Carbohydrate sugars considered in this chapter are those based on sucrose derived from sugar beet and sugar cane and those derived from starch, e.g. glucose syrups in their various forms.

4.1.1 *History*

It would appear that the association between sugars and 'carbonated soft drinks' first occurred in the seventeenth century when lemon juices containing naturally present sugars were added to spring waters. Sugar was also used during the travels of Captain Cook[1] as an addition to lemon juice in order to preserve the juice for long periods. From these origins, the use of sugar to improve preservation and to improve the taste acceptability has rapidly increased. In the UK, the soft drinks trade sector is the largest market for carbohydrate sugars, representing in 1986/87 approximately 21% of the total market. The amount of carbohydrate sugars used in 1986 was approximately 210 000 tonnes, the majority of which was in the form of sucrose-based products. Future usage of carbohydrate sugars is expected[2] to increase to approximately 292 000 tonnes in 1995, even though the percentage share of 'low calorie' carbonated soft drinks, which are entirely artificially sweetened, is expected to reach approximately 18% by the same year. Because of the present European quota arrangements for sucrose and high-fructose glucose syrup, the relative percentage share of these products as used in UK-produced carbonated soft drinks is not expected to alter dramatically. By comparison, a dramatic swing to the use of high-fructose glucose syrup has occurred in the United States and certain South American countries for reasons of pricing.

4.2 Carbohydrate sugars

Carbohydrate sugars used in carbonated soft drinks can be divided into those in a dry, granular form (e.g. granulated sugar (sucrose)) and those in a liquid or syrup form (e.g. liquid sugar – which is a solution of sucrose in water – and glucose-type syrup produced from maize) or, in certain circumstances, wheat (e.g. glucose syrup or high-fructose glucose syrup).

In the UK, a type of granulated sugar is available (mineral water sugar) which, by virtue of its name, could be viewed as the sugar used for production of mineral waters or soft drinks. Although this sugar type is eminently suitable for the preparation of carbonated beverages, its name and indicated use are historical and relate to past times when the quality of standard granulated was deemed unsatisfactory. Nowadays, standard granulated from both beet and cane is of a substantially higher quality and therefore adequate for the production of carbonated beverages. Mineral water sugar, which has undergone additional purification stages to reduce the already minute levels of impurities present, is now used for specific pharmaceutical and crystallisation processes.

4.2.1 Granulated sugar

Granulated sugar is a dry, crystallised disaccharide extracted from sugar beet and sugar cane called sucrose; commercially, however, it is referred to by numerous names – standard granulated sugar, dry sugar or granulated sugar. In Europe and the USA, the quality of granulated sugar is independent of source, being an extremely high purity organic product. However, in some less developed countries the quality of sugar is such that further purification is undertaken by the end-user. This normally takes the form of filtration and carbon treatment of a prepared aqueous solution.

Packaging. Granulated sugar is supplied in various weight packages, including sacks (25 and 50 kg), 1 tonne flexible containers and bulk tankers. The choice of supply depends on the availability and distribution systems in operation in the country concerned.

When a total choice exists, bulk tanker deliveries are most widely received owing to their advantages over sacked products. These advantages are related to convenience of handling, reduced storage space, reduced labour cost and a decrease in sugar contamination and loss, associated with sack opening and emptying. For those locations that require the convenience of 'bulk' deliveries but do not wish to incur the necessary cost of installing bulk tanker reception facilities, 1 tonne flexible containers provide a useful compromise.

Depending on the country, the price of dry sugars, in particular bulk granulated sugar, can be below the price of an equivalent quantity of sugar in commercially available liquid form (i.e. 67% w/w aqueous solution). This price differential can be sufficiently great to be cost effective for soft drinks manufacturers who currently receive liquid sugar to change to dry sugar and dissolve on their own premises (see Section 4.6).

Manufacture. After extracting the sucrose from either sugar beet or sugar cane, the juices so produced containing both sugars and non-sugars are subjected to a series of purification steps which remove the non-sugars and progressively concentrate the sucrose solution. These processes involve precipitation and absorption stages coupled with numerous filtration and evaporation systems. The final purification step involves crystallisation of the pure sucrose crystals in vacuum pans. The resulting mixture of sugar crystals and syrup, known as 'massecuite', is transferred to centrifugal machines where the syrup is spun off and the remaining thin surface film adhering to the sugar crystals is removed by washing with water. The damp sugar crystals are then dried to a moisture content of about 0.02% w/w using a hot-air granulator before being cooled and stored in temperature- and humidity-controlled sugar silos.

The produced granulated sugar is not screened to a particular particle size distribution; the range of crystal sizes is close to the Gaussian or 'Normal Distribution' pattern of spread, and is controlled during the vacuum pan crystallisation stage by skilled operators. However, to remove any over-large lumps due to agglomeration of sugar crystals, a coarse sieve is normally incorporated in the system conveying sugar to and from the storage silos. In addition, a comprehensive system of magnets is employed to protect the final product from chance contamination.

4.2.2 *Liquid sugar*

In the UK, commercially available liquid sugar comprises an aqueous solution of sucrose at a saturated concentration of 67% w/w (67° Brix) at 20°C.

Type of delivery. Although available in drum form from some manufacturers, the majority of soft drinks manufacturers receive their supplies by specially designed bulk road tankers capable of transporting up to 3000 gallons of liquid sugar.

Normally, liquid sugar is delivered at a temperature within the range 45–60°C when its viscosity is 48.4 and 23.9 Centipoise (cP), respectively. However, it is possible for soft drinks manufacturers to receive deliveries that have been cooled to a maximum temperature of 30°C if higher temperatures would cause problems during their filling operations. Even at this temperature the liquid sugar remains relatively free-flowing, having a viscosity of about 114 cP (at 30°C).

Manufacture. Certain types of liquid sugar are produced without undertaking a sugar crystallisation stage. These 'drawn-off' syrups from the manufacturing process are, however, of a colour and flavour unsuitable, without further treatment, for most soft drinks manufacturers.

Because of the high quality requirement of the soft drinks industry, liquid sugar supplies are normally produced by dissolving high-quality granulated sugar in water. This dissolution process is carried out at an elevated temperature to reduce the level of any microflora that may be present. The produced syrup is then normally filtered through a filter-aid based system. Carbon filtration and de-ionisation, using resin columns, are incorporated by some manufacturers in the liquid sugar production unit if the quality of the granulated sugar is insufficient to produce the necessary standard of liquid product.

Treatment of the liquid sugar with ultraviolet radiation is generally undertaken to minimise further the presence of any micro-organisms. This is carried out using an in-line system whereby the liquid sugar passes through a number of narrow annuli which ensures a short path length for the ultraviolet radiation to pass, which is necessary for high absorbing liquids.

Temperature adjustment is then carried out, if necessary, by plate-heat exchangers before final filtration as the liquid sugar is loaded into despatch tankers.

4.2.3 *Glucose syrup: high-fructose syrup*

Glucose syrups and high-fructose syrups can be used as a complete, but more usually partial, replacement of sucrose in the majority of carbonated soft drinks. Used in conjunction with sucrose, syrups with appropriate fructose contents enable sweetness levels to be adjusted according to specific market preferences.

Complete sucrose replacement in carbonated soft drinks has occurred in certain non-European countries with 55% fructose syrup. However, this is not the case within Europe since high-fructose syrup production is governed by the EEC quota system.

Glucose syrups of various types are used exclusively in certain 'health' type soft drinks.

Type of delivery. Glucose syrups, although available in drum containers, are generally supplied in specially designed road tankers. The syrups normally incorporated are: demineralised 95DE (dextrose equivalent, see 'Manufacture' below) syrup; 63DE syrup; high-fructose syrup of 42% fructose; and various blends of the above, with and without sucrose to produce the required level of sweetness, viscosity and mouth feel.

The temperature of delivered glucose syrup depends on the specific type involved. 95DE is delivered at a minimum temperature of 50 °C because of the possibility of crystallisation below that temperature, 63DE at a temperature of 40–45 °C and high-fructose at 28–30 °C.

It is important to note that 63DE glucose syrup, in particular, will increase its solution colour on storage. Consequently, a demineralised form is necessary if the product is to be stored for up to three weeks.

Manufacture. Glucose syrups are manufactured[3] by the acid and enzyme hydrolysis of starch, normally of maize or wheat origin. This treatment breaks down the long-chain carbohydrate molecules into a spectrum of simple and higher sugars. If this conversion is allowed to continue, the end products are dextrose and maltose. However, under controlled conditions syrups of defined composition can be produced.

The extent of hydrolysis is defined in terms of 'dextrose equivalent' (DE); this figure represents the total reducing sugar value of the syrup expressed as a percentage of the reducing sugar value of pure dextrose, calculated on a dry basis.

Additional enzyme treatment enables the dextrose content of syrup to be converted to fructose up to 42%, giving the syrup a greater sweetening power. This level can be further increased by chromatographic enrichment techniques.

4.3 Quality

Carbohydrate sugars are used in carbonated soft drinks not only to provide a level of sweetness to balance flavours and acids present but also to provide mouth feel to the product by increasing its viscosity and dissolved solids content. They also provide an easily metabolised source of energy, a fact utilised in the marketing of certain 'health' related soft drinks. The carbohydrate sugars incorporated therefore require to be of a consistently high quality in their physicochemical properties.

4.3.1 Trade requirement

Carbohydrate sugars are generally supplied to quality specifications agreed with soft drinks manufacturers and can vary depending on the products into which they are to be incorporated and the particular processing techniques involved at manufacturers' premises.

Extraneous matter. The levels of extraneous matter are of particular concern because of (a) their effect on the appearance of the final products and (b) the possibility of loss of carbonation[4] if excessive sites of nucleation are present.

All liquid products supplied (such as liquid sugars and the various types of glucose syrup) can, by virtue of their physical characteristics, be filtered before despatch thus minimising the above problems.

Granulated sugars, although prepared from syrups that have undergone numerous filtration processes, can contain levels of extraneous or water-insoluble matter depending on the country of origin. In the UK these levels are extremely low, typically of the order 6 mg/kg sugar, as determined by a membrane filtration method.[5]

Every precaution is taken to minimise the presence of this water-insoluble matter, which can consist of filter aid or calcium salts from the manufacturing process. It may therefore be prudent for soft drinks manufacturers to consider the filtration of liquids prepared from granulated sugars using a filtration system of adequate porosity commensurate with realistic filtration rates.

Colour in solution. Especially for those carbonated soft drinks which are clear in appearance, carbohydrate sugars of a low solution colour are a prerequisite. The 'National Soft Drink Association'[6] standards of the United States, which in general has been adopted by other countries, state a maximum level of 35 reference basis units for the solution colour of both granulated and liquid sugars. In the UK this standard is obtainable, but in other less-developed countries further treatment of sugars after receipt is necessary in order to conform to this level. The treatment used is normally that of filtration through plate-and-frame filters incorporating carbon sheets.

The production specifications of some international soft drinks companies list the use of these carbon filters for all locations in order to cover the world-wide production of soft drinks, with carbohydrate sugars of differing quality parameters. In those countries whose carbohydrate sugar quality can be guaranteed to be of the highest order, the use of these filters may be negotiable.

Acid floc. Floc formation in carbonated soft drinks is a phenomenon observed in acid solutions and normally appears as a white precipitate. This precipitate can take the form of loosely aggregated particles floating within the solution which, in extreme cases, may take on the appearance of well teased-out cotton wool or as more dense particles that sink through the solution. On shaking the solution the floc usually disappears as the weak forces between the molecules are disrupted.

Floc[7,8] has been shown to be associated with sugars (both beet and cane) and also with polysaccharides present in water[9] as a result of the growth of algae and chemicals from water treatment.

Sugar floc is normally associated with sugars of low quality and these have caused final product problems in some countries. For this reason, the National Soft Drink Association[6] standards include a method for the evaluation of floc-producing substances. This is the Spreckles Qualitative Floc Test Procedure.

Since the National Soft Drink Association makes reference to the testing of sugars from beet for floc evaluation, the sugar beet industry in the UK has made extensive efforts over the years to ensure that all granulated sugars and liquid sugars are free of any substances that could lead to acid floc.

Microbiological. The low moisture content, typically 0.02% w/w, of granulated sugar coupled with its high purity minimises the possibility of microbiological degradation. In solution, carbohydrate sugars generate high osmotic pressure[10] which is a major factor in protecting concentrated solutions from microbiological contamination. However, if granulated sugar is allowed to become damp and syrups are diluted, then an increased risk of microbiological degradation occurs. This can happen if there are deficiences in storage conditions (see Section 4.5).

In the UK, carbohydrate sugar manufacturers' target levels for mesophilic bacteria, yeasts and moulds in liquid products are the standards for 'Bottlers' liquid sugars as defined by the National Soft Drink Association.[6] By comparison the target levels for granulated sugar are twice the National Soft Drink Association standards. This difference relates to the fact that, by virtue of their physical form, liquid products can be treated by filtration and ultraviolet radiation to reduce further any micro-organisms present.

An additional protection operated by some sugar manufacturers is to control[11] the pH of the liquid products to about 8 to reduce even further the chance of spoilage.

In practice, carbohydrate sugar manufacturers undertake regular sanitising programmes of their production and storage systems coupled with comprehensive quality assurance and control, not only to ensure minimal levels of spoilage organisms, but also to ensure freedom from pathogenic bacteria.

4.3.2 *Quality assurance management*

Soft drinks manufacturers in the UK have for some time required guaranteed quality raw materials especially as they introduce Supplier Assurance Systems and Good Manufacturing Practices. Carbohydrate sugar suppliers have responded positively to these requirements by instigating their own quality assurance systems, including certification to British Standard 5750 and the International Standard ISO 9000.

These quality management systems involve documented procedures and instructions which are self and externally audited, concentrating on all key activities affecting quality of products and service.

4.3.3 *Sugar analysis*

The methods generally used in the analysis of sucrose products are those defined by 'The International Commission for Uniform Methods of Sugar Analysis', normally referred to as ICUMSA. Many of their methods[12] have been adopted by the European Economic Community and the Codex Alimentarius Commission. The analytical methods used by glucose syrup manufacturers are generally those standardised by the Corn Refiners Association.[13]

4.4 Transportation and delivery

The mode of transportation of carbohydrate sugars depends on the type of sugar involved and also on the sophistication of the transport system and reception facilities available in any particular country. Less-developed countries normally rely on granulated sugar contained in bags. Ideally this sugar should, for ease of handling, be palletised and transported within covered wagons to reduce contamination of the packaging material which could eventually contaminate the contained sugar.

4.4.1 *Bulk delivery of granulated sugar*

Demand by soft drinks manufacturers for bulk deliveries has grown rapidly over the past years since their use offers a number of important advantages over bagged supplies (Figure 4.1). The advantages relate to: savings in storage space; savings in manpower, both at the reception point and during internal redistribution; improved hygiene and the possibility of automated processing to which the system lends itself.

Bulk deliveries are made in road tankers which are specifically designed to maximise efficient discharge of the sugar and, by virtue of the materials used for construction, to minimise product contamination. Additionally they are designed for ease of internal cleaning, which should be carried out at regular intervals.

Figure 4.1 Pneumatic discharge of bulk granulated sugar. (Photograph taken with kind permission of Coca-Cola and Schweppes Beverages Ltd.)

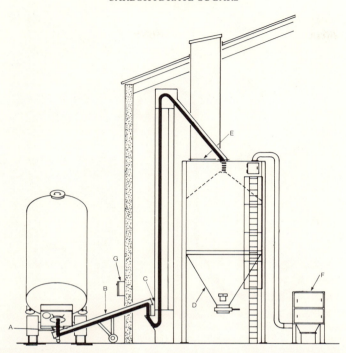

Figure 4.2 Reception facilities for gravity discharge of granulated sugar. A, receiving hopper into which granulated sugar is discharged from the tanker; B, inclined conveyor; C, transfer bin and vertical conveyor; D, storage silo with protective internal coating fitted with inspection door, rodding access adjacent to the outlet of the silo, maximum-level indicator connected to an alarm at the delivery point to prevent over-filling and an excess-pressure valve; E, explosion-relief panel vented to atmosphere; F, fan-assisted dust-collection system; G, indicator panel connected to dust-extraction system and high-level probe.

Granulated sugar can be off-loaded from the road tanker either by a gravity mode or by a pneumatic technique.

Gravity discharge (Figure 4.2) involves elevation of the tanker to allow the sugar to fall from the rear port of the tanker into a receiving hopper and then via a horizontal or inclined conveyor to a transfer bin before being vertically conveyed to a storage silo. Discharge rates as high as 30 t/h can be obtained but can be readily controlled by the tanker driver to suit the conveying capacity of the reception equipment.

Pneumatic discharge (Figure 4.3) involves mixing air, normally from the tankers own air blower, with sugar from the discharge point while the tanker is elevated and conveying the sugar crystal/air mix via a flexible hose to the reception intake pipe. On a world-wide basis a large variation exists in the air volumes and sugar-to-air ratios used, resulting in variable off-loading rates. In the UK the discharge parameters used normally result in a delivery rate of between 10 and 15 t/h.

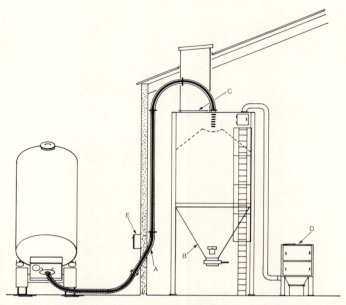

Figure 4.3 Reception facilities for pneumatic unloading of granulated sugar. A, sugar-intake pipe with a smooth and continuous surface; B, storage silo with protective internal coating fitted with inspection door, rodding access adjacent to the outlet of the silo, maximum-level indicator connected to an alarm at the delivery point to prevent over-filling and an excess-pressure valve; C, explosion-relief panel vented to atmosphere; D, fan-assisted dust-collection system; E, indicator panel connected to dust-extraction system and high-level probe.

Gravity discharge involves the provision of mechanical conveying equipment with the result that the initial cost of this system is usually higher than a pneumatic system. Additionally, the running cost of a gravity system is higher because of the increased power and higher maintenance costs required. However, a gravity system normally results in less attrition of the sugar crystal than does pneumatic discharge, especially if a low sugar-to-air ratio is used.

The tanker blower required for pneumatic discharge is generally powered from the tanker engine and can, in certain circumstances, result in problems of noise. This can be virtually eliminated by the installation of a 'land-based' air blower suitably insulated.

With pneumatic deliveries an adequate rate of discharge and a minimisation of sugar crystal attrition is of importance to the user, and are both related to the design of the sugar intake pipe system. The internal surface should be smooth and continuous. The total length should be kept to a minimum and preferably not contain more than three bends which should have a minimum radius of 4 ft (for a 4-in bore pipe). The first section of any pipe, other than very short ones, must be vertical and extend to the maximum height required without involving any 'double' lift sections.

Figure 4.4 Discharge of liquid sugar. (Photograph taken with kind permission of Britvic Soft Drinks Ltd.)

4.4.2 *Bulk delivery of liquid carbohydrate sugars*

The supply of liquid sugars and the various types of glucose syrups (Figure 4.4) is made in specifically designed road tankers. The tank barrels of these vehicles are normally of stainless-steel construction to reduce product contamination to a minimum and to aid internal cleaning. Associated pipework, valves and pumps are of hygienic construction, comprising food-grade material.

Normally liquid tankers are insulated to ensure, as far as possible, that the temperature of the product at the time of delivery meets the user's requirement.

Off-loading is facilitated by positive displacement pumps which are driven by the vehicle, or in some cases by an on-site source of electricity. Delivery rates of about 80 gallons per minute can be achieved but this is dependent on the design of the user's intake pipe system and on the viscosity of the product involved (Figure 4.5).

The majority of the liquid carbohydrate sugar manufacturers clean and sanitise the internal surfaces, pipework and pumps between each delivery. This is normally carried out by the use of steam after cleaning with water to remove all traces of sugar products.

48   CARBONATED SOFT DRINKS

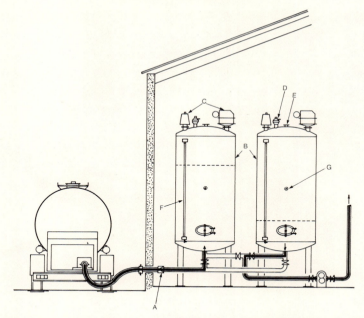

Figure 4.5 Reception facilities for liquid carbohydrate sugars. A, intake pipe connected to the storage tank with a control valve at the delivery point; B, twin storage tank system; C, air-ventilation system designed to prevent microbiological contamination; D, pressure/vacuum relief valve; E, provision to sanitise equipment via sprayballs; F, content-level indicator; G, provision for steam sanitisation.

4.4.3 *Security of delivery*

Owing to a growing concern about the possibility of product adulteration during transportation, a number of carbohydrate sugar manufacturers have introduced a system of security sealing. This involves the placement of numbered seals by key personnel on each entry manway or port. On arrival at the user's premises these seals are inspected to ensure they are intact before being broken under the supervision of local management. The numbers of these seals can also be compared to the log of those applied at the despatching location. This system eliminates the possibility of access to the load for malicious purposes during transportation.

4.5 Storage

The successful storage of granulated sugar depends on controlling the relative humidity and temperature of the storage environment; with liquid products, sanitisation and the minimisation of condensation within storage vessels are the key factors.

4.5.1 *Granulated sugar in bags*

Granulated sugar can, under conditions of high humidity, absorb moisture from the environment forming a thin layer of syrup on the sugar crystals. If the sugar is then subjected to low humidity conditions, crystallisation of the syrup film can result at the point of contact, resulting in the formation of sugar crystal agglomerates. Continuation of this process can eventually lead to 'caking' (lump formation) of the sugar.

The equilibrium relative humidity of granulated sugar can vary depending on its purity, but for high-quality sugars it is generally above 70%. The relative humidity within the storage area should therefore be maintained below 65% within a range ideally of $\pm 3\%$.

Since the migration of any moisture within granulated sugar, which can exacerbate 'caking' problems, is dependent on temperature, the storage area should also be controlled to a constant temperature $\pm 5\,°C$ within any 24-hour period. The actual temperature is not so critical and should be between 10 and 30 °C. In practice, a temperature slightly above ambient is the best choice especially if humidity control is not possible.

Additionally, granulated sugar in bags should be stored in pest-free environments in areas that are free from draughts and direct sunlight and be incorporated in a controlled stock rotation system.

4.5.2 *Granulated sugar in bulk*

The storage silo capacity used to receive the gravity or pneumatically delivered granulated sugar can vary depending, generally, on the load-carrying capacity of the delivery tanker. In the UK, silos of a minimum capacity of 40 tonnes up to about 100 tonnes are now usually installed. The actual size is a compromise between security of supply and an adequate turnover of stocks.

Many different designs are available but generally they are constructed of mild steel, the internal surfaces being coated with a protective layer, such as an epoxy resin paint, to prevent rust and scale contaminating the sugar. The outlet cone should have a minimum slope angle of 55° to prevent bridging of the sugar and as an added precaution rodding access should be provided.

The silo (Figures 4.2 and 4.3) should be fitted with a manway with an external ladder, and facilities for air venting and dust collection. It is necessary to fit an excess pressure relief panel to conform to the regulations prevailing in the country concerned. In normal circumstances, in the UK it is necessary to vent this panel to atmosphere.

Because of the possibility of sugar 'caking' (see Section 4.5.1) the siting of a sugar silo is of paramount importance, a dry atmosphere and minimum temperature variation being required.

Silos can be internally or externally situated. For those inside the factory building it is not advisable to site the silo in such a way that one side is subjected to a temperature variation caused by, for example, adjacent steam or hot water pipes and radiators or any other major source of space heating. Cold draughts from doors sited close to the silo can cause similar problems.

Uneven temperatures of this nature can cause the small moisture content of the sugar to migrate to the coldest part of the mass, resulting in a tendency for the sugar to 'cake'. In some circumstances the silo may require lagging or protecting by an insulated and temperature-controlled housing.

Externally placed silos are normally of special construction, being protected from temperature variations by a surrounding annular air space through which temperature-controlled air passes. Certain designs also incorporate humidity control systems for the air within the silo.

The presence of sugar dust in the granulated sugar, as a result of pneumatic delivery, can have a considerable bearing on its susceptibility to 'caking'. It is therefore imperative that the sugar intake pipe is correctly designed (see Section 4.4.1) to minimise crystal attrition and that a free-standing fan-assisted dust extraction unit is used which removes the produced sugar-dust from within the silo.

In order to ensure trouble-free operation of a bulk silo system, it is essential to adhere to a planned maintenance procedure. The silo should be completely emptied at least once a year. Dust on the internal surfaces should be cleared and any agglomerated sugar removed.

After each delivery, any dust collected in the dust units should be removed and a check made on, at least, a three-monthly basis to ensure that trunking between the silo and the dust unit is clear. The unit itself should be serviced on a periodic basis, dependent on the amount of sugar handled, by the manufacturers to ensure its efficient operation.

4.5.3 *Liquid carbohydrate sugars*

The storage tanks (Figure 4.5) used to hold the liquid sugar or glucose syrup should be of a capacity to receive the maximum delivered load allowed in the country of operation. The tanks can be manufactured from stainless steel, mild steel with a suitable lining or resin-bonded glass fibre. All internal surfaces should be as smooth as possible and all corners rounded. The tanks should be capable of being sanitised by steam or commercially available sterilants. Facilities should be included for the prevention of microbiological infection and the tanks should be capable of being drained completely. It is also advisable to fit level indicators and pressure/vacuum relief valves.

Liquid products may be subject to microbiological deterioration and therefore their storage period should be minimised according to the manufacturers' recommendations.

The growth of micro-organisms in commercially available liquid sugars and glucose is slow, but should condensation take place in the top of the storage tank, a diluted surface layer will be created in which micro-organisms may thrive. This deterioration can be substantially reduced by the following techniques.

(i) Fitting a fan-assisted ventilation system equipped with absolute filters. The filters retain airborne organisms and the fans reduce condensation.
(ii) Storing the liquid product at an elevated temperature. This may, however, lead to colour formation in the liquid if it is stored for more than a few days.
(iii) Circulating the liquid through special ultraviolet lamps. This will reduce substantially any organisms present.

Maintenance of liquid storage plant is essential to ensure trouble-free operation. The system should be kept clean and free from micro-organisms; pumps, meters and valves should be serviced to the recommendations of the manufacturers. Particular attention should be given every month to those valves that are not in constant use, such as drain valves and valves on sight glasses.

On a regular basis, the operation of the various safety valves and other ancillary equipment should be checked. It should be noted that it is possible for a relief valve to become blocked with crystallised product resulting from accidental overfilling of a storage tank.

If ultraviolet lights are incorporated in the system they should be changed after the manufacturer's recommended time interval. The emission of visible light is not indicative of effective radiation. Where special ultraviolet lights are incorporated inside pipelines, they should be withdrawn every three months and any trace of caramel film removed from the quartz envelope that surrounds the light.

Sanitisation of the storage system to prevent a build up of micro-organisms can be carried out using steam or chemical sterilants. With certain types of plastic storage tanks, steam sanitisation is not recommended. Where a chemical sterilant is used, built-in spray balls are recommended.

The tanker delivery pipe and storage tank should be sanitised after each delivery unless the rate of usage is such that weekly sanitisation is found to be sufficient.

Where steam is used to sanitise the delivery pipe it is recommended that the pipe be sanitised immediately after delivery, repeating the process immediately prior to the next delivery. Alternatively, the pipe, if it is a bottom entry pipe, can be washed with hot water immediately after the delivery to remove any traces of sugar and then filled with a suitable sterilant solution. If steam is used to sanitise the storage tanks, care should be taken to isolate any associated equipment likely to suffer as a result.

Sanitisation of storage tanks using chemical sterilants is best achieved by the use of a spray ball assembly. Hot water should be sprayed into the tank to wash away the remaining traces of carbohydrate sugar. The sterilant solution is then sprayed and recycled to provide the manufacturer's recommended contact time and, finally, the tank is rinsed with cold water to ensure complete removal of the sterilant.

It should be noted that with chemical sterilants it is imperative that all traces of carbohydrate sugar be removed from the surfaces to be sanitised otherwise the sterilant will act only on the sugar film and not on the tank surface.

4.6 On-site dissolving of granulated sugar

Although liquid sugars are eminently suitable for internal transportation and metering within a soft drinks factory the high cost of moving large quantities of water from supplier to user may be reflected in the cost of the product. Therefore, it may be of financial benefit to the user to produce liquid sugar on-site from deliveries of bulk granulated sugar. In some areas this change from liquid to granulated sugar is growing in momentum, especially with the larger users of sucrose-based products.

There are several ways of achieving on-site production,[14] the simplest being a batch system involving a sugar silo, simple mixer and a storage tank. For the larger user more sophisticated continuous or high-capacity batch systems are available with output tailored to individual needs.

4.6.1 *Batch dissolving*

In the simple batch system granulated sugar from a storage silo feeds via a screw conveyor a small mixing tank, normally of stainless steel construction. This tank, which is also supplied with hot water, may be mounted on load cells making possible a considerable amount of automation. The resulting liquid sugar is then pumped from the mixing tank to a storage vessel.

The temperature of the hot water used is dependent on the Brix required of the produced liquid sugar and also on an adequate dissolving rate. For Brix levels approaching 67° it is essential that water of about 80 °C is available; for levels around 62 °Brix water approaching a temperature of 20 °C can be used.

It is, however, important to note that lower Brix solutions by virtue of their reduced osmotic pressure are less stable from a microbiological viewpoint. Greater emphasis should therefore be placed on minimising storage time and on in-house quality assurance procedures.

4.6.2 *Continuous dissolving*

The continuous production of liquid sugar[15] is of particular value to those soft drinks manufacturers who use large quantities of sugar consistently through-

CARBOHYDRATE SUGARS

out a long production period. Plant for continuous dissolving is available from a number of manufacturers as unit packages, while specifically designed installations are also obtainable from the large food engineering companies.

Although each type varies slightly in its mode of operation, in essence the granulated sugar from a storage silo and water are fed continuously into a pre-dissolving vessel by feeding mechanisms that are adjusted to provide a Brix level higher than the required final level. The resulting sugar suspension is then pumped to other 'dissolving pipes' where dissolution is effected by a turbulence effect and heat. Part of this flow is recirculated to the pre-dissolving vessel while a portion is fed after de-aeration to a Brix control unit where its concentration is constantly measured (usually by refractometer or densimeter) and adjusted by water addition to the required Brix level.

In those countries where commercially available granulated sugar is too highly coloured to be used directly into carbonated soft drinks, the continuous dissolving plant can be modified to incorporate a decolorisation stage by means of active carbon.

4.6.3 *High-capacity dissolving*

The term high-capacity sugar dissolving refers to the system whereby a tanker load of granulated sugar is directly delivered pneumatically into an enclosed tank containing a predetermined amount of water, the need for an on-site dry sugar storage silo being unnecessary (Figure 4.6).

Figure 4.6 High capacity sugar dissolving plant. (Photograph taken with kind permission of Waters and Robson Ltd, Morpeth.)

The dissolving tanks can be situated internally or externally and can be of vertical or horizontal construction depending on the space available at the soft drinks manufacturer's premises. Dust extraction or containment units are fitted to the dissolving tanks to prevent the small amount of sugar dust created during pneumatic off-loading entering the factory environment.

Dissolution of the sugar crystals is accomplished by either agitators or jet-pump mixers. This latter mode of mixing involves the use of a circulation pump which draws off liquid from the dissolving vessel and reintroduces it via a diffusion jet. The jet of liquid flowing out of the diffusion nozzle at high speed generates reduced pressure at the inlet of the diffuser resulting in considerable turbulence which aids homogeneity and rapid mixing. The main advantages of the jet-mix system are a reduction in dissolving time and a minimisation of entrained air.

As with the other types of dissolving systems the temperature of the dissolving water relates to the final Brix that can realistically be attained. Usually, levels around 62 °Brix are required and this can be achieved in an acceptable time with water at a temperature of about 20 °C.

However, the high-capacity dissolving tank can be fitted with a steam jacket to maintain a high temperature of dissolving water if elevated Brix levels are required.

The operation of this type of dissolving system requires a quantity of water to be added to the vessel before the addition of the granulated sugar such that, after dissolution, a Brix slightly higher than required is produced. After assessing the precise concentration of the liquid sugar produced, dilution with water is then carried out to give the required level. It is prudent to initially 'over-Brix' in this manner because of the practical difficulties involved in adding extra sugar if too dilute a liquid sugar is produced.

It is important to stress that agitators and jet-mix pumps should be in operation before the granulated sugar enters the dissolving vessel and that the vessel should be sanitised between each mixing to ensure freedom from microorganisms.

References

1. 'Soft drinks through the ages', *Soft Drinks Management Intl*, September (1988).
2. 'Non-sucrose sweeteners, major West European markets, update 1986–1990–1995', GIRA S.A. Strategic Agri-Business Consultants, 1239 Collex, Geneva, Switzerland – Report, September (1987).
3. D. Howling, 'The general science and technology of glucose syrups' in *Sugar: Science & Technology*, eds G.G. Birch and K.J. Parker, Elsevier Applied Science Publishers, Barking.
4. 'Loss of carbonation' in *Manufacture and Analysis Carbonated Beverages*, ed. Morris B. Jacobs, Chemical Publishing, Chapter 10, item 7.
5. D. Hibbert and R.T. Phillipson, 'The determination of extraneous, water-insoluble matter in white sugars using membrane filters', *International Sugar Journal*, February (1966) 39.

6. *Quality Specifications and Test Procedures for 'Bottlers': Granulated and Liquid Sugar*, National Soft Drink Association, 1101 Sixteenth Street, N.W. Washington DC 20036, June (1975).
7. M.A. Clarke, E.J. Roberts, M.A. Godsmall and F.G. Carpenter, 'Beverage floc and cane sugar', *Proc. Int. Soc. Sugar Cane Technol., 16th Congress, 1977*, **3** (1978) 2587.
8. F.G. Eis, L.W. Clark, R.A. McGinnis and P.W. Alston, 'Floc in carbonated beverages', *Ind. Eng. Chem* **44** (1952) 2844.
9. 'Snowstorm in a lemonade bottle', *Soft Drinks*, January (1984).
10. W.M. Nicol, 'Sucrose and food technology', *Sugar: Science and Technology*, eds G.G. Birch and K.J. Parker, Elsevier Applied Science Publishers, Barking.
11. D.D. Leethan and F.G. Eis, 'Control of yeasts in sucrose syrup by control of syrup pH', *J. Amer. Soc. Sugar Beet Technol.* **12** (1963) 359.
12. F. Schneider (ed.), *Sugar Analysis ICUMSA Methods*, ICUMSA Publications Dept, c/o British Sugar plc Research Laboratories, Colney Lane, Colney, Norwich NR4 7UB.
13. *Standard Analytical Methods of the Member Companies of the Corn Industries Research Foundation*, Corn Refiners Association Inc, 1001 Connecticut Avenue, N.W. Washington DC, 20036.
14. 'Sugar dissolving made easy', *Soft Drink Trade J.* **33** (No. 11) November (1979).
15. 'Continuous system produces liquid sugar', *Food Engineering Intl*, April (1978).

5 High-intensity sweeteners

K. O'DONNELL

5.1 Introduction

The low-calorie/sugar-free soft drinks market and, therefore, the use of intense sweeteners has grown dramatically in many world markets over the last five years. The major reasons for growth are:

(1) Sweetener development: that is, improvement in the taste quality of high-intensity sweeteners permitted for use in soft drinks and consequently more acceptable low-calorie/sugar-free products.
(2) An increase in consumer awareness of nutrition and 'healthy' eating, making the reduction of sugar intake in the diet desirable for the majority of developed societies.

Saccharin was the first high-intensity sweetener to be marketed, and its usage increased during the First World War owing to a sugar scarcity. Cyclamate entered the UK market during the 1960s and was later controversially banned in many countries as a potential carcinogen.

The 1970 cyclamate ban brought to an end the use of saccharin cyclamate blends in many soft drinks markets. Soft drinks sweetened only with saccharin did not deliver the sweetness taste quality of the blend and this highlighted the need for alternative high-intensity sweeteners.

It was a further 11 years before other high-intensity sweeteners (aspartame, acesulfame K and thaumatin) gained approval for use in foods in major world markets.

5.1.1 Use of intense sweeteners

Use of sweeteners in soft drinks is not restricted to low-calorie or dietetic products. In some countries, particularly where sugar prices are comparatively high, intense sweeteners are used in combination with sugar or glucose syrups to give more cost-effective formulations.

Intense sweeteners provide sweetness, the amount supplied – i.e. the relative sweetness of all intense sweeteners – will depend on application. The values quoted in this chapter are only a guide and demonstrate the wide range of values obtainable under different conditions.

Intense sweeteners do not supply the mouth-feel of sugar and, in some cases, they may supply undesirable side tastes or prove to be incompatible with some flavours. For these reasons, use of intense sweeteners in soft drinks is rarely a case of direct substitution of sucrose in the regular product formulation; more often than not, total reformulation is necessary. It may be necessary to adjust the acidity and use buffers to assist stability of some sweeteners. Some adjustment of the flavour system used is commonly required and the use of gums or small amounts of sugars can improve mouth-feel and control fobbing during filling. Use of ingredients that mask undesirable side tastes may also be required. Increasing the carbonation of low-calorie products may also help mask undesirable side tastes and give the illusion of better mouth-feel.

Sweetness synergy occurs with many combinations of intense (and bulk) sweeteners. The effects can be twofold: a higher perceived sweetness than would be expected from the theoretical sum of the relative sweetness values of the individual sweeteners used and, in some cases, a marked improvement in taste quality of sweeteners that have undesirable side tastes.

The optimum sweetener system will vary depending on the product and will not necessarily be a sweetener blend. However, if a sweetener blend is to be used, a useful starting point often quoted for blends of two intense sweeteners is that sweeteners are used in an inverse ratio to their relative sweetness (to each other), so that each sweetener contributes 50% of the total sweetness. For example, if sweetener A is half as sweet as sweetener B, the sweetener blend would contain twice the amount of sweetener A than sweetener B.

Optimum sweetener blends for three or more sweeteners are not predictable and should be determined by sensory evaluation.

Several intense sweeteners are now approved for use in soft drinks. Four compounds – acesulfame K, aspartame, cyclamate and saccharin – have major importance in the soft drinks market. This chapter will give a brief review of these, together with three other compounds (stevioside, thaumatin and neohesperidin dihydrochalcone) that have limited world-wide approval for use in soft drinks and two other new intense sweeteners – alitame and sucralose – currently seeking approval.

5.2 Current sweeteners

5.2.1 *Acesulfame K*

Acesulfame K (Figure 5.1) is the generic name for the potassium salt of 6-methyl-1,2,3-oxathiazine-4(3H)-one-2,2,dioxide; it is a derivative of acetoacetic acid and was discovered by the German company Hoechst AG in 1967.[1]

Acesulfame K is a white, non-hygroscopic crystalline substance; at room temperature solubility is good (270 g/l) in water, poor in organic solvents, but increases in solvent water mixtures.[1]

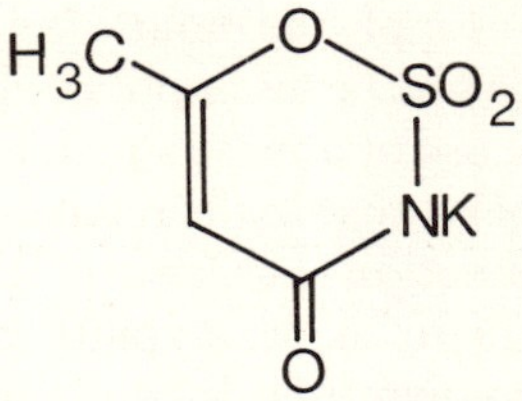

Figure 5.1 Acesulfame K: $C_4H_4N\,SO_4K$ — mol.wt. 201.2.

Application in soft drinks. (a) *Sensory:* As with all intense sweeteners, sweetness potency of acesulfame K relative to sucrose decreases with increasing concentration and varies with the medium in which the sweetener is being tested and the method used for quantifying sweetness.

Values for acesulfame K vary from 110 to 200[2,1] at 10% and 3% sucrose equivalence, respectively. The taste profile of acesulfame K is generally considered to be superior to saccharin. It has a rapid onset time but the sweetness quality is marred by a bitter-astringent aftertaste that is particularly noticeable at higher concentrations. Sweetness quality can be greatly improved by combining with other intense and bulk sweeteners. High levels of synergism (30% and above) reportedly occur with aspartame[3] and, to a lesser extent, with cyclamate, glucose, fructose and sucrose.[1] Very little synergy is reported to occur with saccharin, possibly because they compete for the same sweet receptor site. The aftertaste of acesulfame K can be masked in some cases by the addition of sugar alcohols, maltol and ethyl maltol.[4]

In soft drinks as a sole sweetener, levels of 600–800 and 550–750 mg/l for cola and citrus-flavoured drinks, respectively, are appropriate. Blending with other sweeteners, in particular aspartame, gives a much more acceptable product. In 50:50 combinations with aspartame, taking into account synergy, levels of 160–170 and 140–150 mg/l, respectively, for cola and citrus-flavoured beverages would be appropriate.

(b) *Stability:* Stability of acesulfame K is very good and concentrated stock solutions can be stored and used. In solution, no detectable decomposition occurs at pH 3 at room temperature. Very limited decomposition occurs below pH 3 over extended storage periods.[4]

Heat stability is also good. No detectable decomposition occurs during pasteurisation or UHT treatments.[4]

In general, acesulfame K appears to be non-reactive with other soft drinks ingredients. However, inclusion of acesulfame K adds potassium ions to the beverage and this should be taken into account when selecting clouding agents and stabilisers.[3]

(c) *Analysis:* Qualitative analysis may be performed using thin layer chromatography. HPLC is the main method available for quantitative analysis owing to the low volatility of acesulfame K, detection being in the UV range.[5] Methods using isotachophoretic techniques can be used to determine acesulfame K, saccharin and cyclamate simultaneously.[6]

Metabolism. Acesulfame K is not metabolised and is excreted unchanged from the body primarily in the urine. It, therefore, has a caloric value of zero. Very few micro-organisms have been found to metabolise acesulfame K, indicating that it is also non-cariogenic.[7]

Regulation. A large number of toxicological studies were submitted to the regulatory authorities in order to gain approval for acesulfame K. The toxicity of acetoacetamide (the decomposition product of acesulfame K formed under certain conditions) was also studied and they indicated that both products were non-toxic.[5] The ADI (Acceptable Daily Intake) assigned by JECFA (Joint FAO/WHO Expert Committee on Food Additives) and the FDA (Food & Drug Administration) are 0–9 and 0–15 mg/kg body weight, respectively.[8]

The UK was the first country to approve use of acesulfame K in food and drink with Group A classification in 1983.[9] The FDA gave approval for use in dry mix beverages in 1988. It is approved for use in soft drinks in over 15 countries, with several petitions pending.

Marketing. Acesulfame K is marketed under the brand name Sunett.* Legislative constraints, limited production capacity[10,11] and competition from aspartame, which has better taste qualities, have hindered the development of acesulfame K in the soft drinks market. With capacity problems now overcome and more approvals in different world markets, use of acesulfame K should increase, particularly in areas where aspartame cannot be used. Combination with other sweeteners will take advantage of the improved taste quality and apparent synergism and also assist in keeping within the ADI.

5.2.2 Aspartame

Aspartame is the generic name for N-alpha-aspartyl-L-phenylalanine methyl ester (Figure 5.2). It was discovered as a potential high-intensity sweetener in 1965 by J. Schlatter in the G.D. Searle laboratories.[2]

Aspartame is a white crystalline powder. Solubility in water is 1.0 g/l at 20 °C and this is adequate for most food applications. Solubility increases in acid conditions and with increasing temperatures allowing stock solutions to be made up – however, these solutions should be freshly prepared each day. Aspartame is sparingly soluble in solvents and insoluble in oil.[12]

Application in soft drinks. (a) *Sensory:* Of all the intense sweeteners currently available for use, aspartame has a very similar taste profile to

---

*Sunett is a registered trademark of Hoechst AG.

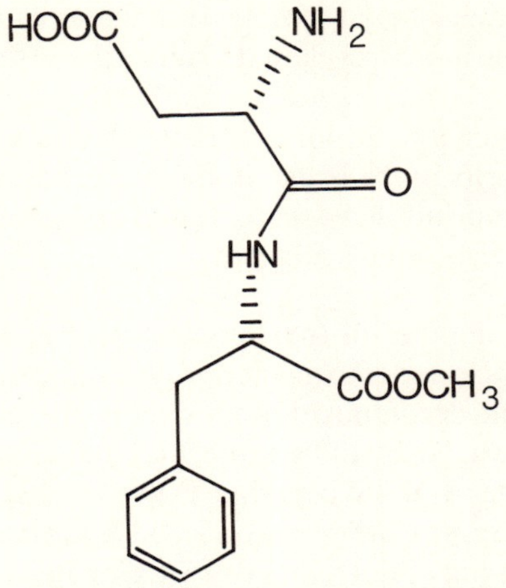

Figure 5.2 Aspartame: $C_{14}H_{18}N_2O_5$ — mol.wt 294.3.

sucrose[12,13] and this has been the overriding factor contributing to its success in the market place.

Relative sweetness values quoted at 4–5% sucrose equivalence in water are in the range 120–215.[14,15] A relative sweetness value of 180 at 10% sucrose equivalence is often used in soft drink formulations. Taste quality of aspartame is a clean sweet taste without the bitter metallic or licorice aftertaste often associated with intense sweeteners – some individuals do, however, notice a slight lingering of the sweet taste. It is synergistic with several other intense sweeteners including saccharin, cyclamates, stevioside,[15] acesulfame K[1] and sugars. Flavour enhancement, particularly with fruit flavours, occurs – most notably with natural flavours.[16,17]

As the sole sweetener, use levels of approximately 500–600 and 400–600 mg/l are appropriate for cola and lemonade beverages, respectively.

(b) *Stability:* As would be expected from a compound essentially made up of two amino acids, aspartame undergoes degradation in solution. Hydrolysis of the ester bond gives the dipeptide aspartyl-L-phenylalanine with the elimination of methanol. At pH 5 and above, the main degradation product is formed by cyclisation to the diketopiperazine (DKP) with the elimination of methanol. DKP may then hydrolyse to the dipeptide which may in turn, hydrolyse to its constituent amino acids, aspartic acid and phenylalanine.[17]

The critical factors that dictate the rate of aspartame degradation in soft drinks are pH, temperature, moisture and time. Fortunately, for the soft drinks manufacturer, the optimum pH range for aspartame stability is pH 3 to 5 with maximum stability at pH 4.3.[17]

The effect of UHT aseptic processes on soft drinks containing aspartame is minimal. Typical aspartame losses would be in the range 0.5–5% for most standard treatments.[19] Therefore, the effect of temperature on stability of aspartame in soft drinks is likely to be a function of storage and distribution temperature.

Stability of aspartame in concentrate and post mix/fountain syrups is generally lower than in the corresponding ready-to-drink product due to the lower pH of concentrates.

There is no direct relationship between the acceptability of an aspartame sweetened product, its perceived sweetness and the actual loss of aspartame.[12] As the concentration of aspartame decreases, the relative sweetness increases, thereby partially compensating for the degradation of the sweetener. Sensory evaluation has indicated up to 40% loss of aspartame before the soft drink is judged unacceptable.[17]

In dry form, when stored correctly, aspartame is stable for several years, making it an ideal sweetener for powdered soft drinks.

The improved stability of aspartame has been the subject of several patents – most of which involve co-drying with various acidulants and/or bulking agents or encapsulation, and are not applicable to liquid systems. However, combinations of aspartame with caramel have been reported to give improved stability and are the subject of one patent application.[20]

(c) *Analysis:* Qualitative and quantitative spectrophotometric analyses can be performed by traditional amino acid detection methods based on the reaction with ninhydrin.[21] Quantitative analysis may also be effected by HPLC.[22-25] Some chromatographic methods allow for the simultaneous analysis of other soft drinks constituents.[26,27] A non-chromatographic method based on a non-aqueous perchloric acid titration may also be used.[28]

Metabolism. Unlike many other intense sweeteners, aspartame is metabolised by the body. It is hydrolysed into the two constituent amino acids and methanol in the gut. These breakdown products are metabolised in the same way as aspartic acid, phenylalanine and methanol from other foods.[7] The aspartame molecule adds nothing new to the food chain.

People with the rare human genetic disease Phenylketonuria have a deficiency in their ability to metabolise phenylalanine and their intake of this essential amino acid must be very strictly controlled from birth to adulthood. Therefore, they must include the phenylalanine content of aspartame in their dietary calculations.[29]

Aspartame is non-cariogenic and has a calorific value of approximately 4 cal/g.

Regulation. The FDA issued approval for the limited use of aspartame in foods and beverages on 24 July 1974. G.D. Searle voluntarily withdrew it from the market shortly afterwards when questions were raised about the validity of some of the toxicological data used to establish its safety. A stay of effectiveness of the aspartame regulation was published in the Federal Register of December 1975.[12]

Further toxicological studies and re-evaluation of the original toxicology data satisfied the FDA that aspartame was a completely safe food ingredient, and in 1981 it gave approval for use in limited food applications. JECFA gave aspartame a comparatively high ADI of 40 mg/kg body weight. The FDA ADI is 50 mg/kg body weight.

Canada was the first country to allow the use of aspartame in soft drinks in August 1981 and in July 1983 the USA followed suit. In the UK aspartame was given Group A classification in the Sweeteners in Food Regulations (1983).[9] It is currently permitted in over 39 countries for use in soft drinks.

Since its approval, aspartame has been the subject of some controversy concerning its safety. Much of the evidence picked up by the lay press is anecdotal. Scientifically controlled studies have consistently failed to produce substantive evidence linking aspartame consumption, even at abuse levels, to adverse health effects,[30] and analyses of adverse reaction reports made by consumers in the USA have not provided evidence suggesting a serious and widespread public health hazard associated with the sweetener.[31,33]

Marketing of aspartame. Aspartame has been marketed in a way unlike any other food ingredient to date. When aspartame was approved for food use in the early 1980s, G.D. Searle had patent protection in several countries (notably the USA, patent expiry 1992; Canada, patent expiry 1989; and the UK, patent expiry 1986). The company chose to market aspartame with a 'branded ingredient strategy' using the brand name 'NutraSweet'.*

The effect of this strategy was that all retail products using only NutraSweet as the sweetening ingredient featured the NutraSweet logo (or the NutraSweet blend logo in the case of the soft drinks using a NutraSweet/saccharin blend) on their packaging. NutraSweet was advertised to consumers in its own right as a healthy, desirable food ingredient. This strategy, although not universally popular with all food and drink manufacturers, proved extremely successful at a time when food additives were under fire from the media and consumer and political pressure groups.

Patent expiry in several markets has seen the emergence of other aspartame manufacturers – some of whom also brand their product, e.g. Sanecta† from The Holland Sweetener Company. It is a little too soon to assess their impact on the market place.

5.2.3 Cyclamate

Cyclamate is the generic name for cyclohexylsulphamate and was discovered by Michael Sveda in 1937 at the University of Illinois; the sodium salt

---

*NutraSweet is a registered trademark of The NutraSweet Company.
†Sanecta is a registered trademark of The Holland Sweetener Company.

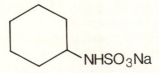

Figure 5.3 Sodium cyclamate: $C_6H_{12}N\ NaO_3S$ — mol.wt 201.2.

(Figure 5.3) is the form more commonly used, although the calcium salt is also intensely sweet.

Cyclamates are white crystalline solids. Solubilities of the sodium and calcium salts are good, 200 and 250 g/l respectively at 20 °C. Solubility in alcohol is quoted at 4 and 20 g/l respectively.[34,35]

Application in soft drinks. (a) *Sensory:* Relative sweetness of cyclamates varies with the type of salt and quoted values vary from 30 to 140.[4,36] In many food systems, values in the range 30–40 are appropriate.

Taste quality of cyclamate as a sole sweetener is not exceptionally good. It has a slow onset time and a detectable sweet/sour aftertaste most noticeable at high concentrations.[15] In combination with other intense sweeteners, its sweetness quality is greatly improved and it is synergistic with saccharin,[10,37] acesulfame K,[1] aspartame.[4,38] The free base cyclamic acid has also been reported to have a flavour-enhancing function at sub-threshold levels.[34]

(b) *Stability:* Cyclamate is stable under conditions likely to be encountered in soft drinks manufacture, i.e. in the pH range 2–7, pasteurisation and UHT heat treatments.[4]

(c) *Analysis:* Cyclamates may be determined by spectrophotometric methods[39] and titration followed by liquid chromatography.[40]

Metabolism. Cyclamates are non-cariogenic. They are also non-caloric, since any metabolism that does occur in some individuals does not release energy.[4] Early studies by Abbott Laboratories on the physiological effects of cyclamate showed cyclamate was poorly absorbed and excreted unchanged in animals.[7]

The majority of people metabolise less than 1.0% of cyclamate intake. However, approximately 47% of the population have the ability to metabolise 20 to 60% cyclamate (via gut microflora) to cyclohexylamine in which form it is excreted.[34,41] Conversion of cyclamate to cyclohexylamine in individuals with the ability to convert is not consistent and may be dependent on induction by chronic administration of the sweetener.[6]

Regulation. In 1958, cyclamate was given GRAS status by the FDA following the Food Additive Amendment to the Food, Drug and Cosmetic Act. In the UK it was first accepted for use by virtue of the Soft Drinks Regulations 1964. The 1960s showed enormous growth in the use of cyclamates – particularly in soft drinks where cyclamate/saccharin blends (usually a 10:1 ratio) were very popular.

In 1969, use of cyclamates in general purpose foods was banned in the USA on the basis of studies linking increased incidence of bladder tumours in rats fed with a cyclamate/saccharin mixture. In August 1970, the FDA banned cyclamate completely from all food and drugs and several other countries followed suit.

The ban on cyclamates was controversial and subsequent studies failed to confirm the original findings.[34] Since 1969, work has been performed to assess the effect of consumption of cyclamates, much of the work supporting the claim that it is not a carcinogen.[42,43] Other work has implicated cyclohexylamine, the principal metabolite of cyclamate, in high blood pressure,[44] testicular atrophy in rats[45] and cancer promotion.[43,46]

Abbott Laboratories have twice petitioned the FDA for reapproval; the first petition in 1973 was denied. The latest petition, backed by the Society of Toxicology, the American Statistical Society,[47] the National Academy of Science and the FDA Cancer Assessment Committee (CAC), was made in 1982. The FDA requested an independent review of the data by the National Academy of Science/National Research Council in 1985. This group concluded that cyclamates may act as promoters or co-carcinogens in the presence of substances such as saccharin, even though they themselves were not carcinogens.[15] The FDA have not yet ruled on the latest petition; however, in May 1989 it was widely reported in the US press that the FDA was expected to lift the 20-year-old ban on the use of cyclamate.[48] This report was denied by the FDA who commented that the results of all the safety studies would not be available until Autumn 1989 and a decision was unlikely before 1990.

Cyclamate is currently permitted for use in soft drinks in over 25 countries; in 1982, JECFA almost tripled the ADI of cyclamate to 0–11 mg/kg bodyweight. If cyclamate is reapproved in the USA, it will put pressure on the UK to also approve. The new EEC Directive on Additives (including sweeteners), includes cyclamates on the positive list. Therefore, restricted reapproval throughout the EEC by 1991 is likely. However, even if it is reapproved, in an age when every new food additive comes under tremendous scrutiny by the media, the marketing of a previously banned sweetener may present problems.

5.2.4 *Saccharin*

Saccharin is the generic name for 1,2-benzisothiazolin-3-one-1,1-dioxide and has been used as an intense sweetener for over a century. It was discovered by Fahlberg and Remsen in 1879 at the John Hopkins University[49] and the first patent for commercial manufacture was granted in 1885.[50] Saccharin use increased during the First and Second World Wars owing to sugar scarcity

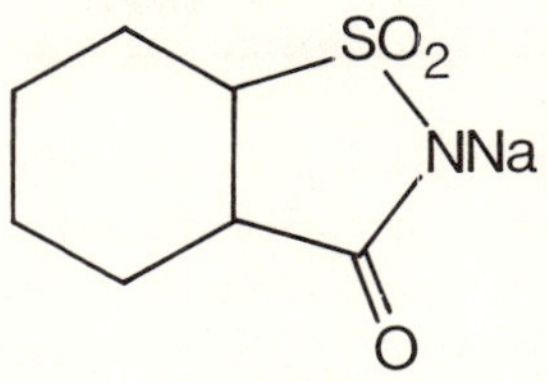

Figure 5.4 Sodium saccharin: $C_7H_4N\ NaO_3S$ — mol.wt. 205.16.

(particularly in Europe), establishing it as an important food and pharmaceutical ingredient.

Saccharin is a white crystalline powder. Solubilities of saccharin, sodium saccharin (Figure 5.4) and calcium saccharin are approximately 3, 700 and 400 g/l, respectively, at 20 °C. Solubilities of the two salts in alcohol are lower: 20 and 200 g/l, respectively.[35]

Application in soft drinks. (a) *Sensory:* Saccharin relative sweetness values in the range 300–700 have been quoted.[50,51] In soft drinks, where the sodium salt is often used, values of the order of 360–500 are usually appropriate.

A significant drawback of saccharin is that its intensely sweet taste is marred by a bitter/metallic aftertaste which is more pronounced at high concentrations. Some individuals are more sensitive to the backtaste than others[52] and much research has been carried out on ways of masking it. Fructose and gluconates have been reported to be effective masking agents,[53] as have tartrates,[54] ribonucleotides,[51] sugars,[55] sugar alcohols,[55,56] and other intense sweeteners.[57] Several sugars, sugar alcohols and intense sweeteners also have the added benefit of being synergistic with saccharin, including fructose,[55] sorbitol, xylitol,[56] aspartame, cyclamate[51] and sucralose.[57]

(b) *Stability:* In dry form, saccharin is stable for several years when stored under the correct conditions. Under processing conditions likely to be encountered in the soft drinks industry, saccharin is completely stable and does not interact with food ingredients. Concentrated stock solutions can be stored.

(c) *Analysis:* Analysis of saccharin in beverages is usually performed by HPLC[58] or spectrophotometric methods.[59] Determination of other soft drinks ingredients is sometimes possible using the same methods.[58]

Metabolism. The consensus of opinion is that saccharin is not metabolised but excreted from the body unchanged primarily via the urine.[41] It therefore has a calorific value of zero.

Regulation. Saccharin has had a controversial history. It was banned for use in food and in drinks in Germany in 1898 and in the USA in 1912. Sugar scarcity in the First World War allowed the reapproval of saccharin in the USA.

More recent controversy regarding the safety of saccharin was initiated in 1970. Research data showed an increased incidence of bladder tumours in rats fed on a saccharin/cyclamate sweetener blend. Subsequent research implicated saccharin as the cause of this effect.[41,60-64] Increase of bladder tumours occurs only at very high sodium saccharin dose levels. The mechanism for the effect is not known and it appears to be affected by the form of saccharin used (the sodium salt being most active).[61] The evidence would also suggest that the effect is species specific and, therefore, the relevance to man has been questioned.[63] Studies in man, including those of groups of people likely to consume larger amounts of saccharin (for example, diabetics), have failed to demonstrate a link between saccharin consumption and bladder tumours.[65] The net result of the controversy has been that the FDA withdrew GRAS status in 1972 and would have banned saccharin for use in food and drink in 1977 had the US Congress not intervened in response to appeals from food manufacturers. The moratorium on saccharin use in food and drinks was extended in 1979, 1981, 1983, 1985 and 1987. It is due to expire or be extended again in 1992. The packaging of food products containing saccharin in the US contain statements warning that saccharin has been shown to cause cancer in laboratory animals. In the UK, MAFF (the Ministry of Agriculture, Fisheries and Food) reacted by assigning Group B status in the Sweetener in Food Regulations 1983.[9]

The SCF reviewed the available evidence on saccharin in 1977 and gave a temporary ADI of 0–2.5 mg/kg body weight with the proviso that the situation was kept under review and reassessed when further evidence became available.[63] The situation was reviewed in December 1987 and the assessment was not changed.

Saccharin is currently permitted for use in soft drinks in approximately 75 countries.

Marketing. Saccharin is presently extensively used in soft drinks world wide either as a sole sweetener or in combination with other intense sweeteners (notably cyclamate and aspartame) in low-calorie and dietetic drinks, or in combination with sucrose in regular drinks, in order to give a more cost-effective formulation.

5.2.5 *Stevioside/stevia*

Stevioside, or stevia, is the name given to a group of sweet diterpene glycosides extracted from the leaves of *Stevia Rebaudiana Bertoni* (Figure 5.5). The plant is native to Paraguay and is now commercially cultivated in South East Asia, including the People's Republic of China, South Korea and Japan, and also in South Africa, South America and the USA.

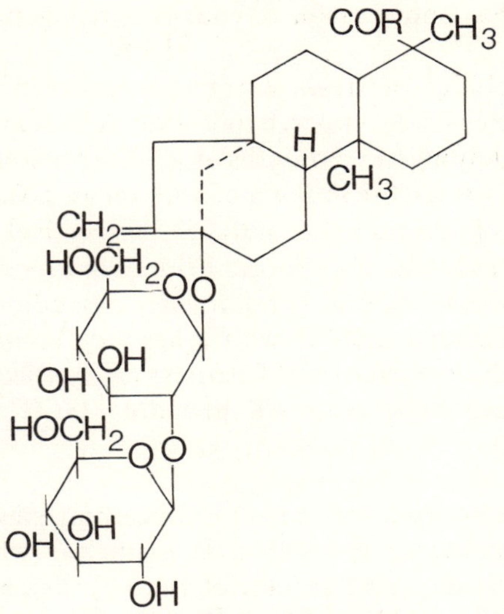

Figure 5.5 Stevioside: $C_{38}H_{60}O_{18}$ – mol.wt. 804.65. $R = \beta - D\text{-glucose}$.

Several sweet diterpene glycosides are present in extracts of *Stevia Rebaudiana Bertoni*. Stevioside and rebaudioside A are the two with commercial importance. The other sweet compounds present in small quantities have been named rebaudioside C–E[66], dulcoside and steviolbioside.[67]

The properties of stevia products are dependent on the composition of the individual extracts and this probably accounts for the conflicting data available.

Pure stevioside is a white hygroscopic powder.[68] Commercial stevia extracts may vary in colour from cream to tan. Solubility of pure stevioside is quoted at 1.2 g/l.[68,69] Rebaudiosides have greater solubility and commercial stevia extracts have very good solubilities: 300–800 g/l.[85,67] Solubility in alcohol is poor.

Application in soft drinks. (a) *Sensory:* Relative sweetness values are dependent on the composition of the stevia extract, values in the range 140–280 times sucrose sweetness have been reported.[14,70]

The taste of stevia extracts is characterised by a lingering sweetness and licorice/bitter offtastes[71,72]; the result of this is that stevia extracts cannot be used as sole sweeteners in most applications. Several methods of improving the taste characteristics have been reported, including increasing the content of the better tasting rebaudioside A fraction or in combination with sugars (notably fructose[70] and lactose[73]), or histidine,[74] chlorodeoxysugars,[57] cyclodextrin,[73] aspartame and cyclamate.[75]

Stevioside also has some useful flavour-enhancing properties with fruit flavours.[69]

(b) *Stability:* Stability of stevia extracts is generally good. Long-term stability tests on stevioside and rebaudioside A in carbonated beverages indicated no degradation over 5 months at 22 °C or below. At higher storage temperatures (37 °C), some breakdown of both rebaudioside A and stevioside does occur over four months (36 and 25% respectively).[75] The extent of degradation is increased in phosphoric acid systems relative to citric acid systems and UV light causes some breakdown of rebaudioside A. It is reported that stevia extracts do not interact with other food components.[4]

(c) *Analysis:* Analysis in soft drinks is most easily effected by HPLC.[75,76] Other methods have been reported, including GLC[77] and colorimetric methods.[78]

Metabolism. Very few data are available concerning stevioside metabolism, and the available data are in conflict. It is unclear whether stevioside is excreted unmetabolised or whether steviol, the aglycone portion, is generated in the gut. The significance of steviol is that, when metabolically activated, it produces a mutagen.[73] Steviol generation has been demonstrated *in vitro*[79] with rat caecal flora and *in vivo* in rats[80] but not in mammalian systems.[73] Metabolism in man has not yet been reported.

Suggestions linking stevioside with anti-hormonal effects and infertility[79,81] have not been supported by the results of controlled experiments.[73]

Regulation. Japan is the main market for stevia, where it has been approved for food use since 1970.[82,83] It is also permitted for food use in Brazil, Paraguay, the People's Republic of China and South Korea. Regulatory approval of stevia products is hindered because there is such a wide variety of extracts of differing compositions available complicating the interpretation of toxicological data. So far, insufficient data have been submitted to MAFF or the SCF.

In the latter's most recent report in December 1987, the conclusion of the review of the available data was that stevioside was not toxicologically acceptable.[73]

Brazil is a producer of stevia, and as part of the Health Ministry's approval of diet drinks in November 1988, the Brazilian Soft Drinks Manufacturers' Association is studying stevia. For the first 8 months of use in diet soft drinks, shelf-life, taste and consumer reaction will be tested.[82]

Until the toxicology issues are resolved, approval in other world markets is unlikely. However, a stevioside producer in the US has predicted FDA approval by 1990.[83]

Marketing. Stevia products are marketed by many companies in Japan. Eleven of the major companies have formed a consortium called Stevia

Konwaki (Stevia Association). This consortium is responsible for much of the toxicology work done so far on stevia products. In Japan, stevia is used by over 50 companies in soft drinks, instant juices and fruit-flavoured drinks – where it has fruit flavour-enhancing properties.

5.2.6 Thaumatin

Thaumatins are a group of intensely sweet basic proteins isolated from the fruit of *Thaumatococcus danielli* (also known as the Katemfe plant).[84] In practical terms, it is perhaps one of the least important of the permitted sweeteners in terms of use in soft drinks in that its taste quality makes it unsuitable for use as a sweetener except in products where a lingering licorice aftertaste can be tolerated. Thaumatins are soluble proteins. There are five distinct proteins designated a, b, c, I and II. Thaumatin I and II are the sweetest and most abundant.

Application in soft drinks.   (a) *Sensory:* Sweetness intensity is quoted as 1300–2000 times sucrose in 5–10% sucrose range.[85] Aluminium ions increase the sweetness of thaumatin by a factor of 2.[86] The taste profile of thaumatin shows a slow onset time and lingering licorice aftertaste.

Lingering aftertaste of thaumatin can be slightly reduced by addition of arabinogalactan, glucuronic acid and several sugars and sugar alcohols. Synergism has been noted with saccharin, acesulfame and stevioside but not cyclamate or aspartame.[86]

(b) *Stability:* Thaumatin is more resistant to denaturation than most soluble proteins. Eight disulphide bridges in its structure confer this extra stability.[86] Thaumatin undergoes denaturation (which may be reversible) when exposed to extremes of pH and high temperatures. The result of denaturation is loss of sweetness.

Thaumatin does interact with some food constituents. The thaumatin molecule is positively charged and may, therefore, form salts with suitable negatively charged compounds, for example, several gums and stabilisers (CMC, xanthan, pectin, locust bean gum, alginates) when they are present in excess. Precipitation may occur with some synthetic colours – particularly in the pH range applicable to soft drinks – although low levels of gum arabic may prevent this.[86]

Metabolism.   Thaumatin is a protein and is digested and metabolised to its constituent amino acid by pancreatic enzymes in a similar fashion to other proteins. Toxicological studies have indicated no adverse effects as a result of consumption.[43]

Regulation.   Thaumatin was first permitted as a natural food in Japan in June 1979. In the UK it has been permitted as a sweetener for use in foods since 1983

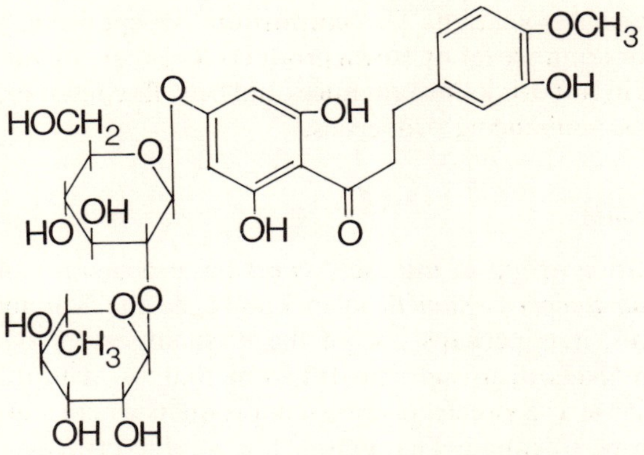

Figure 5.6 Neohesperidin dihydrochalcone: $C_{28}H_{36}O_{15}$ — mol.wt. 612.4.

with group A status. In the US it has GRAS status for use in chewing gum.

Marketing. Thaumatin is marketed by Tate & Lyle who have taken advantage of the increased sweetness caused by aluminium salts. A thaumatin–aluminium product has been given the brand name Talin.* Use of Talin in the UK food industry is generally at sub-threshold levels where the flavour-potentiating effects occur – for example, in chewing gum. Use in soft drinks world wide is very limited.

5.2.7 *Dihydrochalcones*

The dihydrochalcones are a group of intense sweeteners. They are phenolic compounds prepared from the bitter citrus flavanones, naringin and neohesperidin, which are present as major constituents of the peel of some citrus fruits.[87]

The compound neohesperidin dihydrochalcone (NeoDHC) (Figure 5.6) has some commercial importance.

NeoDHC is available as a white to colourless solid. Solubility is not particularly good at 0.5 g/l at 25 °C, although it increases with temperature and, as use level is low, it is sufficient for food applications. Solubility of the sodium and potassium salts is much improved at over 1000 g/l at 25 °C; solubility in ethanol is 20.4 g/l at 25 °C.[88]

Application in soft drinks. (a) *Sensory:* A wide range of relative sweetness values has been quoted for NeoDHC, ranging from 250 (at 5% sucrose) to 1800 at threshold.[87]

*Talin is a registered trademark of Tate & Lyle Plc.

The taste profile of NeoDHC has a slow onset time and lingering menthol aftertaste[72] and this is a major limitation for its use as a sweetener in soft drinks. Gluconates and amino acids have been reported to be effective masking agents. Nearly 250 analogues of NeoDHC have been prepared, only one of which, a homoserine–DHC conjugate, had reduced aftertaste.[79] Synergism and improved taste quality occurs with most other intense sweeteners.[88]

NeoDHC also exhibits other interesting sensory effects: it masks bitterness in fruit juices (particularly grapefruit) when used at low levels (i.e. 6–12 ppm).[87]

(b) *Stability:* Stability in liquid formulations above pH 2 is good with very limited hydrolysis to neohesperidose and the aglycone hesperitin dihydrochalcone which is sweet but not very soluble.[87]

In dry form, NeoDHC is stable for several years.

(c) *Analysis:* Chromatographic methods of analysis have been reported.[89,90]

Metabolism. Metabolism of NeoDHC is through the action of intestinal microflora yielding isoferrulic acid, *m*-hydroxyphenylpropionic and *m*-hydroxycinnamic acid. Excretion studies in rats have indicated that 90% excretion occurs in the first 24 hours, predominantly in the urine.[87]

The caloric value is estimated at 2 cal/g based on the assumption that the sugar residues are hydrolysed and metabolised and the aglycone portion is not metabolised.[87]

Regulation. Much of the toxicology work done on NeoDHC was performed by the US Department of Agriculture, the results of which have indicated NeoDHC to be non-mutagenic, non-carcinogenic and non-cariogenic.[87]

The FDA refused GRAS status for NeoDHC on the basis of insufficient toxicology data.[2,87] SCF have established an ADI of 0–5 mg/kg body weight.[91]

NeoDHC is currently allowed for use in Belgium in sugarless soft drinks and juices (max 50 mg/l)[92] and certain sour beers. It has also been used in Israel in fruit juices.

5.3 Potential new sweeteners

Several new compounds have been discovered to be intensely sweet in recent years. For obvious reasons, details of potential new sweeteners are limited. However, two compounds in an advanced stage of commercial development will be dealt with in this section.

5.3.1 *Alitame*

Alitame is the generic name for L-alpha-aspartyl-*N*-(2,2,4,4-tetramethyl-3-thetanyl)-D-alaninamide hydrate (2:5) (Figure 5.7). Pfizer Inc. patented the sweetener in 1980.[93,94]

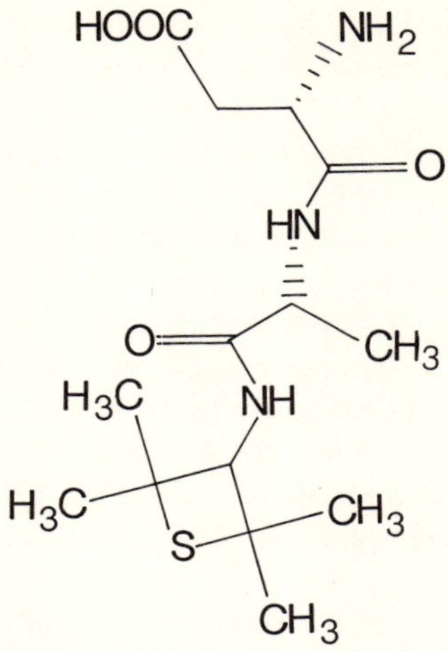

Figure 5.7 Alitame: $C_{14}H_{25}N_3O_4S$ 2.5H_2O — mol.wt. 376.

Alitame is a crystalline, non-hygroscopic powder with good solubility in water: 131 g/l at 25 °C. Solubility in alcohol is also good: 610 g/l in ethanol at 25 °C.[94]

Application in soft drinks. (a) *Sensory:* Relative sweetness of alitame is quoted at 2000 times sucrose (10% equivalent) rising to 2900 times at threshold. Taste quality is said to be good.[94]

(b) *Stability:* Stability of alitame over a broad pH range and at elevated temperatures is said to be superior to aspartame with the half-life being double that of aspartame. Degradation does occur mainly through hydrolysis of the aspartylalanine dipeptide bond to give aspartic acid and alanyl-2,2,4,4-tetramethylthietane amide. All breakdown products are tasteless at low concentrations.[95]

Incompatibility with some food ingredients resulting in off-flavours has been noted, including sodium metabisulphite.[94] It is also reported to be unsuitable for cola beverages.[96]

Metabolism. Alitame is absorbed in man and excreted as a mixture of its metabolites and unchanged alitame.[95]

Caloric value is 1.4 cal/g; this is due to aspartic acid formed during metabolism which is metabolised via normal amino-acid routes.

Regulation. Safety and toxicology studies are complete and petitions for approval for use as a sweetener were submitted to the FDA and MAFF in 1986.[97] Approval is expected within the next 3–5 years.[98]

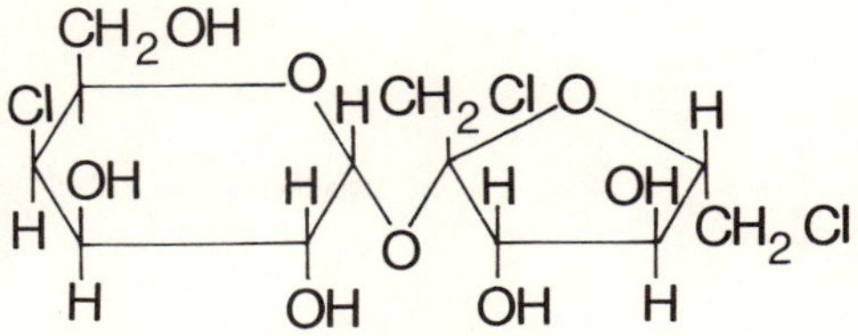

Figure 5.8 Sucralose – mol. wt. 397.6.

5.3.2 Sucralose

Sucralose (Figure 5.8) is the generic name for 4,1',6'-trichloro-4,1',6'-trideoxygalactosucrose (abbreviated to trichlorogalactosucrose or TGS). Sucralose is a derivative of sucrose.[99,100]

Sucralose is a white crystalline powder that is freely soluble in water (280 g/l at 20 °C) and ethanol.[92]

Application in soft drinks. (a) *Sensory:* Relative sweetness of sucralose is quoted at 400–800 where sucrose is equal to 1. Relative sweetness value increases as pH decreases.[101]

Sweetness quality is said to be sucrose-like and can be enhanced in cola beverages by blending with aspartame.[101]

Synergism with acesulfame K, cyclamate, saccharin and stevioside has been noted.[57,103] Sucralose is not synergistic with sucrose and interestingly exhibits negative synergism with aspartame in bipartite blends.[101] High levels of synergism (30–50%) are reported with tripartite sweetener blends of sucralose, cyclamate and aspartame or acesulfame K or saccharin.[103]

(b) *Stability:* Stability of sucralose is very good under most conditions that would exist in food production. For practical purposes, sucralose is stable over the pH range likely to be encountered in carbonated soft drinks. At extremes of pH and at high temperature, limited hydrolysis to the constituent monosaccharides (4-chloro-D-galactosucrose and 1,6-dichloro-D-fructose) does occur.

In solution, sucralose does not appear to interact with food ingredients, with the exception of some iron salts where interaction (possibly chelation) has been reported.[104] Stability of dry sucralose powder is also good when stored correctly. At elevated temperatures, slow decomposition can occur which results in a colour change from white to brown.[104]

Metabolism. Sucralose is said not to be metabolised by mammalian species and is poorly absorbed by the body.

Regulation. The FDA in the US, MAFF in the UK and the HPB in Canada have been petitioned for approval in 1987.[105] A number of regulatory bodies, including JECFA and the SCF, have requested additional information. JECFA have assigned a temporary ADI of 3.5 mg/kg body weight.[106]

Table 5.1 Regulatory status of high-intensity sweeteners in soft drinks in selected world markets

	Acesulfame K	Aspartame	Cyclamate	Saccharin	Stevioside	Thaumatin**	NeoDHC
ADI to JECFA specification	0–9 mg/kg bw	0–40 mg/kg bw	0–11 mg/kg bw	0–2.5 mg/kg bw			0–5 mg/kg bw
Argentine	NP	P	2000 mg/l	150 mg/l	NP	NP	NP
Australia	3000 mg/kg	1000 mg/l	600 mg/l	50 mg/l	NP	P	NP
Belgium	600 mg/l	500 mg/l	NP	125 mg/l	NP	NP	50 mg/l
(Belgium)***	600 mg/l	750 mg/l	400 mg/l	125 mg/l	NP	NP	50 mg/l
Brazil	NP	750 mg/l	1600 mg/l	500 mg/l	750 mg/l	NP	NP
Canada	AP	0.1%	NP	NP	NP	P	NP
Denmark	250 mg/l	500 mg/l	250 mg/l	75 mg/l	NP	P	NP
Finland	NP	500 mg/l	100–400 mg/l*	30–70 mg/l*	NP	NP	NP
France	360 mg/l	600 mg/l	NP	100 mg/l*	NP	NP	NP
E. Germany	NP	NP	450–600 mg/l*	20–60 mg/l*	NP	NP	NP
W. Germany	P	300 mg/l	800 mg/l	200 mg/l	NP	NP	NP
Greece	P	600 mg/l	P	P	NP	NP	NP
Ireland	P	P	P	P	NP	NP	NP
Israel	NP	700 mg/l	193 mg/l	44 mg/l	NP	NP	P
Japan	NP	P	NP	300 mg/l	P	P	NP

HIGH-INTENSITY SWEETENERS

Country						
Kenya	NP	P	NP		NP	NP
Mexico	NP	P	NP	P	P	NP
Netherlands	AP	700 mg/l	P	400 mg/l	P	NP
(Netherlands)***	600 mg/l	750 mg/l	400 mg/l	120 mg/l	P	NP
New Zealand	AP	P	1500 mg/l	125 mg/l	P	NP
Norway	NP	500 mg/l	P	100 mg/l	NP	NP
Saudi Arabia	NP	NP	NP	50–120 mg/l*	P	NP
South Africa	1000 mg/l	1000 mg/l	2500 mg/l	NP	NP	NP
Spain	NP	P	4000 mg/l	500 mg/l	NP	NP
Switzerland	P	P	800 mg/l	200 mg/l	NP	NP
Turkey	P	600 mg/l	NP	P	P	NP
UK	P	P	NP	80 mg/l	NP	NP
USA	P	P	NP	12 mg/fl oz	NP	NP
USSR	NP	NP	NP	P	NP	NP
Yugoslavia	NP	NP	P	180 mg/day	P	NP

Legend:
P   Permitted (includes countries where special permission is required from Government and/or health bodies to market individual products containing high-intensity sweeteners).
NP Not permitted.
AP Approval pending.
*   Maximum level dependent on type of soft drink.
** Thaumatin may be permitted for use in soft drinks as a sweetener or flavour enhancer.
*** New regulations due to be implemented by November 1989.

Regulatory information regarding Acesulfame K kindly supplied by Hoechst AG, Hounslow, Middlesex, UK.
Regulatory information for Aspartame kindly supplied by NutraSweet AG, Zug, Switzerland and The NutraSweet Company, Chicago, USA.

Sucralose is expected to be included in the EEC Directive on Sweeteners which should be adopted by 1992.[106]

Marketing. Sucralose will be marketed in the UK and Europe by Tate & Lyle Speciality Sweetener Division. Johnson & Johnson have obtained a marketing licence from Tate & Lyle to market sucralose in the USA and other selected countries through McNeil Speciality Products Company who are a subsidiary of Johnson & Johnson, Stillman N.J., USA.

5.4 Sweetener approval and regulation

Sweetener approval and regulation varies between countries. The UK is one of the few countries in which general approval for use in food and beverages (excluding products for infants) is given.

In other countries – for example, the USA – approval for use of sweeteners in specific product categories is granted: e.g. table-top sweeteners, yoghurts, soft drinks or dietetic products. Other countries are still more specific and approval for, and registration of, individual brands of product must be sought from the appropriate Food or Health Authority (e.g. Brazil).

Regulatory bodies world-wide have recognised that, as trading between world markets increases, it is logical to instigate moves to harmonise food legislation where possible. The Codex Alimentarius Commission has initiated moves towards world-wide harmonisation of food additive legislation.

In an attempt to harmonise food legislation for countries in the EEC, work is currently under way on a draft proposal for an EEC Directive on Food Additives (including sweeteners), the aim being to have an approved list of sweeteners for Europe. Similar initiatives have occurred in other regions of the world that regularly trade with each other; for example, the Scandinavian countries and Australia and New Zealand. The EEC draft proposal is expected to be put under review in 1989. Agreement to the proposals from regulatory bodies of the countries concerned will be obtained and the Directive is expected to come into operation by 1992.

Table 5.1 shows the current regulatory status for use in soft drinks, at time of going to press, for intense sweeteners in use in several world markets; maximum use levels are indicated where specified. It should be emphasised that this information is only a guide; this is an area of continual change, petitions for the newer sweeteners are pending in several countries and information must be checked before use with the appropriate regulatory authorities.

Several countries require that special permission is sought from the appropriate regulatory and/or health body before a product containing an intense sweetener is allowed to be marketed. Some countries also have special labelling requirements for products containing intense sweeteners and these, too, should be checked with the appropriate regulatory body.

5.5 Future use of intense sweeteners

Aspartame has had a major impact on the low-calorie soft drinks market during the last decade. Trends in markets where alternatives to saccharin have been approved for several years suggest a drift away from saccharin sweetened beverages.

In Europe, the new EEC Directive on Food Additives, including sweeteners, will have an impact on choice of sweeteners available. It would seem likely that the list of approved sweeteners will contain all intense sweeteners currently permitted in any country in Europe – resulting in the reintroduction of cyclamate (albeit probably restricted) in several markets and the approval of acesulfame K, aspartame and possibly sucralose in several new markets.

Local regulations may also have an impact. The forthcoming new soft drinks regulations in the UK are expected to dispense with a minimum sugar content in soft drinks and this may lead to an increase in intense sweetener use.

The increased number of sweeteners available in many markets will present new opportunities for different sweetener blends. Supporters of the multiple sweetener concept claim improved taste quality and a reduction of the chance of exceeding the ADI of intense sweeteners with lower values. The disadvantage, in an additive-conscious market, is an increased number of additives on the ingredient list.

Stevioside and thaumatin are unlikely to have wide use in soft drinks outside specific markets, notably Japan. Use of NeoDHC is extremely limited and there is no evidence to suggest usage will increase.

Limited information is available about the sweeteners currently seeking approval. Their effect on the soft drinks market will be dependent to a large extent on their sensory characteristics in comparison with currently permitted sweeteners and their cost effectiveness.

Aspartame successfully commanded a large price premium over saccharin on its approval for use – because its taste characteristics were vastly superior. The newer sweeteners are unlikely to offer significant, if any, taste benefits over aspartame or currently available sweetener blends. They would, however, offer stability benefits but, judging by the extensive use of aspartame in soft drinks, the soft drinks manufacturer can live with stability limitations. In order to have a significant impact on the market, new sweeteners must be able to offer equivalent sweetness quality at a competitive price.

It seems likely that, in the foreseeable future, use of intense sweeteners in soft drinks will continue to increase, as will the choice of intense sweeteners available to the soft drinks formulator.

References

1. G.W. Von Rymon Lipinsky and B.E. Huddart, *Chemy Ind.* **11** (1983) 427.
2. J.D. Higginbotham in *Developments in Sweeteners*, Vol. 2, eds T.H. Grenby K.J. Parker and M.G. Lindley, Elsevier Applied Science Publishers, Barking (1983).
3. G.W. Von Rymon Lipinsky, Hoechst AG: Personal communication (1988).
4. A.I. Bakal, *Chemy Ind.* **18** (1983) 700.
5. G.W. Von Rymon Lipinski in *Alternative Sweeteners*, eds L. O'Brien Nabors and R.C. Gelardi, Marcel Dekker Inc., New York (1985).
6. H. Klein and W. Stoya, *Ernahrung* **II** (5) (1987) 322.
7. A.G. Renwick in *Developments in Sweeteners*, Vol. 2, eds T.H. Grenby, K.K. Parker and M.G. Lindley, Elsevier Applied Science Publishers, Barking (1983).
8. Hoechst AG–Sunett PR Information Sheets (1988).
9. 'Sweetener in Food Regulations', *Statutory Instrument* No. 1211, *Food Composition and Labelling*, HMSO, London (1983).
10. G.W. Von Rymon Lipinsky, *Food Marketing and Technology* (1987) June.
11. A. Woollen, *Soft Drinks Management Int.* (1988) June, p. 18.
12. A. Ripper, B.E. Homler and G.A. Miller in *Alternative Sweeteners*, eds L. O'Brien Nabors and R.C. Gelardi, Marcel Dekker Inc., New York (1985).
13. N. Larson-Powers and R.M. Pangborn, *J. Food Sci.* **43** (1978) 47.
14. A. Tunaley, D.M.H. Thomson and J.A. McEwan, *Int. J. Food Sci. Technol.* **22** (1987) 627.
15. R. Franta, B. Beck, F. Katz, N. Primack, R.D. Varvil and F.A. Voirol, *Food Technol.* **40** (1) (1986) 116.
16. R.E. Baldwin and B.M. Korschgen *J. Food Sci.* **44** (1979) 938.
17. B.E. Homler in *Aspartame – Physiology and Biochemistry*, eds L.D. Stegink and L.J. Filer, Marcel Dekker Inc, New York (1984).
18. *The NutraSweet Technical Information Bulletin*, The NutraSweet Company, USA.
19. W.H. Shazer, A. Kedo, L.M. Hill and L. Metcalf, Presentation to Society of Soft Drink Technologists, USA, 30 March (1988).
20. Pepsico, UK Patent Application GB 2104369A (1982).
21. O. Lau, S. Luk and W. Chan, *Analyst* **113** (5) (1988) 765.
22. L. Fox and G. Anthony, The NutraSweet Company, *Standardised High Pressure Liquid Chromatography Procedure for Alpha-APM Determination (HPLC)*, Searle Laboratories Analytical Research Laboratory Progress Report No. 7320015.
23. K. Tamase, *J. Food Hyg. Soc. Japan* **26** (5) (1985) 515.
24. W.S. Tasang, M.A. Clarke and F.W. Parrish, *J. Agric Food Chem.* **33** (4) (1985) 734.
25. A. Kedo and A. Fox in *Proc. 44th Annual Meeting of the Society of Soft Drink Technologists*, ed. by SSDT, Brentwood (1988).
26. H.J. Issaq, D. Weiss, C. Ridlon, S.D. Fox and G.M. Muschik, *J. Liq. Chromat.* **9** (8) (1986) 1791.
27. Anon, *Soft Drinks Trade Jl* **39** (12) (1985) 487.
28. Technical Information, *Alpha-APM Determination by Perchloric Acid Titration*, The NutraSweet Company, USA.
29. A.E. Harper, in *Aspartame – Physiology and Biochemistry*, eds L.D. Stegink and L.J. Filer, Marcel Dekker Inc., New York (1984).
30. A. Leon and D. Hunninglake, Paper given at 72nd Annual Meeting FASEB 1–5 May (1988).
31. P.J. Janssen and C.A. Van den Heyden, *Toxicol.* **50** (1988) 1.
32. International Food Information Council, *Aspartame Safety Issues – A Scientific Report* (1986).
33. Council Report, 'Aspartame Review of Safety Issues', *J. Amer. Med. Ass.* **254** (3) (1985) 400.
34. R.W. Kasperson and N. Primack, in *Alternative Sweeteners*, eds L. O'Brien Nabors and R.C. Gelardi, Marcel Dekker Inc., New York (1985).
35. Martindale, *The Extra Pharmocopoeia* (28th edn), J.E.F. Reynolds, The Pharmaceutical Press, London.
36. K.M. Beck in *Kirk-Othmer Encyclopedia of Chemical Technology* (2nd edn), Vol. 19, p. 598, Wiley (1969).
37. O. Kirk, *Encyclopedia of Chemical Technology* (34th edn), Vol. 22, p. 449, Wiley Interscience (1983).

38. G.D. Searle, British Patent 1256995 (1971).
39. G. Kruger, *Lebensmittel Ind.* **27** (6) (1980) 264.
40. A. Herrmann, *J. Chromat.* **280** (1) (1983) 85.
41. A.G. Renwick, *Food Chemy* **16** (3/4) (1985) 281.
42. F. Coulson, E.W. McChesney and L. Goldberg, *Food Cosmet. Toxicol.* **13** (1975) 297.
43. FACC Report on the Review of Sweeteners (1982).
44. M. Eichelbaum, J.H. Hengstmann, H.D. Rost, T. Brecht and H.J. Dengler, *Arch. Toxicol.* **31** (3) (1974) 243.
45. P.L. Mason and G.R. Thompson, *Toxicol.* **8** (1977) 143.
46. R.M. Hicks et al., *Chem Biol. Int.* **11** (1975) 225.
47. W.T. Miller, *Food Technol.* January (1987) 116.
48. M. Gladwell, *The Washington Post*, 16 May (1989).
49. C. Fahlberg and I. Remsen, *Berichte* **12** (1879) 469.
50. US Patent 319082 (1985).
51. A. Bakal, *Food. Technol* **1** (1987) 117.
52. L.M. Bartoshuk, *Science* **205** (1979) 934.
53. US Patent 3743518, *Chem. Abstracts* 79:64836 (1979).
54. British Patent 1239518, *Chem. Abstracts* 75:117353 (1975).
55. L. Hyvonen, R. Kurkeki, P. Koivistoinen and A. Ratilainen, *J. Food Sci.* **43** (1978) 251.
56. A. Askar, F.R. Hassanien, M.G. Abdel Fadeel, A. El-Saidy and M.S. El-Zoghabi, *Alimenta* **24** (1985) 37.
57. Tate & Lyle Plc, European Patent 0064361 (1986).
58. J.T. Hann and I.S.A. Gilkison, *J. Chromat.* **395** (1987) 317.
59. A.G. Ramappa and A.N. Nayak, *Analyst* **108** (1289) (1983) 966.
60. J.M. Taylor, M.A. Weinberger and L. Friedman, *Toxicol. Appl. Pharmacol.* **54** (1980) 57.
61. G.J. Walter and M.L. Mitchel, in *Alternative Sweeteners*, eds I.L. O'Brien Nabors and R.C. Gelardi, Marcel Dekker Inc., New York (1985).
62. R.W. Morgan and O. Wong, *Food Chem. Toxicol.* **23** (1985) 529.
63. *Report of the Scientific Committee for Food on Sweeteners*, Commission of the European Communities (1984).
64. C. Verbanic, *Chem. Bus.* **1** (1986) 29.
65. B. Armstrong, *Brit. J. Prev. Soc. Med.* **30** (1976) 151.
66. O. Tanaka, *Trends in Analyt. Chem.* **1** (11) (1982) 246.
67. Stevia Corporation Ltd, Technical Information 'Sato Stevia' (1986).
68. *The Merck Index* (9th edn), ed. M. Windholz, Merck & Co Inc., New Jersey (1976).
69. J.D. Higginbotham, *International Sweeteners Report* (4) (1986) 13.
70. F.J. Pilgrim and H.G. Schutz, *Nature* **183** (1959) 1469.
71. G.A. Grosby and J.R. Wingard in *Developments in Sweeteners*, Vol. 1, eds C.A.M. Hough, K.J. Parker and A.J. Vlitos, Elsevier Applied Science Publishers, Barking (1979).
72. K.J. O'Donnell, PhD Thesis, University of Reading (1983).
73. K.C. Phillip, in *Developments in Sweeteners*, Vol. 3, ed. T.H. Grenby, p. 21, Elsevier Applied Science Publishers, Barking (1987).
74. Ajinomoto Co., Japanese Patent 8111772 (1969).
75. S. Chang and J. Cook, *J. Agric Food Chem.* **31** (1983) 409.
76. H.C. Makapuga, N.P.D. Nanayakora and A.D. Kinghorn, *J. Chromat.* **283** (1984) 390-S.
77. M. Kobayashi, S. Horikawa, I.H. Degrandi, J. Ueno and H. Mitshuhashi, *Phytochem.* **16** (1977) 1405.
78. Morita Koga and Ku Kogyo, Japanese Patent 61202667 (1986).
79. G.E. Dubois, P.S. Dietrich, J.F. Lee, G.V. MuGarraugh and R.A. Stephenson, *J. Med. Chem.* **24** (11) (1981) 1269.
80. K. Nakayama, D.D. Kashara and F. Yamamote, *J. Food Hyg. Soc. Japan* **27** (1) (1986) 1.
81. Anon., *Scrip* **197** (1984) 22.
82. Anon., *Soft Drinks Management Int.* Dec. (1988) 12.
83. K. Weisberg, *Beverage World Int.* August (1987).
84. J.D. Higginbotham, US Patent 4011206 (1977).
85. J.D. Higginbotham, in *Developments in Sweeteners*, Vol. 1, eds C.A.M. Hough, K.J. Parker and A.J. Kitos, Elsevier Applied Science Publishers, Barking (1979).
86. J.D. Higginbotham, in *Alternative Sweeteners*, eds L. O'Brien Nabors and R.C. Gelardi, Marcel Dekker Inc., New York (1985).

87. R. Horowitz and B. Gentili, In *Alternative Sweeteners*, eds L.O'Brien Nabors and R.C. Gelardi, Marcel Dekker Inc., New York (1985).
88. P.J. Pratter *Perfumer and Flavourist* **7** (15) (1980) 12–4 + 16–8.
89. J.F. Fisher, *J. Agric. Food Chem.* **25** (3) (1977) 682.
90. R. Schwarzenbach, *J. Chromat.* **129** (31) (1976) 9.
91. Report of the Scientific Committee for Food on Sweeteners, December (1987).
92. Anon., *Food Eng. Int.* June (1986) 28.
93. Pfizer Inc., European Patent 34876 (1980).
94. Pfzier Inc., US Patent 4517379 (1985).
95. Pfizer Chemicals, Alitame Technical Summary – Publicity material (1987).
96. Landell Mills Commodities, Presentation given at Annual Meeting of ISA, Madrid, 17–18 May (1988).
97. Anon., *Food Chem. News* **28** (31) (1986) 50.
98. Pfizer, Personal Communication (1988).
99. L. Hough and S.P. Phadnes, *Nature* **263** (1976) 800.
100. L. Hough and J. Elmsley, *New Scientist* **1573** (1986) 41.
101. Tate & Lyle Plc, UK Patent Application GB 2153651A (1985); published 29 August, Application No. 8503284.
102. Tate & Lyle Plc, US Patent 4495170 (1985).
103. Tate & Lyle Plc, UK Patent Application GB 2154850A (1985); published 18 September, Application No. 8503285.
104. M.J. Jenner, Paper given Beh'r Seminar: Artificial Sweeteners and Sugar Substitutes, 8–9 May (1988).
105. Anon, *Food Engineering* January (1988).
106. L. Yeomans, Tate & Lyle, Personal communication (1988).

6 Flavourings and emulsions

G. HOPKINS

6.1 Flavourings

Although flavourings are normally used in extremely small quantities in a carbonated soft drink, their impact can make the difference between a tasty product and one that is bland and uninteresting. If follows, therefore, that many flavourings are highly concentrated and their application dose rate must be optimised with great care.

6.1.1 Legislation

The use of flavourings is controlled in most countries by the Food Regulations, which should be checked very carefully. Flavourings are normally classified into three categories:

- *Natural flavourings,* in which the components are obtained by an appropriate physical process (including distillation and solvent extraction) or an enzymatic or microbiological process from material of vegetable or animal origin, either in the raw state or after processing for human consumption by traditional processes of food preparation (including drying, torrefaction and fermentation).
- *Nature identical flavourings*, which are produced by chemical synthesis or isolated by a chemical process, and are chemically identical to a substance naturally present in material of vegetable or animal origin.
- *Artificial flavourings*, which are produced by chemical synthesis but are not chemically identical to a substance naturally present in material of vegetable or animal origin.

To illustrate this, consider bitter almond oil or, as it is commonly known, benzaldehyde (Figure 6.1). When it is extracted totally from almonds, it is *natural*. When it is produced by either the oxidation of benzene carbinol or by chlorinating toluene to produce dichlorotoluene and then saponifying with lime water, it is *nature identical*.

In some instances, the use of an artificial ingredient may be preferable, because it contains no impurities. Analysis can be undertaken with such accuracy that traces of undesirable components can easily be detected. Some

Figure 6.1 Bitter almond oil or benzaldehyde.

of these impurities may be harmful and too costly to remove. However, if the trace components are not harmful, the material may be used.

Although many flavourings will contain ingredients from two or even all three of these categories, it is normal for the status of a flavour to be given as that of the lowest percentage component. This can often mean that a flavouring that contains 99.5% of natural components and only 0.5% artificial ones will be given the status *artificial*.

6.1.2 *Creation*

The creation of new flavourings is a skilled job and can only be undertaken by someone with several years' experience. A flavourist will have a wide range of raw materials available, arranged on an 'organ' (Figure 6.2) from which the appropriate ingredients will be selected. A flavouring will often consist of over 25 individual ingredients. For example, raspberry can be broken down to several basic types – top notes, fruity, green, berry, background, woody/pippy, and sweet – the whole of which will combine to give a full round flavouring.

6.1.3 *Production*

The production of flavourings can be as simple as mixing two or three ingredients together. In most cases, however, the use of sophisticated and specialised equipment is necessary. Below is a brief description of some of the various techniques used by the flavour industry.

Distillation. A typical example of this method uses soft fruit. The fruit is crushed and at the point of fermentation pure alcohol is added. The flavour and aroma pass into the solvent. The alcoholic solution is then filtered and by fractionated distillation the alcohol is recovered (for further use) leaving the concentrated flavouring behind. This is done under vacuum conditions so as not to harm the delicate flavour which may be damaged by excessive heat.

Extraction. Probably the best-known extraction process is in the manufacture of separation flavourings. In this method, a citrus oil can be washed with a mixture of solvent and water to extract the oxygenated (flavour-containing)

FLAVOURINGS AND EMULSIONS

Figure 6.2 A flavourist's 'Organ'. (Courtesy of IFF.)

compounds from the insoluble terpenes. The resulting soluble flavouring can be boosted with other ingredients to give the desired finished flavouring.

Lemon oil	80 g
Isopropyl alcohol	257 g
Water (de-ionised)	235 g

This mixture is stirred well and allowed to separate. The alcohol and water mix is then run off. To the remaining lemon oil add

| Isopropyl alcohol | 257 g |
| Water (de-ionised) | 235 g |

Stir well and allow to separate. Once again run off the alcohol and water mix and combine this with the previous mix. Now add

| Lemon oil, terpeneless | 12 g |
| Citral | 5.2 g |

Stir well and filter clear. This will produce a good-quality lemon flavouring that can be used at a dose rate of 0.6 g/l ready-to-drink beverage.

Maceration. As the word implies, this method involves soaking the raw material. It is commonly used for citrus peel, herbs, such as basil and mint, or spices such as ginger or chillies. The material is ground and placed in a tank with solvent to extract the flavour. A period of time may be allowed to elapse for full extraction, and this may vary from a few hours to several months. Eventually, the liquid is separated by decantation and is filtered. The use of ultra-waves can speed up this process and is now being used successfully on large-scale production.

Carbon dioxide extraction. Carbon dioxide can exist as a solid, liquid or gas, and is easily isolated in any of these forms. At normal atmospheric temperature and pressure, the solid form becomes gas without passing through the liquid phase. However, when solid carbon dioxide is heated under pressure, the liquid form is produced. The use of liquid carbon dioxide is becoming increasingly important for the extraction of delicate natural products. Although the cost of high-pressure equipment is expensive, carbon dioxide has several advantages over other liquid solvents:

CO_2 is non toxic
CO_2 is not flammable
CO_2 does not cause environmental problems
CO_2 is relatively cheap.

Natural products usually contain a large number of different chemical compounds which, with different solvents, can be extracted to a greater or lesser extent depending on their solubility in that solvent. The use of carbon dioxide can be very important with delicate compounds as extractions are normally carried out at low temperatures ($-20\,°C$ to $+20\,°C$) thus ensuring that the material is not damaged by excessive heat.

The use of materials produced by carbon dioxide extraction is growing, enabling the flavour industry to develop flavourings more true to nature than previously.

6.2 Emulsions

An emulsion can be described as a dispersion of one liquid in another in which it would not normally be miscible. Most emulsions used in carbonated soft drinks consist of two phases, which are homogenised under positive pressure. The following indicates some of the ingredients used in beverage emulsions.

Oil phase: Oils – essential or otherwise
Flavour chemicals
Oil-soluble colours
Weighting agents
Anti-oxidants

Water phase   Gums
               Water-soluble chemicals
               Modified starches
               Pectins
               Preservatives

Although a wide range of materials is used to produce these emulsions, the ingredients are normally controlled by the Food Regulations which specify not only the flavourings but also the emulsifiers and stabilisers that can be used.

The main use of emulsions in carbonated soft drinks is as a clouding agent. When considering the organoleptical features of a soft drink, the visual appearance is extremely important as the 'eye appeal' can often be the determining factor in deciding which drink is actually purchased. The main requirements of an acceptable emulsion are:

Ease of use   The emulsion must disperse readily when mixed into the syrup.

Taste   The emulsion should not interfere with the taste or aroma of the drink.

Stability   The finished drink should be stable and have a shelf life of six or twelve months, as appropriate, without ringing or sedimentation.

Legal   The emulsion must comply with the current regulations of the country(ies) in which the drink is retailed.

6.2.1 Manufacture

Although there are many machines that can be used for the manufacture of emulsions, it is the opinion of this author that the best and most reliable results are obtained with an APV Gaulin Homogeniser (Figure 6.3). This machine is described in APV's literature as a simple three-throw positive displacement pump which operates via a belt-drive from an electric motor. This drive is transmitted to three pistons reciprocating in a cylinder block. On the suction stroke of each piston, liquid is drawn into the pump cylinder via a suction valve; on the discharge stroke the liquid is fed to a two-stage homogenising valve inside which the liquid impinges on impact rings set at right angles to the direction of flow, causing a high degree of homogenisation.

In a typical process the following procedure is followed:

Oil phase: Ester gum   64 g
               Orange oil (Brazil)   63 g
               Orange oil (Florida)   25 g
               Mandarin oil (Italy)   2.5 g
               Bulylated hydroxyanisole   q.s

Figure 6.3 APV Gaulin homogeniser. (Courtesy of IFF.)

Blend the oils, mix in the BHA anti-oxidant and then dissolve the ester gum, gently warming the oil blend if necessary. Direct heat to the oils must be avoided to ensure the flavour is not damaged.

Water phase:	Spray dried gum arabic	155 g
	Water	688.5 g
	Sodium benzoate	1 g
	Citric acid	1 g

Dissolve the gum arabic in hot water using a high-speed stirrer. Sodium benzoate is then added and finally the citric acid. The whole mix is then sieved or passed through a centrifuge to ensure no particulate matter remains. Deaeration may also be necessary. Cool to a maximum of 20 °C.

When both phases are prepared, the oil phase is added slowly to the water phase while stirring fast enough to draw the oil phase into the mix, but at the same time ensuring that air is not incorporated.

Stirring is continued for a short period to reduce the size of the oil droplets. Finally, the mixture is passed through a two-stage homogeniser at pressures sufficient to reduce the particle size to less than 1 μm.

An emulsion of this type will have a dose rate of about 1.5 g/l ready-to-drink beverage or less if fruit juice is included in the recipe.

6.3 Application of flavourings and emulsions

The dose rate of flavourings in a carbonated soft drink is normally extremely low. Nevertheless, the taste, aroma and appearance of the drink are critical to its success in the market place.

6.3.1 Selection

The type of flavouring material used will depend upon the finished drink required. If cloudiness is needed, an emulsion will be used either as a neutral cloudifier or in combination with flavouring oils and, possibly, chemicals to give a flavoured emulsion. Where cost is not a factor, or legal reasons dictate, a natural flavouring will give a more 'fruit-like' character, but this must be weighed against the problems of oxidation and spoilage, which are less likely in artificial flavourings.

The regulations of the country in which the drink is to be marketed must also be considered. This applies to the flavouring ingredients and the solvent system used. Most flavour companies will furnish, on request, full details of the status and legality of the flavourings they supply.

Finally, the storage conditions of the flavourings will have to be considered. There is no point in selecting a flavouring that must be stored at 4 °C to maintain its character if such storage conditions are not available at the factory. Normally, flavourings should be kept cool and in the dark, and under these conditions will maintain a good quality over a six- to twelve-month period. Subjecting flavourings to heat (either by placing near a steam pipe or in a sunny area) can cause a considerable amount of evaporation, especially when isopropyl alcohol or ethyl alcohol is used as the solvent. Equally, unsealed containers can be spoiled by airborne contamination, and even air can cause oxidation and give off-tastes to the drink.

6.3.2 *Methods of use*

The accurate measurement of flavourings is very important as, once a flavouring has been standardised in a beverage, subsequent production must not vary in the smallest detail. In the author's opinion, this can only be achieved by measuring the quantity of flavouring by weight and not by volume. The variations that can occur in a solvent-based flavouring where isopropyl alcohol has been used as the solvent can be quite large owing to the expansion of the low specific gravity material when subjected to the common temperature variations often found in the laboratory and syrup rooms. Such variations will increase (or decrease) the dose rate of that flavouring, and because the flavouring is so concentrated the effect on the drink can be greatly out of proportion to the actual volume error.

Emulsions should always be diluted with an equal amount of water before being added to the syrup. This avoids the possibility of gelling the gums, which, apart from making the drink look unappetising, can also cause a rapid deterioration of the cloud, giving a clear end product.

6.4 Evaluations

Having selected the appropriate flavouring material and developed the carbonated soft drink so that the taste, aroma and appearance are acceptable, a series of evaluations must be undertaken to ensure the full stability of the drink over the required shelf-life.

Process evaluation. With the many and varied process conditions being used in the drinks industry today, it is surely necessary to subject any new drink to the conditions that will be in use during manufacture. The effect of heat, either by flash pasteurisation or in-can pasteurisation, can lead to flavour losses and these must be compensated when finalising the formulation. The homogenisation of fruit-containing beverages will have a marked effect on the appearance of that drink and must be taken fully into account before the stability of the drink is accepted.

In-pack storage. The shelf-life of a carbonated beverage can be dramatically affected by the container into which it is poured. Plain glass and PET bottles allow ultraviolet rays to penetrate the drink, and these can damage the flavouring, even giving variations in the stability of the same flavouring when packed in both containers. The effect of light is easily demonstrated by allowing a bottle of Tonic Water to stand in direct sun light for a few days. The loss of bitterness is due to the degradation of quinine by light and is a well-known phenomenon.

Cans obviously protect the product from the light, but because carbonated

Figure 6.4 The light box. (Courtesy of IFF.)

soft drinks are acidic, the lining of the can must be compatible with the drink otherwise contamination may occur.

Shelf testing. After the newly-developed carbonated soft drink has been found compatible with the process conditions and the packaging has been selected, some form of shelf test must be undertaken to ensure that the drink will be stable over a period of time. In the fast-moving world of beverages, it is not always possible to evaluate a product over, say, twelve months, so an accelerated storage test has to be undertaken. This can be achieved by use of a light box (Figure 6.4), a simple device that should be large enough to accommodate the number and size of packs that need to be evaluated at any one time. Daylight tubes are used for the light source, and their position should affect both the top and the bottom of the beverage container. A thermostatically controlled fan ensures that overheating does not take place. The temperature in the box can be adjusted to suit those found in the country for which the beverage is designed. Light boxes of the type illustrated have been used by International Flavours and Fragrances over many years to ensure the stability of the flavourings offered to customers.

Acknowledgement

I should like to thank my colleagues for their help with this chapter.

7 Acids, colours, preservatives and other additives
B. TAYLOR

7.1 Introduction

The commercial success of a soft drink formulation depends upon a number of factors. A strong, well-placed advertising campaign will bring the consumer to purchase the new product but, thereafter, the level of repeat sales will reflect the degree of enthusiasm with which the new drink has been received.

Taste panelling and market trials are also preliminaries to a successful launch, yet continuity of sales will ultimately depend upon the product itself, primarily its appearance and taste, as assessed by the consumer, and then, perhaps, the reproducibility of quality in both manufacture and storage – these latter being the major concerns of the producer and soft drinks retailer, who must maintain a regular turnover to survive.

It is hardly surprising that the development of a new drink product can take many months, while all aspects of its appearance, organoleptic properties and stability are tuned to requirements. In the final analysis, organoleptic properties are paramount, and the aroma, taste and mouth-feel must be complementary in their contribution to the resulting drink. However, the immediacy of colour and its importance to the success of the product cannot be underestimated.

In recent years, the use of synthesised ingredients has frequently been under attack by the media and, as a result, market forces in many countries have initiated a rapid move in the direction of natural ingredients.

We have seen an influx of various natural colour extracts to the food industry which, being largely pH dependent and light sensitive, have found limited use in soft drinks. A few have found acceptance, but even so are still open to scrutiny in terms of adverse metabolic effects. Many have no recommended ADI (Acceptable Daily Intake in mg/kg body weight) values, while others have values allocated which are not far removed from those of the synthetic colours they have replaced (cf. Table 7.4).

Preservatives also show signs of being phased out, as improved methods of pasteurisation and aseptic filling are devised. The ability of carbon dioxide to act as a preservative places carbonated drinks in a strong position for future development.

A typical carbonated soft drink comprises carbonated water, sugar, citric

ACIDS, COLOURS, PRESERVATIVES AND OTHER ADDITIVES   91

Table 7.1 Components of a carbonated soft drink

Component	Typical use level	Contribution
Water (quality to meet rigid requirements)	Up to 94% v/v	Bland carrier solvent for other ingredients (must be chlorine free)
Sugars	7–12% m/v	Sweetness, body, balance to flavour
Carbon dioxide	0.3–0.6% m/v	Provides mouth-feel and 'sparkle' to drink
Acids (e.g. citric)	0.05–0.30% m/v	Sourness, sharpness background to flavour; increases thirst quenching effect
Flavours Nature identical and artificial Natural	0.1%–0.28% m/v Up to 0.5% m/v	Provides flavour character and identity to the drink
Emulsion (flavour or otherwise)	0.1% v/v	Produces cloud; opalescence in drink to replace or enhance cloud effect from natural juices (e.g. crush)
Colour (natural or synthetic)	0–70 ppm	Standardises and identifies colour tone of drink
Preservatives	Statutory limits generally apply, eg. SO_2 70 ppm, benzoic 160 ppm in UK	Restricts microbial attack and prevents destabilisation
Anti-oxidants (e.g. BHA, ascorbic acid)	Less than 100 ppm subject to user country legislation	Limits flavour and colour deterioration owing to oxidation
Quillaia extract (saponins)	Up to 200 ppm (UK), up to 95 ppm (USA)	Heading foam e.g. in shandies, cream soda, and colas
Hydrocolloids (gums), e.g. carageenans, alginates, polysaccharides, etc.	0.1–0.2% GMP minimum amount required to create desired effect	Mouth-feel, shelf-life stability, viscosity
Vitamins/minerals	ADI applies	Nutritional requirements

acid, flavouring, acidity regulators (e.g. sodium citrate), colouring, preservative and artificial sweeteners, if used (Table 7.1). The flavour component is presented against a finely tuned backcloth of the other ingredients, providing the right degree of sweetness, bitterness, sourness, and acidity (pH) to enhance drink palatability.

7.2 Acids

Following water and sugar, the acid component is third in terms of concentration (Table 7.2). Its presence tends to be taken for granted, yet, without its contribution, the other formula components are left lacking in character. Because of the general tartness or sourness in taste, acidity is useful in modifying the sweetness of sugar. It will increase the thirst-quenching effect of the drink by stimulating the flow of saliva in the mouth and also, because of a reduction in pH level, tends to act as a mild preservative. While the majority of soft drinks contain acids, it is the carbonated drinks that have the additional

Table 7.2 Acidulants used in beverage formulation

CITRIC ACID		
2-Hydroxy-1,2,3-propane tricarboxylic acid		
$HO \cdot OC \cdot CH_2 \cdot C(OH), (CO \cdot OH) \cdot CH_2 \cdot CO \cdot OH$	mw 192.1	mp 153–154 °C
pK_1, 3.08, pK_2 4.74, pK_3 5.4		
TARTARIC ACID (d-tartaric)		
2,3-Dihydroxy butan-dioic acid		
$HO \cdot OC \cdot CH(OH) \cdot CH(OH) \cdot CO \cdot OH$	mw 150.1	mp 171–174 °C
pK_1 2.98, pK_2 4.34		
PHOSPHORIC ACID		
Orthophosphoric		
H_3PO_4	mw 98	mp 42.4 °C
pK_1 2.12, pK_2 7.21, pK_3 12.67.		
LACTIC ACID (DL-Lactic)		
2-Hydroxy propanoic acid		
$CH_3 \cdot CH(OH) \cdot CO \cdot OH$	mw 90.1	mp 18 °C
pK_1 3.08		
ACETIC ACID		
Ethanoic acid		
$CH_3 \cdot CO \cdot OH$	mw 60	mp 16–18 °C
pK_1 4.75		
MALIC ACID (d-malic)		
2-Hydroxy butan-dioic acid		
$HO \cdot OC \cdot CH(OH) \cdot CH_2 \cdot CO \cdot OH$	mw 134.1	mp 98–102 °C
pK_1 3.4, pK_2 5.11		
FUMARIC ACID		
trans-Buten-dioic acid		
$HO \cdot OC \cdot CH = CH \cdot CO \cdot OH$	mw 116.1	mp 299–300 °C
pK_1 3.03, pK_2 4.44.		

pK values taken from CRC, *Handbook of Chemistry & Physics*.

ACIDS, COLOURS, PRESERVATIVES AND OTHER ADDITIVES

effect of dissolved carbon dioxide. Not officially recognised as an acid addition, the presence of carbon dioxide under pressure certainly provides that extra sparkle to mouth-feel, flavour and sharpness (or bite) to the drink, so it has been included here under the identity given to its soluble form.

7.2.1 Carbonic acid

The solution of carbon dioxide in water exploits weakly acidic properties. Neither liquefied nor dry gaseous carbon dioxide affects dry blue litmus indicator paper, but if the paper is moistened it will provide an acid reaction in contact with the gas. There is little doubt that in solution some of the gas forms carbonic acid by combination with water.

$$H_2O + CO_2 \longrightarrow H_2CO_3$$

Carbonic acid, however, is unstable and has never been isolated. It is dibasic and forms two series of salts: the carbonates and the bicarbonates. By acting carbon dioxide upon a solution of sodium hydroxide, for example, the normal sodium carbonate is first formed:

$$2NaOH + CO_2 \xrightarrow{pH > 10} Na_2CO_3 + H_2O$$

With an excess of the gas being passed through the solution, sodium bicarbonate then results:

$$Na_2CO_3 + H_2O + CO_2 \xrightarrow{pH < 8} 2NaHCO_3$$

Potassium and sodium carbonates can be used in the production of 'dry' carbonated drink mixes, where a blend of sugars, fruit acid crystals, spray-dried flavourings and other additives such as stabilisers is formulated to produce a drink which, when dissolved in water, has a carbonation level of about $1-1\frac{1}{2}$ volumes carbon dioxide. In its more regular role, during the production of carbonated drinks, carbon dioxide is introduced as part of the bottling sequence, being dissolved under pressure before or after dilution of the bottling syrup with water. The practicalities of carbonation are dealt with in Chapter 11, and it will be appreciated that the level of dissolved gas needs to be carefully controlled to maintain the flavour balance of certain drink products. Measured in volumes of dissolved gas per unit volume of water at a specified temperature and pressure (usually 'Volumes Bunsen' at 0 °C and 1 atm), the average level employed is in the region of three volumes although extremes of perhaps one volume and six volumes are sometimes encountered where highly specialised flavoured products are required.

7.2.2 Citric acid

This is by far the most widely used acid in fruit-flavoured beverages. It has a light fruity character that blends well with most fruits and, in fact, is found as a major constituent in many of them, e.g. unripe lemons contain 5–8% of the acid. It is also the chief acid constituent of currants, cranberries, etc., and is associated with malic acid in apples, apricots, blueberries, cherries, gooseberries, loganberries, peaches, plums, pears, strawberries and raspberries, with isocitric acid in blackberries and with tartaric acid in grapes.

It was originally obtained commercially from lemons, limes or bergamots by pressing the fruit, concentrating the expressed juice and precipitating citric acid as its calcium salt by running in, with constant stirring, a slurry of chalk and water. The crude calcium citrate was then filtered off, filter pressed and washed prior to treatment with sulphuric acid to yield the free citric acid, which was then filtered from the precipitated calcium sulphate, and finally isolated by concentration of its solution by boiling, from which crystals of the monohydrate formed.

It was noted at the time of Dr Martin's Treatise on Industrial and Manufacturing Chemistry (1917) that a known organism existed – *Mucor Piriformis* (C. Wehmer, German Patent 72,957) – that could ferment sugar directly into citric acid. Owing to the low market prices of Sicilian lemon juice, no wide technical application of this early enzyme process had been made. However, citric acid is now produced by the action of specific enzymes upon glucose and other sugars.

Citric acid is a white crystalline solid and can be purchased in its powdered form or as the monohydrate. This latter state is more convenient in terms of storage, as it does not have a tendency to absorb moisture, as does the anhydrous form.

7.2.3 *Tartaric acid*

This acid occurs naturally in grapes as the acid potassium salt and, during fermentation of grape juice, will be seen to deposit from solution as its solubility decreases with increasing alcoholic content of the wine. The acid can be obtained in four forms: dextro, laevo, meso-tartaric and the mixed isomer-equilibrium, or racemic acid. Commercially it is usually available as the dextro-tartaric acid. The acid possesses a sharper flavour than citric and, as such, may be used at a slightly lower rate to give an equivalent palate acidity. (Note that palate acidity is a purely subjective measurement and it is generally agreed that a number of acids may be used at a concentration different to that indicated by their chemical acid equivalent (see Table 7.3).)

Tartaric acid may be isolated from the crude deposit of tartrates obtained from the wine fermentation process in a similar manner to that originally used for citric acid – by leaching the deposit with boiling HCl solution, filtering clear and re-precipitation of the tartrates as the calcium salt. Further treatment with sulphuric acid is used to liberate the acid, which can then be purified by crystallisation.

Tartaric acid (dextro form) exists as a white crystalline solid mp 171–174 °C.

Table 7.3 Palate acidity* equivalents of various acids

Acid	Concentration (g/l)**
Acetic	1.00
Ascorbic	3.00
Citric	1.22
Lactic	1.36
Malic	1.12
Phosphoric	0.85
Tartaric	1.00

*Palate acidity, while a subjective measurement, can give a guide to equivalent sourness of the acid type.
**The above concentrations in water were considered to be equal in palate acidity following a series of taste trials carried out in the laboratories of Barnett & Foster International.

If used in beverage production, the acid must be perfectly pure and guaranteed for 'food' use. It has disadvantages in that its salts are of a lower solubility than those of citric, particularly the salts of calcium and magnesium. When using hard water, it is therefore advisable to use citric acid to avoid unsightly deposition of insoluble tartrates.

7.2.4 Phosphoric acid

The acid is derived from mineral and not vegetable sources although occurring naturally in some fruits, e.g. limes, grapes, in the form of phosphates. It is used in some beverages as a substitute for, or in addition to, citric and tartaric acids, having a sharper and drier flavour than either of the above acids. Its taste is of flat 'sourness', in contrast with the sharp fruitiness of citric acid, and it seems to blend better with most non-fruit drinks. In the UK, it is not allowed in drinks claiming the presence of fruit juices and comminuted fruits. Its main use is in cola-flavoured beverages, where its special type of acidity complements the dry, sometimes balsamic, character of the cola drinks.

Pure phosphoric acid is a colourless crystalline solid (mp 42.35 °C) but is usually used in solution as a strong, syrupy liquid, miscible in water in all proportions. It is commercially available in concentrations of 75, 80 and 90%. The syrupy character is the result of hydrogen bonding, which occurs at concentrations greater than 50%, between the phosphate molecules. It is corrosive to most construction materials and rubber-lined steel or food-grade stainless steel are recommended for holding vessels.

7.2.5 Lactic acid

Sometimes used for the acidification of beverages, lactic acid possesses a smoother flavour than any of the foregoing acids. It is supplied commercially as an odourless and colourless viscous liquid and is obtained from the fermentation of sugars by lactic acid bacillus.

7.2.6 *Acetic acid*

As in the case with phosphoric acid, under UK legislation this acid is limited to use in non-fruit juice drinks and really only qualifies where its vinegary character can contribute to a suitable flavour balance. Pure glacial acetic acid is a colourless, crystalline solid of mp 16 °C and is one of the strongest of the organic acids in terms of its dissociation constant and displacing carbonic acid from its carbonates.

7.2.7 *Malic acid*

This is the natural acid found in apples and other fruits. A crystalline white solid (mp 100 °C), it is highly soluble in water. Being less hygroscopic than citric acid it possesses improved storage and shelf-life properties.

Malic acid is slightly stronger than citric in terms of perceived palate acidity and imparts a fuller, smoother, fruity flavour. It is, of course, first choice for apple-flavoured drinks.

Unlike tartaric, its calcium and magnesium salts are highly soluble and the acid presents no problems in hard-water areas.

7.2.8 *Fumaric acid*

Not permitted under UK soft drinks legislation, fumaric acid is widely used in other countries as an acidulant, notably in the US market.

In terms of equivalent palate acidity it can be used at a lower rate than citric acid and typical replacement can be employed at two parts fumaric per three parts citric in water, sugar water and carbonated sugar water. Its main drawback is a reduced solubility compared with the citric acid and special methods need to be employed in getting it into solution.

7.2.9 *Ascorbic acid*

This acid (known as Vitamin C) is not only used as a contributory acidulant but rather as a stabiliser within the soft drinks system and its anti-oxidant properties improve the shelf-life stability of the flavour component in many cases.

As will have been gleaned from Chapter 6, many of the ingredients used in flavourings are susceptible to oxidation, particularly the aldehydes, ketones and keto-esters. Ascorbic acid shields these from attack by itself becoming preferentially oxidised and lost, leaving the flavour component unaffected.

It should be noted, however, that while a browning inhibitor in unprocessed fruit juices, the effect can later be reversed should the juice be subsequently heat treated (pasteurised) when the ascorbic acid present can itself initiate a chemical browning reaction. Another disadvantage of ascorbic acid is its effect upon some colours in the presence of light.

7.3 Colours

The sensory perception of colour will influence the taster's reception of the drink. It has been generally demonstrated that the colour can far outweigh the flavour in the impression made upon the consumer. Both quality and quantity of colour are of importance and certain colours provoke, or perhaps complement, a particular taste. Reds will favour the fruitiness of soft drinks, e.g. blackcurrant, raspberry, strawberry, etc. Orange and yellow tend towards the citrus flavours. Greens and blues reflect the character of peppermints, spearmint and cool flavours, sometimes herb-like and balsamic and the browns align with the heavier flavours, e.g. colas, shandies, dandelion and burdock.

Having briefly considered the influential qualities of colour, let us return to practicalities. How do we create the illusion of colour in a soft drink and what factors will govern the choice of colouring agents?

There is little doubt that in the early years many questionable practices were involved in beverage production and there is an interesting reference in *Skuse's Complete Confectioner – A Practical Guide to the Art of Sugar Boiling in all its Branches*. This book, published *c*. 1890, contained information on cordials and other beverages and, under its section on flavours and colours, the author felt it necessary to point out the dangers of using certain colours such as sulphate of arsenic, iodide of lead, sulphate of mercury, carbonate or sulphate of copper and seriously admonished the used of chrome yellow (lead chromate) by certain confectioners who were partial to using a little chrome yellow for stripes in sweets. Such colours were officially banned from food use in 1925.

Today, the use of food colouring is carefully controlled under various legislations, with an ongoing programme of toxicological studies where there is suspicion of harmful or allergic effects.

Both the EEC (European Economic Community) and the FDA (Food and Drug Administration of the USA) have published permitted lists that are under regular review. Most concern has been expressed over the azo colours as certain people can demonstrate an allergic reaction to some of them. Toxicological and allergic reactions have been reported most frequently with Sunset Yellow (E110) and Tartrazine (yellow) E162.

It has been found by experience that a number of food colours give a broadly satisfactory performance in soft drinks and carbonated beverages.

The colour properties can be affected by a number of soft drink ingredients and good storage stability is required in the presence of acids, flavouring compounds and, where necessary, the preservative. The colour component must also be stable in the presence of light. It is well known that the combination of ascorbic acid and light has a detrimental effect on many colours. While it can be said that the colours permitted for soft drinks have a reasonably good all-round performance, there is no substitute for storage trials in new product development to ascertain the real behaviour in the finished beverage.

Table 7.4 Colours used in soft drinks

Colour	E No.	CI No.	Colour stability	
			Fruit acids	Light
NATURAL				
Annatto	160b	75120	M	FG
β-Carotene	160a	40800	G	L
Curcumin	100	75300	G (pH 3)	L
Riboflavine	101	—	M	G
Riboflavine-5-phosphate (sodium salt)	(101a)*	—	G	G
Xanthophylls	161b	75135	G	G
Canthaxanthin	161G	40850	M	G
Red sandalwood	—	75540	G	G
Anthocyanins	163d, e, f	—	G	M
Hibiscus	—	—	G	M
(Grape Juice Blue) Anthocyanins	163d, e, f	—	Unstable (reddish)	M
Carmine	120	75470	G	VG
Beet-Red	162	—	G	M
Copper-chlorophyll	141	75810	FG	FG
Chlorophyll	140	75810	M	FG
Caramel	150	—	G	VG
ARTIFICIAL				
Quinoline Yellow	104	47005	VG	G
Tartrazine (FD & C Yellow No. 5)	102	19140	VG	G
Sunset Yellow (FD & C Yellow No. 6)	110	15985	VG	G
Ponceau 4R	124	16255	G	G
Carmoisine	122	14720	G	G
Amaranth	123	16185	VG	M
Patent Blue V	131	42051	Unstable (greenish)	M
Indigotine (FD & C Blue No. 2)	132	73015	M	L
Brilliant Blue FC (FD & C Blue No. 1)	133*	42090	VG	G
Food Green S	142	44090	G	L

Legend: M, Moderate; G, Good; VG, Very Good; FG, Fairly Good; L, Low.
E Numbers are in accordance with guidelines of EEC re food additives.
CI Number: Colour Index number allocated by Society of Dyers & Colourists (Bradford, England) and American Association of Textile Chemists and Colourists (Mass, USA).
FD & C Numbers: These colours are permanently listed for food use and subject to certification in USA.
*Under consideration by EEC for 'E' prefix.

The colours most commonly encountered in the soft drinks industry are Tartrazine, Sunset Yellow FCF, Carmoisine, Green S, Chocolate Brown HT, the caramels and the nature-identical carotenes (Table 7.4).

Amaranth, previously widely used, has lost ground since its exclusion from the US permitted list by the FDA of America in January 1976. Although Amaranth (E123) is still permitted in Europe, there has been a tendency towards the use of Carmoisine instead. Tartrazine and Sunset Yellow are being replaced more frequently by Quinoline Yellow, the slight differences in colour tone being compensated for in terms of intensity by altering dosage rates.

The consumer, also being remarkably tolerant, tends to demonstrate the fact that, unless one can make immediate comparisons – and batches of beverages using different colour types are not presented side by side – then the changeover to a new colour will have little effect on sales. In recent years there has also been an increase in the usage of natural colour extracts within the regulatory lists. Curcumin, carotenoids, (caramel) flavenoids, anthocyanins and chlorophyll have all been produced where necessary in water-soluble forms (emulsions, salts, etc.) with varying success.

During the eighteenth century there was little need for rigid laws controlling additives in food or drinks. Until the Industrial Revolution, food had been produced in Britain for the immediate needs of the local communities, and trade was restricted likewise to the immediate area. Producer and consumer were often neighbours with a high level of trust between them. However, the result of the new industrialisation changed all that.

Between 1834 and 1856 it was discovered that aniline, produced from coal tar (a by-product of coal gas manufacture), could, in conjunction with other agents, provide a wide range of vivid and fast colours. The patent (No. 1984, August 1856), taken out by a young chemist, William Henry Perkin, for a mauve colour produced from aniline, opened the door to a succession of new dye-stuffs from coal-tar products. These transformed the textile industry which had hitherto relied upon natural colouring extracts. Low cost and bright hues ousted the use of natural colours and had a marked effect upon world trade. In 1868, alizarin, the colouring principal of the madder root (*Rhubia tinctoria*), was prepared synthetically and, during the same period, natural indigo was also being displaced in commerce by the artificial version.

Textile manufacture was not the only use to which the new products were directed and a selection of them soon became available for food use, often with dire results to the consumer.

In 1925 the compounds of arsenic, antimony, cadmium, etc. (referred to earlier), were finally officially banned from use as legislation began to take hold. Even so, it appears that little was done until the early 1950s to regulate the use of food colours other than to ban from use those colours that had become obviously unsuitable for consumption, usually at the behest of interested parties following the outbreak of poisoning owing to excessive use of a particular additive. In 1954, a list of acceptable food colours was drawn up (hitherto only negative lists had been available) and subsequently, in 1957 and 1973, the list as we know it today was drawn up of both natural and synthetic colours.

In line with greater concern over the food we consume, there is a greater regard to the toxicological effects of food additives in general and, accordingly, not only do we consider the suitability but also the Acceptable Daily Intake (ADI). This is expressed in milligrams per kilograms of body weight as the amount of food additive that can be taken daily in the diet, without risk.

Within the EEC, the allocation of ADI values is the responsibility of JECFA (the Joint Expert Committee on Food Additives), which comprises experts representing the World Health Organisation (Geneva) and the Food and Agricultural Organisation of the United Nations, (often referred to as WHO/FAO).

Control of food additives in the USA comes under the auspices of the FDA, who have devised a permitted list of additives. The EEC and FDA lists, while subjected to a similar degree of toxicological testing, may differ in content. For instance, Amaranth, permitted by the EEC list, was de-listed by the FDA in 1976.

The subject is controversial and it is often difficult to identify the actual number of persons showing the allergic reaction, as the offending substance may only show adverse effects when in combination with a food or beverage to which the person is also allergic. The major deterrent is the list of ingredients on the label, which enables those who are allergic to identify the substance and avoid intake.

7.4 Preservatives

A preservative can be defined as any substance that is capable of inhibiting, retarding or arresting the growth of micro-organisms or any deterioration of food due to micro-organisms or of masking the evidence of any such deterioration, etc.

Defined maximum levels for named preservatives are given according to the food substrate; in the case of soft drinks for consumption without dilution, the UK regulations read as follows (see also Table 7.5):

		mg/kg	E No.
	Sulphur dioxide	70	E220
	Benzoic acid	160	E210
or	Ethyl 4-hydroxy benzoate	160	E214
or	Propyl 4-hydroxy benzoate	160	E216
or	Sorbic acid	300	E200

(Schedule 2, *Preservatives and Food Regulations 1979*, SI 1979, No. 752 as amended.)

Table 7.5 Preservatives used in soft drinks

Preservative	E No.	Alternative form used at equivalent level	E No.
COOH-C₆H₅ (benzene ring) mp 122 °C **BENZOIC ACID**	E210	Sodium benzoate Potassium benzoate Calcium benzoate	E211 E212 E213
COOR-C₆H₄-OH structure where R = —CH$_3$ (methyl) mp 126 °C —C$_2$H$_5$ (ethyl) mp 116 °C —C$_3$H$_7$ (propyl) mp 97 °C Methyl-4-hydroxybenzoate Ethyl-4-hydroxybenzoate Propyl-4-hydroxybenzoate **PARABENS**	E218 E214 E216	Sodium salt (of E218) Sodium salt (of E214) Sodium salt (of E216)	E219 E215 E217
$CH_3 \cdot CH=CH_2-CH_2=CH \cdot COOH$ 2,4-Hexadienoic acid, mp 133 °C **SORBIC ACID**	E200	Sodium sorbate Potassium sorbate Calcium sorbate	E201 E202 E203
SO_2 **SULPHUR DIOXIDE**	E220	Sodium sulphite Sodium hydrogen sulphite Sodium metabisulphite Potassium metabisulphite Calcium sulphite Calcium hydrogen sulphite	E201 E222 E223 E224 E226 E227

While not added specifically as a preservative, carbon dioxide contributes to the inhibition of micro-organic growth and, coupled with other factors (e.g. pH), contributes well to the micro-stability of the drink. It is deemed to be effective at over 2.5 or 3.0 volumes CO_2 and for this reason the incidence of micro-damage in carbonated beverages is less of a problem than with the non-carbonated varieties.

While preservatives can prevent spoilage of food products, they should not be considered as a panacea for all micro-organic deterioration in foods.

Certain strains of yeast moulds and bacteria are able to survive and some can indeed thrive in the presence of preservatives. The old maxim of 'prevention is better than cure' applies equally to soft drinks as to other food products. It is therefore essential that all raw materials used in the formulation are carefully checked for microbial activity *before use* and are deemed to fall within realistic levels. Equally, all processing plant, machinery, or containers likely to come into contact with the product should be subjected to thorough cleaning (sanitisation) before use. Everything must be done to prevent the yeast organism from growing, which it can at an alarming rate (according to strain). Under favourable conditions a typical rapidly growing yeast strain can double its numbers every 30 minutes. In 12 hours 1 yeast could become 16.7 million at this rate, providing no inhibitory factor is present.

7.4.1 *Micro-organisms and soft drinks*

Because of their utilisation of sugars to produce alcohols and carbon dioxide, yeasts are of more immediate concern. Yeasts (which are in fact classified with the fungi and are unicellular for most of their life cycle), moulds and bacteria can bring about deterioration in flavour, producing taints, off-notes, differences in mouth-feel, etc. Most yeasts can grow with or without oxygen, whereas some bacteria cannot survive in it. Most yeasts thrive at temperatures between 25 and 27 °C; some can survive over 70 °C and others can exist, apparently quite comfortably, at 0–10 °C.

Bacteria exhibit similar characteristics, with the optimum growth temperature being 37 °C.

Micro-organisms require nutrient media to enable growth. The range depends upon the individual organisms but apart from water, as the environmental necessity, typical requirements are sources of carbon (carbohydrates), nitrogen (amino acids), phosphorus (phosphates), potassium, magnesium, calcium (mineral salts) and traces of other minerals, e.g. sulphur, iron, cobalt, and even vitamins.

Because of the obvious link with protein formation during cell growth, the presence of combined nitrogen is of particular importance. Also, where introduced to beverages via fruit pulp or caramel (colouring), there will be greater susceptibility to spoilage by certain micro-organisms.

It is useful to consider the individual status of a bottled drink. It exists as a unique system that can enhance or inhibit the growth of micro-organisms. Micro-flora, which may be present, will enter a dormant stage while the chances of survival are 'assessed'.

Following this lag stage, during which specific micro-flora adapt to their new environment and start to grow, there is a burst of activity (dependent upon the species) during which the population will double itself repeatedly at a steady rate. Waste products and diminishing nutrients will act to slow down this growth and eventually bring it to a standstill when, perhaps, the nutrient materials are exhausted and the death rate is seen to increase.

However, in practical terms the drink, while not necessarily a health hazard, no longer exhibits those qualities intended by its formulation.

7.4.2 Sulphur dioxide

Because of the ease with which gaseous SO_2 can be produced, it was one of the earliest chemical compounds manufactured and used by man. Homer (c. 800 B.C.) refers to the use of burning sulphur in fumigation, Pliny (c. A.D. 100) reports the fumes to be used in purifying (or bleaching) cloth and, in early times also, sulphur dioxide was used as a preservative for wine by burning sulphur prior to sealing wine into its barrels or storage jars. It is one of the most versatile agents used in food preservation and well known for its microbiological effect upon bacteria, moulds and yeasts.

Seldom used as the gaseous form nowadays, it is generally employed in the form of a sulphur dioxide generating salt. For example, sodium metabisulphite is converted as indicated in acid medium:

$$\underset{(190)}{Na_2S_2O_5} + H_2O \longrightarrow 2NaHSO_3$$

$$2NaHSO_3 + H^+ \longrightarrow 2Na^+ + H_2O + \underset{(64)}{SO_2}$$

i.e. 190 parts metabisulphite produce 64 parts SO_2.

The microbiocidal effect increases as the pH falls below 4.0, and hence it is ideally suited for most soft drink formulations. However, the efficacy of its preservation action is impaired by the tendency to react with many fruit components of soft drinks to form organic sulphites, in which state the sulphur dioxide is said to be 'bound'. It is therefore necessary to analyse for 'free' as well as total SO_2 because its preservative activity is mainly due to the 'free' SO_2 and legislation, prompted by the safe working requirements, refers only to maximum *total* concentrations (i.e. 'free' plus 'bound' SO_2).

The disadvantages associated with SO_2 are: (1) its use can be detected by some tasters, even at low concentrations, as an unpleasant backnote or taint; and (2) it has a tendency to evoke allergic responses from some individuals.

The FAO/WHO committee have recommended an ADI of not more than 0.7 mg/kg body weight.

7.4.3 *Benzoic acid and benzoates*

Benzoic acid occurs naturally in a number of fruits and vegetables and is also found freely in some resins, chiefly in gum benzoin (from *Styrax benzoia*) and in coal tar. Commercially available benzoic acid is produced by chemical synthesis.

A white powdery solid (mp 122 °C), and only sparingly soluble in water at normal temperature, it is added into the drink formulation as the sodium salt, which is soluble, taking care to avoid local precipitation of the free acid owing to inadequate mixing in low pH (acid) solution.

It is the free or undissociated form of benzoic acid which exhibits preservative action, and hence its use is only effective when low pH values are encountered, ideally below pH 3, at which point the degree of dissociation of the acid has reduced to less than 10%.

Benzoic acid is often used in conjunction with sulphur dioxide where it is claimed that the 'joint' performance is better, owing to a synergistic effect. The UK preservative regulations allow for mixed preservatives, providing the concentrations of each, when expressed as a percentage of the respective legal limit, will give a cumulative figure of no more than 100.

The maximum ADI for benzoic acid, as recommended by the Joint WHO/FAO Expert Committee on Food Additives (JECFA), is 5 mg/kg body weight.

Benzoic acid is generally considered to exhibit an inhibitory effect on microbial growth, whereas sulphur dioxide shows selected microbiocidal activity. Allergic responses to benzoic acid have been reported, particularly among children known to be hyperactive towards some other agents, e.g. Tartrazine.

7.4.4 *Esters of para-hydroxy-benzoic acid*

These preservatives, often collectively referred to as parabens, are permitted for use in soft drinks in a number of countries in the form of the methyl, ethyl and propyl esters and their sodium salts. A maximum level of 160 mg/kg under the UK regulations has been set in soft drinks for consumption without dilution. Parabens demonstrate a greater tolerance to increased pH and are more effective than benzoic acid at values above pH 3.

The anti-microbial activity increases with chain length from methyl to propyl, but this effect is somewhat offset by a decrease in solubility and hence the lower esters appear to be more frequently used in aqueous food preparations such as soft drinks.

Extensive test work has shown that the parabens do not accumulate in the human body but are excreted mainly unchanged. Some allergic responses have been reported and the ADI has been recommended at 10 mg/kg body weight.

7.4.5 *Sorbic acid and sorbates*

Sorbic acid appears naturally in a number of fruits and vegetables, notably in the juice of unripe mountain ash berries (from *Sorbus aucuparia*) where it occurs together with malic acid. In common with benzoic acid and some other microbial inhibitors, sorbic acid and the sorbates show reduced activity with increase in pH; however, the upper limit of effectiveness is considerably higher than for benzoic acid, at around pH 6 to 6.5. It is undissociated form that inhibits microbial growth. Recommended ADI is 25 mg/kg body weight.

Sorbic acid is permitted in the UK at levels not exceeding 300 mg/kg in soft drinks for consumption before dilution. Because of poor solubility of the free acid it may be introduced in the form of its sodium, potassium or calcium salts, taking care (as in the case of the benzoates) to avoid local precipitation of free acid owing to low pH conditions, before complete dispersion has been achieved. Sorbic acid may be used in admixture with other preservatives.

7.5 Other additives

7.5.1 *Emulsifiers*

The function of an emulsifier is to enable, and maintain, a uniform dispersion of oil droplets within the aqueous phase. Concentrated emulsions are used to impart both cloud (neutral emulsions) and flavour (flavoured emulsions) characteristics to the drink and are usually formulated to be used at a rate of about 0.1%.

In its simplest form the emulsion system consists of an oil phase, usually a citrus essential oil, suitably prepared and 'weighted' with an oil soluble gum/resin component. This is dispersed and homogenised into an aqueous solution containing gum arabic or other similar acting hydrocolloid. Droplet size is standardised in most cases between 1 and 2 μm in diameter to give optimum stability and cloud effect. The weighting or clouding agent appears to interact with the hydrocolloidal phase providing the oil droplet with a stable surface layer, and this limits the tendency for coalescence to occur.

Since 1970, when brominated vegetable oil was withdrawn from UK Soft Drinks Legislation, a great deal of work has been carried out in developing an alternative to what had been a most stable and well-tried cloudifier system during a permitted use period of some 30 years.

The various alternatives have included sucrose esters (e.g. SAIB – sucrose diacetate hexa-isobutyrate), rosin esters (e.g. ester gum-glyceryl ester of wood rosin), protein clouds, benzoate esters of propylene glycol and glycerol, colophony, titanium dioxide, waxes and gum exudates. A number of these have won prominence in some countries but no one system has been universally accepted.

Despite toxicological considerations taking priority, there is often the problem of background flavour from a particular emulsifier system which can seriously limit its use unless some form of masking effect can be incorporated; rosin esters and gum exudates are prone to background flavour.

7.5.2 Stabilisers

While stabilisers can be used in the preparation of emulsions (as referred to above) they are also utilised, where appropriate, in soft drink formulations to impart stability to natural clouds (e.g. dispersions of fruit solids) and to improve mouth-feel characteristics of the drink by increasing viscosities.

Although there are over 50 E-coded substances with stabilising properties for food use, perhaps no more than ten are used to any regular basis in soft drinks; these include the alginates, carrageenans, vegetable gums, pectin, acacia, guar, tragacanth, xanthan and carboxy methyl cellulose. Also included, but without an E number coding, is extract of Quillaia which is permitted specifically in soft drinks and, apart from emulsifier properties, is valued primarily for its foaming qualities.

7.5.3 Saponins

Saponins are well known for their foaming abilities and are used in shandy, ginger beer, cream soda, cola formulations, etc., to improve heading foam characteristics.

Saponins are contained in most natural plant material but notably in Quillaia bark and the Yuccas. Of the latter species, two varieties are used in the USA for water extraction; these are the Mohave Yucca (*Yucca mohavensis*), and the Joshua Tree (*Yucca brevifolia*). Permitted limits are quoted in terms of weight of dry extract.

7.5.4 Anti-oxidants

These additives are used to off-set deterioration as a result of oxidation of flavour and colour components in the presence of dissolved oxygen. Ascorbic acid has already been referred to under acids and is noted for its use in the aqueous phase.

Most vulnerable to oxidation are the oil-based flavours (e.g. citral in lemon oil) which are present in the form of an emulsion where the actual emulsification process may have introduced air into the flavour and an oxidation pattern initiated before addition of the emulsion to the beverage.

The oil phase can be protected from oxidation by use of an oil soluble anti-oxidant such as BHA (butylated hydroxy anisole) or BHT (butylated hydroxy toluene) added before emulsifying; 1000 ppm is the typical usage level in essential oils. As the use of the flavour emulsion will be in the region of 0.1%, the level of anti-oxidant in the drink will be no more than 1 ppm (or 1 mg/kg), which will comply safely with an ADI of 5 mg/kg body weight for the additive. At these levels the anti-oxidant is being employed as a processing aid in stabilising the added beverage ingredient, i.e. the emulsion.

Both BHA and BHT are subjected to restricted use in many countries on health grounds and, increasingly, are being replaced by the more acceptable natural (or nature-identical) anti-oxidants such as: ascorbyl palmitate and its sodium and calcium salts (6-O-palmitoyl-L-ascorbic acid), natural extracts rich in tocopherols (vacuum distilled soya bean oil, wheat germ, rice germ, cottonseed oil, etc.) and synthetic, alpha-, gamma- and delta-tocopherols.

Ascorbyl palmitate synthesised from ascorbic and palmitic acids is oil-soluble and, when used in conjunction with tocopherol (Vitamin E), exhibits enhanced anti-oxidant properties.

7.6 The safety of food additives

To ensure the carbonated product reaches the consumer in a satisfactory state, it is necessary to create a balanced blend of component parts that will remain predictably sound during retail shelf-life for at least twelve months, under recommended storage conditions. To some, 'component parts' are synonymous with 'additives', a term which often provokes concern, but it must be emphasised that food additives receive official approval only if toxicological testing in the light of current knowledge yields no signs of harmful effects, nor any reason to suspect that harmful effects may arise.

The safety of food additives is tested according to guidelines issued by the joint committee of JECFA, WHO and FAO. Knowledge of health safety is gained primarily in animal experiments. In later stages of testing, humans may also be included in the studies to ascertain that their physiological reactions are similar to those found in animals.

Because of potentially longer term ingestion, the standards applied to food additives in feed trials are significantly higher than those applied to pharmaceuticals where, to treat successfully some conditions of illness, certain side effects caused by the treatment can be tolerated. The ADI is an estimate of the amount of a food additive that can be consumed safely over an entire lifetime. ADI levels are set by JECFA after considering the results of various feeding trials. As a general rule, the ADI value is set at one hundredth of the intake, which produces virtually no toxicological effects in long-term animal feeding trials.

8 Syrup room operation

M.J. TURNER

8.1 Introduction

It is possibly true to say that the design of any syrup room depends upon the desired end product. This chapter attempts to detail some of the equipment and materials that are necessary for a modern syrup room operation. In designing a syrup room, there are many considerations to be taken into account. These may be grouped as follows.

Hygiene-related. The design of infrastructure, tanks and pipework must ensure sterility and safeguard the end product from future spoilage. The choice of detergent and type of CIP (Clean-in-Place) system is important.

Product-related. To do with the raw material, its storage, handling and treatment, all of which affect the future quality of the drink.

Process-related. Plant and equipment. From the humble origins of the syrup rooms of the 1930s and 1940s (when syrup batches of 100 and 250 litres were commonplace and 1000-litre batches were unusual), the concept has evolved to the present-day syrup room being considered as a 'mini-factory' within the main soft drinks complex – clinically designed, staffed and operated by qualified technicians assisted by modern plant and sophisticated instrumentation.

Present developments include the automated and computerised systems currently available, and the latest advances where a multiple component mixing plant prepares finished product as opposed to the conventional preparation of syrup. These areas, together with the storage and handling of raw materials, are discussed to give an appreciation of the size, complexity and importance of any syrup room operation.

8.2 Syrup room design

8.2.1 *Wall finishes*

An elaborate interior is not necessary, but a critical area of any syrup room is the finish on the walls as these are constantly exposed to harsh cleaning compounds as well as the acid components of the product; for wall surfaces to last, they must be impervious to this kind of treatment. With continual use of water and inevitable sugar deposition, these surfaces are extremely prone to mould growth. Ceramic tiling has generally been used but tends to be expensive and spray tile finishes may be used as an alternative; this finish is applied directly to masonry blocks and consists of several layers of epoxy enamel paint covered with a glaze coating.

8.2.2 *Floors and drainage*

A surface coating capable of resisting strong acid and alkaline solutions is necessary; unprotected concrete surfaces offer little resistance to these types of solutions. Quarry tiling is an excellent surface but, again, can be very expensive; several less costly products have now been introduced, based on epoxy finishes that have been developed to resist the corrosive effects of specific compounds. Proper drainage must be installed to allow fast flow of waste to reduce contact time of corrosive products on floor areas, and drains should always be designed to allow adequate cleaning.

8.2.3 *Ceilings and lighting*

A concealed or dropped ceiling system is generally used and is supplied as lightweight panels; the surface should be resistant to the type of products used in a syrup room since vapours are often carried upwards and deposits can occur. Fluorescent lighting systems are most commonly used and should be waterproof to reduce the effects of corrosion and prevent ingress of insects.

8.2.4 *Heating, ventilating and air conditioning*

Environmental conditioning is preferable but is not installed in many factories in the UK. The advantages of these types of systems are:

1. They reduce relative humidity which, in turn, reduces condensation on equipment and piping that could support mould growth.
2. They reduce airborne contamination by filtering out dust, etc.
3. They reduce ambient syrup temperatures which could affect filling performance.
4. They improve equipment performances by permitting them to operate at cooler temperatures.

8.3 Syrup room equipment

8.3.1 *Storage, mixing tanks and systems*

Tanks are used for a number of purposes and can vary from a small vessel for dissolving purposes to large storage tanks. Some factories are equipped with

Figure 8.1 Syrup room tank layout.

syrup tanks of capacities greater than 30 000 litres, although many factories still use tank sizes of 5000 to 20 000 litres and consider such batches economical. The number and sizes of tanks are determined by many factors:

- Number of hours in a working day
- Line filling speed
- Variety of products and flavour changes
- Filtration requirements of syrup
- Sterilisation requirements between critical flavours
- Ageing or maturation time of syrups
- De-aeration of syrups after mixing

Traditional syrup rooms had tanks with open tops and the only form of agitation was manual, often using a wooden paddle. Modern syrup room tanks are now designed to be fully enclosed and are fitted with manholes, inspection lamps, agitators, high- and low-level probes and full CIP spray-ball assemblies; level indication is often by sight glass or the use of load cells. The general layout of syrup room tanks and pipework is shown in Figures 8.1 and 8.2.

Mixing systems. The correct amount of agitation in a tank is very important; the agitation should be smooth since violent agitation leads to aeration which

Figure 8.2 Syrup room tanks and associated ingredient feed pipework.

is difficult to remove and causes problems at the filler. A comprehensive range of mixers is now available and can vary from propeller-type mixers to high shear mixers. The positioning and type of mixer to be used should be carefully selected depending upon the size of tank and the nature and viscosity of product to be mixed.

Mixers can be attached to the tops of tanks or by more permanent mountings at the bottom or side entries. Variable-speed motors can be fitted to ensure that excessive aeration does not occur when low levels of syrup are mixed in large tanks. One of the major problems encountered in mixing syrup is the exclusion of air; it is good practice when preparing syrup to first add water to the tank.

The agitators should not be set in motion until they are well covered with liquid. All ingredient pipework should be directed to run additions down the side of the tank instead of allowing them to splash into the bulk of the liquid. Some tanks have inlet points of the bottom feed type and these help in reducing splashing, foaming and aeration. Syrup batches should generally be allowed to stand to de-aerate before filling to allow any entrapped air to escape; it is usual to allow at least one hour minimum or two hours ideally for sugar-based syrups.

High-shear mixers. High-shear mixers are often employed in the soft drinks industry to disperse stabilisers, e.g. xantham gum and sodium carboxymethylcellulose; these can be very difficult to dissolve unless the correct technique is employed. The high-shear mixer should be started in water and the powder added as quickly as possible; the use of a venturi to pre-wet the particles is particularly useful. The powder is sucked down into the disintegrating head and dispersed before the viscosity has developed. It is important that the powder should never be added slowly since it will mean that the last of the powder is added to a mix that is already of high viscosity and will inevitably form lumps that will be difficult to disperse.

Liquid jet mixers. These offer an alternative to conventional agitators and have been used successfully for the production of finished syrup. The system provides a homogeneous mixture with very little entrained air; this is due to the short mixing time as the mixing process starts while the vessel is being filled. The application for liquid jet mixers is determined by the viscosity of the liquid to be mixed; generally these mixers can be used where centrifugal pumps are capable of transferring the liquids to be circulated.

Jet mixers are normally installed at the lowest point possible in a tank to ensure adequate and efficient blending in the event of a low liquid level; the

Figure 8.3 Jet mix nozzle viewed from inside tank.

SYRUP ROOM OPERATION

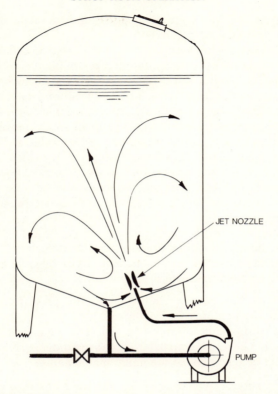

Figure 8.4 Circulation of liquid in a tank using jet mixer system.

installation of a jet mixing nozzle in the base of a mixing tank can be seen in Figure 8.3 and the circulation of liquid is illustrated in Figure 8.4.

The jet of liquid flowing out of the diffusion nozzle at high speed generates a reduced pressure in the inlet cone of the diffuser which causes a liquid stream to be sucked out of the vessel and carried along; the diffusion jet mingles with the drawn-off liquid, increasing its velocity. The turbulence in the diffuser produces a homogeneous liquid mixture and the entire contents of the vessel are blended in a short time without creating circular movement.

8.3.2 *Pipework, fittings and connections*

The majority of pipework is stainless steel and if suspended ceilings are used the pipework can be run above the ceiling. All pipework to the syrup-manufacturing area should be marked or colour coded; this helps to eliminate mistakes and promotes safety.

Various fittings and valves are used to connect pipes; e.g. bends, tees and reducers; sight-glasses and instrument ports; and valves for directing and regulating the flow and the pressure.

Permanent joints are sometimes used by welding or expansion. Where disconnection of pipework is required the pipe coupling is in the form of a threaded union which has a seal or gasket in between. Swing bends are often used where a piping run needs to be switched from one line to another, e.g. connecting CIP line to the syrup line. All pipe connections should be tightened firmly to prevent air being sucked into the system and to prevent leakage of syrup. Sight-glasses are often located in the pipeline where a visual check on the ingredient is required. Connections may be provided to allow the mounting of measuring instruments such as thermometers, pressure gauges, conductivity probes (to detect liquid). Sampling ports can be included, but it is essential that the sampling port does not allow contamination to occur to the ingredient.

Valves. Many designs of valves are available for process systems and can vary from manual operation to air-operated types. The choice of valve is important for a number of reasons:

- Valves should be leaktight against known line pressures.
- Air valves should close on air failure.
- Diaphragms should be of approved food grade material.
- Valves should be resistant to harsh corrosive products, should be reliable and should be designed for minimum maintenance.

Shut-off and *change-over* valves have distinct positions. A *regulating* valve allows the passage or flow of liquid to be controlled gradually and is used for the fine control of flow and pressure at various points in the piping system.

Check valves are fitted where it is necessary to prevent product from flowing in the wrong direction; the valve is normally kept open by the flow of liquid in the right direction. If the pressure on the downstream side of the valve becomes greater than the flow pressure, the valve disc is forced against its seat by the back pressure and the valve is effectively closed against reversal of the flow.

Pressure relief valves are used for regulating the pressure of the product in the piping. If the pressure is too low the spring presses the plug against its seat; when the pressure reaches a certain level, the force of the plug overcomes the spring and the valve opens. Spring tension is adjusted to set the desired opening pressure.

8.3.3 Ingredient flow

Pipework should be designed to prevent any pockets occurring along the line where the product or cleaning detergent can collect; this could lead to microbiological spoilage or contamination of the syrup by a detergent. It is also desirable to ensure that low points are designed into pipework so that full drainage may be obtained via a valve at that point.

It is essential to design the pipe sizes to give adequate flow but ensure that mechanical bruising of the product does not occur. Flow resistance will occur in pipework due to friction in straight pipe lengths, changes in direction of flow due to bends, valves and other fittings. The flow resistance is expressed in terms of the column or 'head' of water necessary to compensate for loss of pressure due to resistance.

Traditional syrup rooms had tanks on a floor above the filling lines and the pressure arising from the head of syrup was usually sufficient to transport it to the production line at the required flow rate. In modern factories, which are designed on high output, it is more difficult to gravity feed and so pumps are therefore used to generate the pressure required.

The resistance to the flow of a liquid results in a loss of pressure and the component is therefore said to cause a pressure drop in the pipe. The pressure drop is measured in terms of head and is equivalent to the resistance of the ingredient, the size of pressure drop being governed by the velocity of the flow, i.e. flow rate and size of pipe. If the velocity exceeds a certain value (dependent upon the nature of the liquid) then not only does the frictional head increase (requiring a more powerful pump) but the flow in the pipe becomes turbulent and disturbed, producing a possible adverse effect on the syrup.

8.3.4 *Pumps*

Nowadays, with high throughput operations, it is necessary to convey large quantities of liquid through pipelines often with large numbers of bends, valves, etc., and through pasteurisers, homogenisers and other associated equipment which could all possibly contribute to pressure drop. Pumps need to be fitted at various points in the line to convey the liquid and compensate for the loss of head; however, pumping agitates the product and it is essential that the correct pump is chosen. Many different types are available from the range of centrifugal, diaphragm, peristaltic, gear and positive displacement pumps.

Centrifugal pumps. These are often used since they can be manufactured to a sanitary design, are suitable for CIP and are not capable of producing accidental over-pressure. A centrifugal pump comprises an impeller rotating in a casing, a delivery chamber and an electrical drive. Liquid entering the pump at the centre of the casing is carried round by vanes. If the liquid has to be pumped up to a tank at a higher level, the pump discharge pressure must be sufficient to raise the liquid to that height. This type of head is known as a static delivery head. The pump characteristics supplied by a manufacturer usually relate to water; the viscosity of a product makes a difference. When highly viscous liquids are pumped, the pressure losses in the pump are higher and so the energy of the product leaving the impeller will be lower than for water.

During pumping, liquid is carried from one side of the pump to the other, creating a partial vacuum in the space once occupied by the liquid. This space on the suction side is then refilled with more liquid.

Cavitation on a pump can occur when the pressure at the suction falls below the saturation pressure of the liquid (varies with temperature) and dissolved oxygen comes out of solution; this can be avoided by reducing the pressure drop on the suction line, e.g. large pipe size, fewer valves, raising liquid level above pump inlet, etc.

Flow controllers are often used to maintain a constant flow rate; an effective method of flow control is to vary the speed of the pump which can be achieved mechanically, hydraulically or electrically. The centrifugal pump can be used to handle a wide range of liquids provided the viscosity is not too high; the pump is not self-priming and the suction line and pump casing should be filled with liquid before switching on.

Positive-displacement pumps (rotary). There are many types of self-priming positive-displacement pumps. The rotary pumps work on the principle of two synchronised driven lobed rotors which have a very close clearance but do not actually touch each other. As the rotors turn, the volume between the lobes at the suction port increases and the partial vacuum created causes liquid to enter the pump. The liquid is carried in the space between the lobes and the pump casing to the outlet; as the volume between the lobes is reduced, the pressure increases and the product is discharged.

In order to prevent excessively high pressures, positive-displacement pumps usually have some form of relief valve which automatically returns some of the liquid to the inlet if the pressure becomes too great. Flow is normally regulated by varying the speed of the pump. Positive-displacement pumps are generally used for handling high-viscosity liquids.

8.3.5 *Measurement of liquid*

The contents of tanks may be measured using meters, dipsticks, sight-glasses or load cells. The use of load cells is a common method for monitoring ingredients; these, however, have some disadvantages in that all additions need to be converted to weight.

Reliable sanitary meters have been developed for measuring liquids. Accurate meters are the most practical method of measuring quantities of liquid into a tank. Figure 8.5 shows the installation of meters for measurement of sugar syrup, glucose and citric acid; they have all been installed with ease of access for cleaning and maintenance.

Registering meters are normally fitted with shut-off valves which automatically close at a predetermined setting, shutting off the flow of liquid. Meters do have an advantage in that multiple additions may be made simultaneously, thus saving time and often aiding mixing. In the absence of meters, the dipstick

Figure 8.5 Meters installed for measurement of sugar, glucose and citric acid.

method of measuring liquid is reliable and has been widely used for many years. These are made from stainless steel and have a hooked end to facilitate hanging the strip from a specially marked spot on the side of the tank. With the use of dipsticks, each tank must be calibrated individually. Using an accurate meter or calibrated measure, water is introduced into the mixing tank: the dipstick is then calibrated to this level and marked. The mark is checked by repeating the process several times, being certain to hang the dipstick from the same point each time. Once the calibration has been accurately established, the level is scribed permanently into the stainless steel.

A sight-glass, mounted on the side of a syrup tank with sanitary mountings, is another means of measuring contents in a tank. The sight-glass should be accurately calibrated and marked. Unless careful cleaning is employed, sight-glasses can offer hiding places for contamination.

8.3.6 *Filtration of ingredients*

A number of ingredients in syrup manufacture – and, indeed, in certain instances, the syrup itself – require some form of filtration, which could range from a simple mesh filter to a more sophisticated plate-and-frame or cartridge-type system. Microfiltration is nowadays a safe and efficient method of removal of unwanted particles and other turbidity components. Where high carbonation mixer drinks are concerned, it is preferable to ensure all active centres are removed to aid good filling: the syrup itself can be filtered typically through a nominal 50 μm cartridge for this purpose.

In contrast to the conventional filter plate systems the modern cartridge housing concept has numerous advantages.

- Completely enclosed, sanitary and leak free
- Quick change out of filter media
- Short cleaning and sterilisation times

Some cartridges can be cleaned in forward flow with hot water and significantly reduce costs by prolonging cartridge life. Typical cartridge systems feature a polypropylene centre core, outer cage and end caps and either a cotton wound cartridge or often a charged nylon membrane. The components are thermo-welded to eliminate the use of glue or resins which may impart unwanted off-flavours.

8.3.7 *Ultraviolet sterilisation*

Ultraviolet sterilisation of liquids is a method of purification by means of UV radiation.* UV light at a wavelength of 254 nm has the ability to kill micro-organisms that come into contact with it. The light rays penetrate the outer membrane of the organism, reach the central DNA and destroy it.

Figure 8.6 UV lamp installation.

*Bronzepalm, High Wycombe, Bucks, England.

In operation, the liquid enters tangentially giving full swirl throughout the period within the reactor, exposing the micro-organisms to the UV source for the maximum time. UV light does not affect the critical taste, odour and colour of the liquid, nor does it change its chemical structure. The UV light source is mounted inside a protective quartz envelope which allows the UV light unrestricted passage; this is totally enclosed by a stainless-steel outer casing containing inlet and outlet ports. The electrical circuitry is mounted separately in a control panel. Figure 8.6 shows how a UV lamp is connected to conventional pipework; the system occupies very little space. UV equipment has a proven record in the treatment of sugar syrup solutions and is often a fundamental piece of equipment to be designed into any syrup room operation. Very often pipework runs can be extremely long in modern factories and are therefore prone to bacterial spoilage. UV is a means of ensuring sterility of liquid sugar and water prior to syrup manufacture.

It is essential that the UV lamp is sited in the correct location, i.e. either immediately after a storage tank or, if long pipe runs exist, then immediately prior to batch mixing.

8.3.8 *Pasteurisation*

Fruit-containing or beer-containing syrups will often require pasteurisation; this may be achieved by the use of a flash pasteuriser. The layout of a pasteuriser unit can be seen in Figure 8.7. The syrup is heated to a preset

Figure 8.7 Layout of pasteuriser unit and control panel.

temperature, held at that temperature for a prescribed time and then cooled below the pasteurising temperature. A simple heat exchanger would have a pre-heating section to raise the temperature of the syrup to the required level, and regeneration and cooling sections which are constructed so that the hot outgoing syrup heats the cold incoming syrup, thus saving energy.

The heat-treatment stage is carried out in plate heat exchangers; plates are packed together in each stage to form a section and are sealed by the use of gaskets round the edges. The syrup is introduced through a corner hole of the plate into the first section and flows vertically through the channel; it leaves at the other end through a separate corner passage which takes it past the next passage and then into the plate. The syrup therefore flows through alternate plates.

The heating or cooling medium is also introduced at the other end in the same way through alternate plates; each syrup channel has channels for

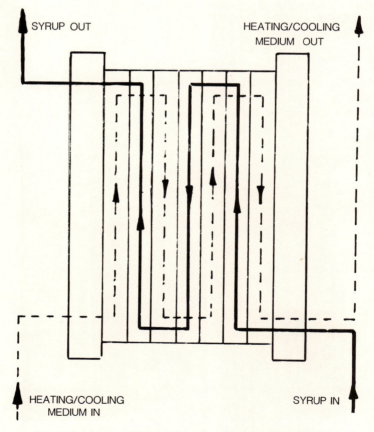

Figure 8.8 Passage of liquids through a plate-type heat exchanger.

heating and cooling on both sides of it. The passage of syrup through a set of pasteuriser plates is illustrated in Figure 8.8. The correct holding time is usually achieved in a holding section of tubes built onto the heat exchanger. Since the holding tubes are sized to give a holding time at a specified flow rate, it is essential that accurate flow-rate control is installed.

Constant temperature is maintained by a temperature controller acting on a steam-regulation valve; should the temperature deviate from the preset limits a flow-diversion valve should operate to recirculate the syrup back to the balance pan.

8.3.9 *Homogenisation*

Homogenisation is a process that traditionally was used in the dairy industry; with the increasing demand to present a more uniform product to the consumer it has been necessary to evaluate the use of homogenisation of syrup in the soft drinks industry. This has become more widespread with recent changes in the use of certain colour systems, e.g. β-carotene; the colour requires homogenisation into the syrup to reduce or prevent neck ringing in the bottle.

Figure 8.9 Homogeniser unit and pipework.

The principle components of a homogeniser are: a high-pressure pump, a back-pressure device and the homogeniser head. The pump is driven by a powerful electrical motor through a crankshaft and connecting-rod transmission which converts the rotary motion of the motor into the reciprocation action of the pump pistons. Figure 8.9. illustrates a homogeniser and associated equipment. The pistons run in cylinders in a block and are often cooled by the use of water; these are illustrated in Figure 8.10.

With the use of any homogeniser it is essential that air is not allowed to enter the system; air mixed with the product causes shocks to the mechanical components and is very damaging to the homogeniser. An adequately sized supply tank is always used and a positive liquid level maintained above the homogeniser inlet.

The heart of any homogeniser is the valve, which is normally manufactured of a single strand of fine stainless-steel wire formed into a compact, resilient valve element; the product flows through the labyrinth of minute passages formed between the wire loops and folds. Here the product is subjected to hydraulic shear forces, turbulence and innumerable changes in flow direction. These forces applied over a relatively long flow path produce a maximum homogenising effect with the lowest operating pressures; by replacing the valve on a scheduled basis, repeatability of homogenisation is assured.

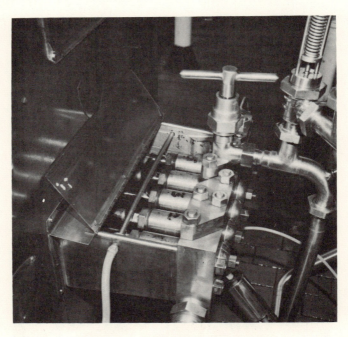

Figure 8.10 Pistons on an homogeniser.

8.4 Syrup room materials storage and handling

The type of storage and degree of control for any raw material will depend to a large extent on the effect the ingredient has on the finished product. Certain ingredients require controlled temperature conditions due to problems with stability, microbiological spoilage or crystallisation.

8.4.1 *Sugar*

Sugar was traditionally made available in bagged form but is now more widely delivered by road tanker as bulk granulated or dissolved in water to give a 67 °Brix solution. Bagged sugar should be stored off the floor on pallets with spaces left for air circulation; bulk granulated is normally stored in hoppers, which are temperature controlled to avoid caking of the sugar. Filtration is important to remove any debris and is achieved using paper filters in a horizontal or vertical plate type filter or cartridge system. Filtration systems used in the soft drinks industry are covered in Section 8.3.6.

Liquid sugar of 67 °Brix will not support the growth of most micro-organisms: however, dilution can occur at the top of stored sugar syrup due to condensation which then allows growth of micro-organisms. Some storage tanks are fitted with UV lamps in the headspace of the tank, but UV systems in line after storage now have a more proven record of effective sterilisation of sugar solutions; the use of in-line UV systems has been covered in more detail in Section 8.3.7

8.4.2 *High-fructose glucose (corn) syrup*

High-fructose glucose syrup has been used as a partial or full replacement for sugar for many years and is supplied in liquid form having a range of around 70° Brix. It has a high viscosity and a tendency to crystallise and therefore storage tanks invariably require some form of thermal jacket dependent on their location.

8.4.3 *Acids*

The main acid used in soft drink production is citric acid and is generally supplied in paper sacks or fibre drums as a powder. Anhydrous citric acid is fairly stable to air and heat but is somewhat hygroscopic; it should be stored in a well-sealed container in a dry, cool place. The powder is normally dissolved in water to manufacture a 50% w/v monohydrate solution. Citric acid can also be supplied as a 50% w/w anhydrous solution either by road tanker or in Rotoplas (1000-litre PVC containers). It is recommended that the solution should be stored for no longer than four weeks at ambient temperature due to possible microbiological spoilage; crystallisation may occur after prolonged storage at temperatures below 20 °C.

Phosphoric acid is also supplied in drums or road tankers as required; it should be stored at room temperature and protected against light and heat. Although phosphoric acid is liquid when packed, strengths above 84% may crystallise in cold weather. If this occurs the acid should be remelted slowly by storing in a warm place.

Ascorbic acid is also used but mainly for its benefits of enrichment of Vitamin C or as an anti-oxidant. It is generally supplied in fibre-board containers and should be stored in well-closed, non-metallic containers protected from light in a dry, cool place; ascorbic acid is fairly stable to air if protected from humidity but is somewhat sensitive to heat.

8.4.4 *Sweeteners*

The three sweeteners generally used are: sodium saccharin, acesulfame K and aspartame. Saccharin is normally supplied as the dihydrate in fibre-board drums: it should be kept in a tightly closed container in a cool, dry place. It should be stored away from oxidising materials, heat and sources of ignition. Saccharin will decompose on heating, especially in the presence of acids, and therefore it should never be boiled.

Acesulfame K is normally supplied in fibre-board drums and should be stored in a tightly closed container in a cool, dry place. It is readily soluble in water but the solution may often be slightly turbid; impurities or insoluble particles should not be visible.

Aspartame is also normally supplied in lined drums and should be stored in a tightly closed container in a cool, dry place. It is sparingly soluble in water but the solubility increases with increasing temperature and lowering pH; it is generally dissolved by sprinkling the powder into a dilute citric acid solution.

8.4.5 *Preservatives*

Sodium benzoate is supplied in paperwall sacks or fibre-board drums and should be stored in cool, dry conditions; it is readily soluble in water.

Sodium metabisulphite is also supplied in fibre-board drums or paperwall sacks and should be stored in cool, dry conditions in airtight containers; it must not be allowed to make contact with, or be stored near, oxidising agents since it will give off SO_2 gas.

8.4.6 *Flavourings*

Most flavours are supplied in glass jars or polythene containers and storage conditions will vary dependent upon flammability and composition. Certain flavours will develop haze or separation if stored at excessively low temperatures while excessively high temperatures will often reduce the shelf-life of the material. Materials with a flashpoint below 32 °C are considered highly flammable and must be kept away from sources of ignition.

8.4.7 Colours

These can be either natural or synthetic and are normally supplied in polyethylene drums or vacuum-packed polythene bags; they are usually in powder form and are readily soluble in water. However, with recent changes in colour systems, it has meant that to achieve good solubility of colours it has been necessary to homogenise certain colour systems into the syrup (see Section 8.3.9).

8.4.8 Fruit juices and comminuted bases

These fruit products can be supplied from many countries and may be preserved or non-preserved, of different strengths and in a variety of packages, from steel drums to Rotoplas to road tanker. Storage conditions of the fruit material would tend to be determined by the type of product supplied. Non-preserved products may require chill or frozen storage, whereas preserved products in Rotoplas could often be stored at ambient; however, care must be taken not to store Rotoplas in direct heat since extremely high temperatures can be generated within the Rotoplas.

8.5 Syrup room CIP systems and detergents

To obtain the required degree of sterility all cleaning operations must be performed according to a carefully worked-out procedure; the sequence of operations must be exactly the same every time. A CIP (or Clean-in-Place) system is usually designed into modern equipment; this enables cleaning to take place without the need to dismantle and physically clean pipelines, pumps and tanks.

The design of any CIP circuit is very important to ensure adequate spraying from spray balls to all areas of the tank. Considerable savings can be made since these systems are designed to optimise the use of water, steam and detergents; there is little manual effort involved with most computerised systems. However, despite the luxury of a CIP system it is still vitally important to inspect all areas and recognise that manual cleaning may still be necessary; agitator inlets and ledges on sight-glasses are particular areas endangered by microbiological contamination and it is essential that, if they are not adequately reached by the spray balls, they are physically cleaned. Syrup residues on pumps and dispensing pistons and soiling residues behind seals are all sources of contamination, and in certain instances can only effectively be removed by manual cleaning.

8.5.1 *Design of a CIP unit*

The CIP unit* should consist of two sections, one for detergent/sanitiser and one for recovered water; if there are likely to be problems with the fresh-water supply, a third section should be added to act as a fresh-water break tank.

The size of the tank in which the CIP solution is stored depends on the maximum volume of the liquid in circulation. When considering vessel cleaning, the volume in the CIP delivery and scavenge lines has to be calculated and the amount of CIP liquid in the vessel has also to be considered; when cleaning pipelines the total volume in the circuit has to be calculated. When calculating the size of the recovered water section the pre-rinse times and flow rates must be taken into account; this applies for vessels and pipelines. Burst rinsing (intermittent bursts of water) can be used on vessels to reduce the amount of recovered water required.

The detergent section should have a facility for heating and should be insulated. It is always preferable to design CIP pipework to be separate from syrup or other ingredient lines; where the CIP liquid and the ingredient are only divided by a valve, there is always the risk of cross contamination and well-designed systems provide safeguards to make this impossible.

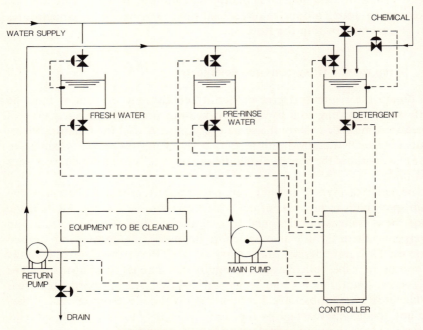

Figure 8.11 Simple three-tank CIP system.

*Ciptek Ltd, Wolverhampton, England.

The CIP control system should be microprocessor-based and will perform two basic functions: it will check that the various conditions required prior to cleaning are met and it will control the sequences that make up the cleaning programme. Figure 8.11 illustrates a simple three-tank CIP system.

The conditions to be met before the programme can be activated are generally as follows:

- the sections are at high level
- the heating section is at correct temperature
- detergent is at correct strength
- there are no fault conditions.

The vessel cleaning programme would normally comprise a three-stage cleaning sequence as follows:

Stage (i): *Removal of product residues*
All residues should be removed from tanks, pumps and pipework by rinsing with water; this initial pre-rinse should always be carried out immediately after the tank is emptied to prevent residues from drying out and sticking to the surfaces, which would then make the tank more difficult to clean. The pre-rinsing will be carried out using recycled final rinse water from Stage (iii) and it is beneficial to warm this pre-rinse water to make it more effective in removing stubborn residues.

Stage (ii): *Cleaning with detergent*
A number of variables need to be controlled to ensure adequate results:

- Concentration of detergent
- Temperature of detergent
- Mechanical effect on the cleaned surface
- Duration

The detergent strength should be determined before cleaning starts and adjusted if necessary; the strength should be checked against the manufacturer's instructions to ensure the correct dose. Increased concentration does not always improve the cleaning and can often result in foaming and considerable expense.

The detergent temperature should be controlled since most blended detergents have optimum temperatures; when using chlorinated blended detergents the temperature must never exceed approximately 40 °C since chlorine is liberated and can corrosively attack stainless steel.

The conductivity of the returning CIP liquid should be checked to ensure that the return valves change over at the conductivity set point; the correct concentration is obtained by the use of conductivity probes which introduce fresh detergent to maintain an adequate solution strength. In any CIP system the detergent feed pumps need to be designed to produce sufficient velocity to

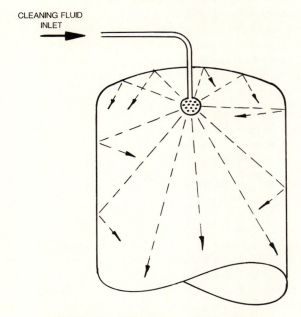

Figure 8.12 Spray ball in a vertical tank.

create turbulence from the spray balls and maintain an efficient scouring action.

Spray balls can be of a static design in a tank or in some instances have a rotating jet device. The spray balls are drilled to provide 360° coverage to ensure adequate cleaning. Figure 8.12 illustrates how a CIP system incorporating spray balls needs to be designed specifically to ensure that every area is covered by the spray. The spray balls are prone to blocking with water scale, rubber from pump impellers, fruit material, etc., and should be regularly inspected and cleaned. The circulation time must be controlled to obtain the maximum cleaning and scouring effects.

Stage (iii): *Rinsing with water*
After the detergent wash, the systems should be flushed with water to remove all traces of detergent and the rinsing water directed into the pre-rinse tank; all drain valves, sample taps, etc., should be opened towards the end of this operation to ensure that every section of pipework has been flushed clean and fully drained. Simple quantitative checks should then be carried out using indicators or paper test strips to confirm the absence of detergent before re-using the system.

Cleaning procedures can be only used as guidelines and will vary dependent upon the type of product being produced. In some instances it is necessary to adopt a five stage cleaning process, i.e.

1. Removal of syrup residues
2. Clean with alkaline detergent
3. Flush with water
4. Sanitation with recommended sterilising solution
5. Flush with water.

Most syrup-room vessels have dished bottoms which may not drain well; care should be taken to ensure that each part of the cleaning cycle is completed and the system drained before stepping to the next stage as this will prevent unncessary dilution and wastage of detergent.

When using flow plates to direct the flow of both product and CIP solutions, it is recommended that the distance from the vessel outlet to the flow plate should be as short as possible. Pipework circuits would also be cleaned from flow plates by setting up a continuous route using a return line, as shown in Figure 8.13. Route-proving can be achieved by sensing the positions of the various link pieces on the flow plates and comparing the route with the programme selected. To clean pipelines a combined scrubbing effect and chemical action is required. To achieve the scrubbing effect the cleaning solutions are pumped around at a velocity that induces turbulent flow. The measurement of turbulence is given as a *Reynolds Number* and flow is said to be 'laminar' up to a Reynolds Number of 2000 and 'turbulent' over 4000.

Figure 8.13 Flow plate for syrup feed and CIP return.

Experience has shown that 50-mm diameter pipework with the usual bends, tees, valves, etc., is cleaned satisfactorily at a flow velocity of 2 metres per second; this gives a Reynolds Number of approximately 100 000.

The verification of any cleaning process should be carried out in two ways:

1. Regular inspection of internal surfaces of tanks and pipes.
2. Adequate microbiological programme at strategic points.
3. Maintenance of detailed records of inspections and findings.

8.5.2 *Rate of flow in pipelines for CIP*

When liquid flows through a round pipe, the speed measured at different points across the diameter is not the same; the liquid in the centre tends to move faster than the liquid towards the wall of the pipe and the liquid in contact with the wall moves slowest of all. At very low flow rates the motion can be thought of as a series of tubes, one fitted inside the other, rather like a car radio aerial; each tube slides slowly inside the other. The effect is not noticeable when looking at the same liquid but it can be seen quite clearly when a mixture of water and a coloured liquid is used. When we refer to *velocity* of a liquid we really mean the *average speed*; as the velocity increases, the tubes vibrate and the first signs of turbulence become noticeable – the tubes in the centre cores dissolve and mingle in a spiral eddy, one with the other. As the velocity is increased, more and more of the tubes intermingle until the flow is really turbulent.

The steady, slow-speed streaming is referred to as *laminar* flow and the intermingling, eddy-type as *turbulent* flow; there is an intermediate state between the two known as *critical* flow.

8.5.3 *Calculation of Reynolds Number*

The nature of flow, whether laminar, critical or turbulent, depends on the pipe diameter, density and viscosity of the liquid and the velocity of the flow. A mathematical equation expresses these variables to calculate the *Reynolds Number*:

$$R_n = \frac{DV\rho}{\mu}$$

R_n = Reynolds Number
D = internal diameter of pipe (m)
V = mean velocity of flow, (m/s)
ρ = Weight density of the liquid (kg/m^3)
μ = Absolute viscosity (kg/ms)

To clean pipes there must be *turbulent* flow.

Table 8.1 Disinfectants and their applications

Impurity	Active chlorine	Hydrogen peroxide	Peracetic acid	Quaternary ammonium compounds
Bacterophages	+ +	+	+ +	−
Small virus	+ +	+	+ +	−
Large virus	+ +	+	+ +	+ +
Gram + ve bacteria	+ +	+ +	+ +	+ +
Gram − ve bacteria	+ +	+ +	+ +	+
Spore-forming bacteria	+	+	+ +	−
Yeasts	+ +	+ +	+ +	+ +
Moulds	+	+	+	+

+ + = quick killing;   + = killing;   − = non-killing.

8.5.4 Choice of detergents

The requirements of a disinfectant, both in its application and its microbial activity, are so diverse that it is impossible to have a single universal product. Table 8.1 shows the properties and applications of a small selection of well-known disinfectants currently in use in the soft drinks industry.* These are dealt with in more detail below.

Active chlorine. Active chlorine has been used for many years in the food industry because of its broad bactericidal spectrum and economic advantage; generally the effect of active chlorine is best in a neutral or weakly acidic solution (pH 5–7) but alkaline products have an excellent effect against all groups of micro-organisms. The question of the corrosiveness of hypochlorite solutions on stainless steel results in certain reservations as to their use, especially for soaking utensils or leaving in machinery overnight; levels of 5–10 ppm chlorine are known to attack stainless steels over a period of time.

Working mechanism: The microbiological effectiveness of a chlorine-based product is primarily based on irreversible oxidative action on the cells of micro-organisms.

Hydrogen peroxide. Disinfectants based on hydrogen peroxide have been used for years in the pharmaceutical industry and now more recently in areas of aseptic filling. They are generally used at higher temperatures since, when used cold and at low concentrations, a longer contact period is necessary because the active oxygen is not liberated so rapidly.

Working mechanism: The microbiological effect of hydrogen peroxide relies on its oxidation of the biologically active systems of the microbial cells to be destroyed.

P_3 Disinfectant Guide, Henkel Chemicals, England.

Peracetic acid. This type of product has been available since the beginning of the twentieth century, but was not used readily due to difficulties in handling. Products based on hydrogen peroxide/peracetic acid combinations are now available and are used successfully.

Unlike many other products, not only are vegetative bacteria of all kinds, yeasts and moulds destroyed with low concentrations at room temperature, but also endospores of the normally hard-to-destroy bacteria, such as types of bacillus and clostridium, are eliminated.

Working mechanism: Peracetic acid does not only react with the protein content of the micro-organism's cell walls, but also through wall penetration; the product has an oxidative destructive action on all proteinous components of the cell, destroying these together with the enzyme systems and thus killing the cell.

Quaternary ammonium compounds. These types of detergents (known as QACs) may be used under various pH conditions ranging from weakly acidic to fairly alkaline. The bactericidal spectrum of the QAC is not as broad as that of the halogens or the peroxide-based products, in that they do not kill spores. The use of QAC-based products has lost its attractiveness over recent years mainly because residues of QACs on various surfaces have been thought to have caused problems; most of the QACs are high foaming and difficult to rinse away completely. Surface residues have been thought to be responsible for the precipitation of ingredients in certain soft drinks.

Working mechanism: A QAC has an ability to lower surface tension and this, together with the negatively charged condition of the active molecules, accounts for its microbial effect; the QAC causes an inactivation of the enzyme and a denaturing of the cell protein.

8.6 Automation and computerisation in syrup rooms

Automation and mechanisation in syrup rooms have removed the heavy manual labour-intensive jobs; increased production capacities have meant a greater demand on the syrup-room operative to manufacture the syrup as quickly as possible. Complex formulations have required the addition of many more ingredients and all have to be added at exactly the correct stage of manufacture. Any malfunction in a process or any wrong decision made by an operator could have a serious effect on product quality.

More remote-controlled facilities have been installed over the years such as central electrical panels to control pumps, agitators, valves and the replacement of manual valves with air-operated valves; ingredients, instead of being dissolved separately, are now dissolved and stored as bulk solutions. The syrup-room operator's role has changed from mixing ingredients in a bucket to controlling the entire process from a central panel. An automatic control system needs to be in communication with each controlled component within the process; the operation of valves and pumps can be achieved through a computer interface which receives instructions and relays signals in the correct form to appropriate items of plant.

8.6.1 Typical system description

The computer or microprocessor can be divided into sections, each section having a specific role to play within the total operation. A programmable control system consists in principle of a central unit, a memory unit and a unit for input and output signals; the central unit has an arithmetic/logic unit that can perform calculations and a controller that performs the operations in the correct order.

There is usually a mimic or flow diagram which is linked via an interface detailing the positions of all valves and pumps in relation to the various manufacturing vessels; a typical mimic diagram and central control unit is shown in Figures 8.14 and 8.15. Each valve and pump on the mimic diagram would be denoted as energised or running by the illumination of an LED (Light Emitting Diode); if the system is operating manually the LEDs will generally flash to indicate the status. Swing bends are sensed by the use of

Figure 8.14 Mimic diagram of a central control unit.

Figure 8.15 Central processor and mimic diagram.

proximity switches that would also be denoted on the mimic diagram. LEDs also denote high level, mid-level and low level on tanks where probes are fitted. The process routes are colour-coded, e.g white, sugar syrup; blue, acid; green, fruit material; red, CIP liquid.

To supply or receive information, a printer terminal or video screen must be connected to the computer. When accessed, the micro will print all the information such as date, time, batch number and the ingredient quantities (those required and those actually measured) for each batch manufactured. The print-out would show the formulation or recipe number selected, request how many batches are required and also through which weighing vessel the ingredients would be dispensed. During normal running the micro would print all start and stop sequences, faults and alarm conditions.

The role of the operator in an automated process control system will be:

- to key in required production requirements
- to ensure all swing bends are in place
- to supervise the process
- to deal with any malfunctions that occur.

The operation of the system would be controlled by a number of simple key entries.

The automated CIP sequences would be similarly controlled with a number of in-built safeguards. For example, all feed and return swing panels must be correctly positioned for CIP before the micropressor will initiate the clean; all manufacturing vessels would have LEDs showing whether the interlock on the tank door was made and whether or not the tank was empty.

8.6.2 Typical operating sequence for syrup manufacture

1. Ensure the swing panel is in the correct position after CIP (LED will be lit).
2. Key in the recipe number and batch factor (number of batches)
3. Select the route for filling the chosen vessel.

Note: The teletype would confirm these entries and would not carry out the sequences until confirmation is received.

4. The teletype would print that a particular sequence had been started and LEDs would light, showing that those particular valves had been energised and opened.
5. The route would now be selected and ready for the first ingredient to be pumped into the tank.

Note: Each ingredient would have already been pre-dissolved and stored in bulk tanks; the ingredients would then be dispensed into a weighing vessel (small tank on load cells) before the main tank to ensure the correct quantity was added.

6. The process would continue until the liquid made contact with the high-level probe (LED will light) at which time the ingredients will have been added and the batch topped up with water to the final volume.
7. The batch would mix for a pre-set time allowing the operator to take a sample to carry out any necessary quality control checks.

Note: Safeguards can be incorporated to ensure that the syrup cannot be allowed to go on forward flow unless the operator has keyed in a sequence.

8. A continual batch-making sequence can be selected to ensure continuation of filling in the same tanks (assuming they are clean and sterile).

The production line would automatically be presented with completed batches of syrup in the sequence selected with very little manual operation by the syrup room operator.

The manufacture of syrups and receipt of raw materials – e.g. liquid sugar, glucose, citric acid, etc. – would all be controlled by the main console. At the end of each day or week the computer would produce a hard copy of all the ingredients used compared to the actual recipes entered; this will enable the syrup room operator to monitor stocks of raw materials, yields and accuracy of all metering units.

8.7 Multiple component mixing plant

It is now possible to purchase a purpose-designed multiple component mixing system for full automatic production of all kinds of soft drinks. The individual components of the beverage are directly processed according to the volumetric batch principle; this eliminates the concept of preparing the conventional syrup and reduces the need for large, complex syrup rooms and associated equipment. The use of computerised systems ensures full automatic production to provide uniform product quality. A microprocessor supervises and controls the individual functions and simultaneously monitors and records all operating data; microbiological safety is ensured by the design of an integrated CIP system.

The beverage is produced from the liquid-based components and carbon dioxide, as opposed to the standard procedures, i.e. the manufacture of syrup is eliminated. The example quoted is based on a unit called the Mixomat M.*

8.7.1 *Construction*

The machine comprises five major construction units, i.e. control panel, de-aeration station, dosing station, mixing station and carbonation station. Figure 8.16 shows the overall size of the installation.

Figure 8.16 'Mixomat M' installation. (Photograph courtesy of Seitz Enzinger Noll.)

*Seitz Enzinger Noll (AG) Ltd, West Germany.

Figure 8.17 Control panel of 'Mixomat M'. (Photograph courtesy of Seitz Enzinger Noll.)

8.7.2 Control and operation

The control panel (illustrated in Figure 8.17) accommodates the microprocessor, a mimic flow diagram and a monitor; the monitor screen displays all data entered, any operational faults occurring, and the Brix and temperature readings of each batch. Data are therefore recorded to document the beverage production.

De-aeration. A vacuum de-aeration system is used to de-aerate the water. A water-sealed vacuum pump creates a vacuum of 95–98% in the vacuum tank; as the water enters, it is sprayed into this vacuum and almost completely de-aerated; the de-aerated water is then pumped out of the vacuum tank to the dosing station.

Dosing. The beverage components are pre-dissolved in small concentrate component tanks; sufficient component is dissolved for a number of hours or multiple batches of production. The components are pumped by high-viscosity, positive-displacement pumps to a dosing station; the computer can be programmed to water rinse after each cycle dependent upon the recipe requirements. The order of dosing is important to ensure that ingredients that are not compatible with each other are added at the appropriate time.

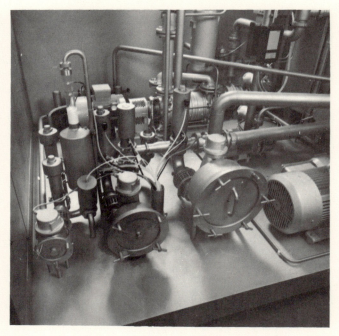

Figure 8.18 Rotary piston meters on 'Mixomat M'. (Photograph courtesy of Seitz Enzinger Noll.)

The dosing assembly consists of three dosing lines – for water, sugar and concentrates. Rotary piston meters (illustrated in Figure 8.18) register the volumes and pass this information to the microprocessor which, by means of valves, controls the quantities required for the particular recipe entered.

Mixing. The dosed beverage is passed to one of two batch tanks, where it is mixed by an agitator. After the mixing time has elapsed, the Brix is checked either manually or by incorporation of an automatic refractometer or density meter; the beverage is then released for the next stage. The batch mixing tanks operate in an alternating manner so that during the mixing process one tank is in readiness for filling.

Carbonation. The carbonating pump sucks the homogeneous beverage from the mixing tank and conveys it to the carbonating venturies where, according to the desired CO_2 content, the correct amount of CO_2 is added to the mix, controlled by the corresponding feed pressure of the gas. The carbonated liquid enters the storage tank below tank level to prevent turbulence and uncontrolled gas pick-up. By means of CO_2 over-pressure, the finished beverage is then transferred to the filler.

8.8 Future developments

The development of computerised systems has raised the soft drinks industry into the modern automated industrial world; this is especially so in the *syrup room operation* where consistent quality of batches to exact recipes is essential. Preparation of syrup with the minimum amount of aeration is of paramount importance to ensure high efficiencies on high-speed filling lines.

Manpower has been minimised which, in turn, has reduced costs and automation has reduced human errors. Raw materials can now be dispensed more accurately and handled in a safer and more hygienic manner.

The introduction of Clean-in-Place systems has ensured that sterilisation routines can be carried out to the exact requirements of the pre-set programmes. Longer production runs will continue to be necessary; it is therefore essential that where cleaning time is reduced, the CIP is still effective ensuring that microbiological quality continues to improve.

The multiple component mixing plant is the latest development in an obvious deviation from traditional syrup room operations; it has its benefits and its limitations. The amount of room it occupies is very minimal compared to the large number of tanks required in most general syrup rooms; these new systems are likely to find a place where minimal flavour changes occur and space is at a premium.

The development of programmable control systems has opened up greater possibilities for the future. Process control, utility supply and material handling will all continue to be automated and this will make it possible to calculate and control raw material costs and syrup yields very accurately. This type of information will be available for management to assist in production planning and give exact information on raw materials to guide purchasing requirements.

9 Containers and closures

I.S. ROBERTS

9.1 Introduction

The development and growth of the carbonated beverage industry has to a great extent been characterised by an ever-increasing diversity in the size and form of the packages used to distribute the industry's products to the ultimate consumer. A multi-million dollar industry involving several large multi-national corporations is dedicated to the manufacture and supply of the specialised containers and closures to the carbonated beverage industry.

It is very difficult to generalise on a global scale about the types of packaging used by the industry, because the mix of packages used in different markets around the world varies tremendously, owing to local customs, practice, legislation and economic circumstances. To exploit fully every sales opportunity, it is usually desirable to market a range of different package types and sizes, each of which is suited to a particular pattern of use. In many countries, most notably the USA, the packages used by the industry are almost exclusively designed to be used once only. In other parts of the world, containers (and in some cases their closures) are used many times, with a washing procedure immediately before each refilling. This type of returnable, re-usable packaging was once the norm throughout the industry, but demands for more convenient one-trip packaging, in a variety of forms and sizes, have resulted in the diversity of package mix that we see in the global industry today.

9.1.1 Basic package types

There are six basic types of containers and four types of closures used in the carbonated beverage industry:

Containers: Returnable, refillable glass bottles
                    Non-returnable, non-refillable glass bottles
                    Non-returnable, non-refillable plastic bottles
                    Non-returnable, non-refillable metal cans
                    Non-returnable, non-refillable plastic cans
                    Returnable, refillable plastic bottles

Closures: Lever-off crown corks
Metal screw-threaded resealable closures
Screw-threaded resealable plastic closures
Easy-open can ends

9.1.2 The function of packaging

Before describing in detail the various container technologies employed by the carbonated beverage industry, it is useful to consider briefly the basic functions of a package. This discussion applies to all packaging for all types of product, but the emphasis will fall more heavily on certain functions than others, depending upon the particular product sector for which the package is used. These are the fundamental functions of any package:

- *Contain* a particular volume of product.
- *Protect* a product from the environment and in some cases to protect the environment from the product.
- *Preserve* a product in an acceptable form or condition until the product is used or consumed.
- *Dispense* the product in desired quantities.
- *Inform* those handling or using the package about the product and its use, hazards, ingredients, shelf-life, etc.
- *Sell* the product in a competitive environment.

In considering the functions of a package, it must not be forgotten that its success in executing these functions will take into account its compatibility with the product, filling and handling systems, distribution systems, disposal methods and cost.

9.1.3 Pressurisation

For carbonated beverage packaging, containment takes on a special emphasis; unlike many other product categories, carbonated beverages are products that are maintained in an unnatural state of having an excess (above equilibrium) of carbon dioxide dissolved in the liquid. The natural tendency for all systems to reach an equilibrium state causes carbonated beverages to try to expel the excess gas into the atmosphere. Packages for carbonated beverages must therefore have the ability to contain liquid and the gas pressure necessary to retain the excess carbon dioxide dissolved in the product. Among food packages, this requirement to contain a gas pressure above atmospheric pressure is unique and dictates some of the special features that carbonated beverage packages must have.

In common with all food products, the beverage container must be capable of providing protection from the range of environmental hazards that could render the product unfit for human consumption. However, unlike most food products, carbonated beverage packaging must ensure that the pressure inside the sealed container does not contribute to the type of catastrophic failure that could cause bodily injury. A great deal of effort is expended by the industry to ensure that the packages it uses are fully capable of containing product-generated pressure so that consumers are not exposed to dangers resulting from package failure.

9.2 Glass bottles

Glass containers were first produced several thousand years ago in ancient Egypt and have been in continuous use throughout the ages since. It is not known exactly when glass bottles were first used for carbonated beverages, but it is known that they were being used for packaging champagne early in the eighteenth century. When Jacob Schweppe started in business as a manufacturer of mineral water in the 1780s, earthenware bottles were initially used; these were soon replaced with glass bottles which were impermeable to gases.

This remarkably long history of glass packaging is due in part to the inert nature of the material. Unlike many other packaging materials, glass will not interact with the packaged product; it will not absorb or allow permeation of any of the product's constituent parts, nor will it impart any 'off-taste' of its own to the product. There are several stories of products packaged in glass being discovered after many tens of years of storage and still being fit for consumption.

Until the end of the 1950s, glass bottles were the only package type used for carbonated drinks, and despite the phenomenal growth in the use of metal cans and plastic containers in the 1960s, 1970s and 1980s, glass containers still predominate in many parts of the world.

9.2.1 *Raw materials*

Despite its long history as a packaging material, the composition of glass has changed remarkably little over the centuries. The main constituents of modern container glass are as follows:

- Silica (SiO_2) accounts for about 70% by weight and is supplied to the melting furnace in the form of high-quality sand.
- Soda (Na_2O) accounts for about 15% by weight and is supplied in the form of soda ash (Na_2CO_3).
- Lime (CaO) accounts for about 10% by weight and is supplied in the form of limestone ($CaCO_3$).

Other minor constituents are included in the glass, either as impurities in the main raw materials or as deliberately added components that are used to control or modify the properties of the finished glass.

One very important raw material in glass production is 'cullet'. This is simply crushed glass derived from substandard containers that have been rejected from the production process or from used bottles that have been collected after use. The use of cullet greatly assists in the melting of other constituents, reducing energy consumption.

The raw materials for glass production are blended together in the correct proportions in modern plants automatically and continuously. The resulting mixture is known in the industry as the 'batch'.

9.2.2 *Glass production*

To produce the large quantities of glass required for a modern bottle-making plant, the 'batch' is fed to a large furnace in a continuous stream, while molten glass is drawn off from the other end of the furnace. The furnace itself can best be described as a small swimming pool with a domed roof, constructed from specially made blocks of heat- and corrosion-resistant refractory material. During steady-state operation, the furnace is filled to a depth of between four and six feet with molten glass and the fuel used to melt the glass is injected into and burned in the space between the glass surface and the furnace roof. The temperature required to melt and fuse the batch materials into molten glass is about 1500 °C.

Most modern furnaces use oil or gas as their source of thermal energy and, in many cases, the burners are designed to be capable of alternating between both fuels. Molten glass is electrically conductive and, in some furnaces, electrodes are installed below the glass surface to allow an electrical boost of the glass temperature.

Glass production on a large scale for container manufacture is, of necessity, a continuous process because of the very high temperatures involved. Therefore, once a new furnace goes into production, it normally remains in operation continuously for its entire life, which can be up to eight or ten years if it is well managed and maintained.

9.2.3 *Container formation*

One glass-melting furnace will normally provide molten glass for several bottle-forming machines. Molten glass leaves the furnace and flows along individual channels, leading to each bottle-making machine. During this transfer, the temperature of the glass is adjusted to the correct temperature for bottle formation.

At the end of the feeder channel is located equipment which accurately controls the flow of molten glass through an orifice. Immediately below this orifice is a pair of metal shears that cut the flow of glass into 'gobs', which then fall by gravity down inclined metal chutes into the moulds on the bottle-forming machine. The weight of each 'gob' of glass is controlled by

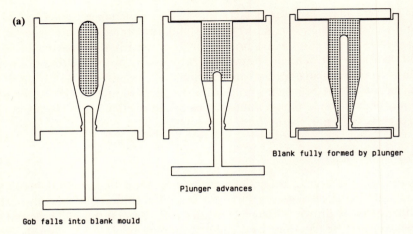

(a)

Gob falls into blank mould

Plunger advances

Blank fully formed by plunger

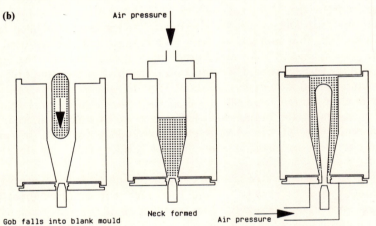

(b)

Air pressure

Gob falls into blank mould

Neck formed

Air pressure

Blank blown

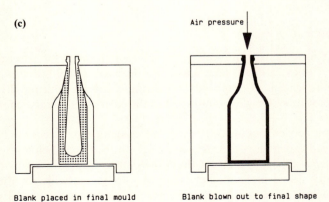

(c)

Air pressure

Blank placed in final mould

Blank blown out to final shape

the temperature and composition of the glass and the speed at which the shears operate.

Modern container-manufacturing machines all operate on the same principle of a two-stage forming operation. The gob of glass from the feeder falls under gravity into the 'blank' or 'parison' mould where it is either blown with air into a bubble that conforms to the internal shape of the mould or it is pressed by a metal plunger to take up the shape of the mould. The 'blank' or 'parison' is then mechanically transferred, held by the neck, to a second mould where air pressure is used to blow the blank out to the final bottle shape, dictated by the shape of the mould.

In the case where the blank is pressed by a plunger to the shape of the mould, the overall two-stage process is called 'press and blow' formation. Traditionally, this process is used for jars and other wide-mouthed containers, but is now being used increasingly for narrow-necked bottles because of its ability to produce greater consistency in bottle-wall thickness. The bottle-forming process is illustrated in Figure 9.1.

In the case where the blank and the finished bottle are both formed by air pressure, the process is called 'blow and blow' forming. Traditionally, this process has been used for all carbonated beverage bottles, but it does have the disadvantage in that control of glass thickness, important in providing a strong bottle, is more difficult than in press-and-blow forming.

There are, of course, some variations to the processes described above. In some types of machines, vacuum applied to the outside of the blank is used to assist the blow and in some cases replaces blowing altogether.

A complete bottle-forming machine will consist of a number of sections, each of which is equipped with blank moulds and finished bottle moulds with a mechanical transfer mechanism between the moulds (Figure 9.2). In most modern machines, each section will carry two, three, or even four blank and finished bottle moulds and the gob feeder is designed to provide two, three, or four gobs at the same time. The number of individual sections on a machine will vary up to ten or even twelve. Each section is out of phase with all of the other sections to allow the gob feeder to supply each section with gobs, in turn.

The bottle moulds themselves are produced from fine-grained cast iron. The rough mould castings are machined accurately to produce a bottle of the desired shape and size for a given glass weight. The bottle-forming surfaces are polished to ensure a smooth surface for the finished bottle and any desired lettering or decoration that is required on the finished bottle is machined into the mould surface.

Figure 9.1 (a) 'Press and blow' blank forming. (b) 'Blow and blow' blank forming. (c) Bottle blowing.

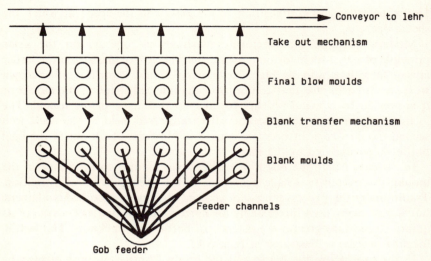

Figure 9.2 Schematic layout of a six-section double-gob bottle-making machine.

Upon completion of the moulding cycle, during which heat is continually being removed from the glass by the moulds, the finished bottle is taken out of the mould by a pair of take-out fingers and placed onto a moving conveyor. At this point the outer surface of the glass has cooled to around 300 °C and is rigid, but the inner parts are still a good deal hotter and soft. If the bottles were to be allowed to fully cool naturally, the hotter parts would contract more than the cooler outer skin, resulting in a build-up of stresses within the glass. Such tensile stresses on the outer surface of the glass would cause containers to be extremely fragile and unserviceable. The bottles must therefore be annealed to relieve these stresses.

This process is achieved in an annealing oven or lehr, through which the bottles pass prior to final inspection and packing. In the lehr, the temperature of the bottles is raised to around 550 °C, at which point the outer skin softens slightly, allowing stresses to relax out. The temperature is then lowered at a carefully controlled rate through the remainder of the lehr so that all of the glass cools uniformly and further stresses are avoided.

9.2.4 *Glass bottle strength*

Glass is an extremely strong material when it is first formed – stronger, in fact, than steel. Unfortunately, much of this strength is lost through the formation of minute flaws in the regions close to the surface of the material. These flaws result from contact with the mould surface and other equipment used in bottle manufacture and, in practice, a bottle's strength may be reduced by a factor of

10 between its initial formation and the point at which it goes into service. Further reductions in strength result from contact with other bottles, filling and handling equipment and various other objects that the bottle touches during its life.

For the carbonated beverage industry, the maintenance of an adequate minimum bottle strength is of critical importance and bottles must be sufficiently strong to withstand the internal pressure generated by the carbonated product. There are many factors that influence bottle strength – weight, design, quality of annealing, impurities in the glass, surface damage, etc.

The measure of bottle strength most frequently used in the carbonated beverage industry is internal pressure resistance (IPR) – for obvious reasons. The IPR of a bottle can be measured accurately, using standardised equipment and conditions and is routinely measured both by the bottle maker and filler. Bottles are pressurised until they burst and a read-out of IPR (corrected to a standard period of time of pressure application) is taken. In general, single trip (or one-way) bottles are expected to have a minimum IPR of 13.8 bar on delivery to the filling plant and new returnable refillable or multi-trip bottles are expected to have a minimum IPR of 15.5 bar. In practice, these levels are regarded as minima and the mean levels are usually much higher. Multi-trip bottles are specified with higher IPR to ensure that adequate strength is maintained throughout the life of the container.

9.2.5 *Surface coatings*

The strength of the bottle cannot be increased by the use of external coatings, but the resistance to surface damage can be greatly enhanced, which results in better maintenance of a bottle's initial strength.

Surface coatings may be applied to the external surface of the glass, either when it is still hot, before annealing, or when it has been annealed. Various types of hot-end and cold-end coatings are used but all have the objective of increasing surface lubricity and avoiding damage to the glass surface.

On some types of bottles, plastic coatings or sleeves are applied immediately before packing. Some returnable bottles are coated with a tough polyurethane-based clear coating which sticks securely to the walls and base of the container in a continuous film. Such films are designed to contain fragments and thus prevent possible injury should the glass break due to an impact. Of course, these coatings do provide excellent surface protection for the glass, which is able to maintain its original strength much better than if the coating were absent.

Non-returnable bottles often have shrunk-on plastic wraparound labels. In protecting the glass surface from damage, these sleeves allow some reduction in glass weight without jeopardising the in-service strength of the bottles. They also provide excellent surfaces for decorative labelling.

9.2.6 Future developments

Glass-container usage for carbonated beverages, while still predominant in many countries, is declining rapidly in popularity where alternatives are economically and technically viable. This trend is largely due to the weight and fragility of glass and the environmental hazards presented by broken bottles, which are often thoughtlessly discarded by unthinking or uncaring consumers.

A major international co-operative research programme has been initiated by major glass-bottle producers around the world, with the aim of enhancing the strength of glass very considerably so that much lighter containers of great strength may be produced. It is hoped that this programme produces rapid results before alternative materials replace glass bottles altogether. No other material used in beverage packaging has the ability to totally resist product interaction and thus maintain the original taste and carbonation of the product.

9.3 Plastic bottles

The carbonated beverage industry has been relatively slow to adopt plastics for its primary packages. The reasons for this lie in the properties of the available plastics and how these materials interact with the products.

Unlike glass, plastics are not perfect gas barriers and they have a tendency to creep under applied stress. Early attempts to produce a plastic beverage container failed because the package distorted severely due to the internal pressure from the beverage and because the dissolved carbon dioxide in the product slowly permeated through the walls of the container leaving behind a flat, uncarbonated liquid.

In the 1940s it was discovered that by stretching certain amorphous polyester plastic materials at a temperature just above their softening point (glass-transition temperature), a degree of crystallinity or molecular orientation could be induced in the material, giving it enhanced physical properties. This discovery gave rise to the production of high-performance textile fibres. In the 1960s, the process of molecular orientation was further developed in order to create orientation in two perpendicular directions (biaxial orientation). This gave rise to the production of high-performance polyester films which were widely adopted for use as packaging films and tapes and for production of recording tape and photographic films.

Finally, in the early 1970s, the Du Pont Company patented a process for inducing biaxial orientation in a three-dimensional bottle structure (Figure 9.3). Bottles made using this process were found to have previously unattainable levels of dimensional stability, clarity and barrier to gases. The carbonated beverage industry at last had access to a plastic container that had

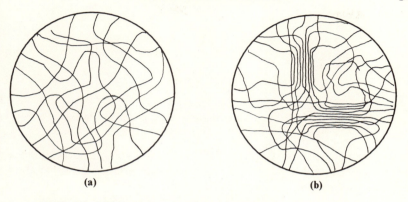

Figure 9.3 PET structure (a) amorphous material (b) biaxially oriented.

the right combination of properties to be a technically and commercially viable alternative to traditional packages. The lightness, resistance to breakage and the absence of any potential for corrosion offered by plastics are highly desirable properties for a carbonated beverage container.

9.3.1 Raw material

By far the most widely used plastic for beverage containers is PET (polyethylene terephthalate). This is a saturated polyester that is produced from the reaction of ethylene glycol with either terephthalic acid or dimethyl terephthalate. The resulting polymer molecule is a long chain consisting of repeated units of the basic ethylene terephthalate monomer group (Figure 9.4). The total length of the polymer chain may vary, depending upon the length of time that the polymerisation reaction is allowed to proceed and, in general, the longer the polymer chains, the stronger the bottle. However, longer chains take longer to produce and high molecular weight polymers are thus more expensive to produce. For a typical carbonated beverage bottle, the polyester chain length will be around 130 monomer units, giving a molecular weight of about 25 000.

Copolymers, in which a small proportion of the terephthalic acid is replaced with another dibasic acid, are used extensively in bottle production. The effect of copolymerisation is to modify the moulding behaviour and properties of the material to suit particular applications, and the number of copolymer variants is endless.

Another thermoplastic that is often used for beverage container production is PVC (polyvinyl chloride). Like PET, this material shows enhanced properties through molecular orientation brought about by stretching –

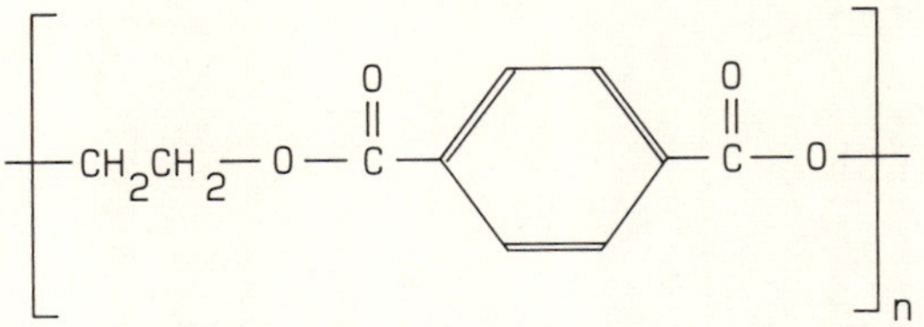

Figure 9.4 Poly(ethylene terephthalate) monomer unit (n = approx. 130).

though not to the same degree as PET. However, PVC is not as good a gas barrier as PET and is not widely used in bottle production other than for very low carbonation products and mineral waters.

9.3.2 PET bottle production

The PET bottle production process is similar in many ways to glass bottle production in that it is a two-stage operation. In stage one, raw material is converted into a parison or pre-form and in stage two, the pre-form is converted to the finished bottle. In some manufacturing equipment the two stages are combined into a single machine, while in other systems the two stages of manufacture are performed in separate pieces of equipment which can be in totally separate locations. However, in both types of operation the principles of bottle formation are identical.

Pre-form manufacture. The process most frequently used for pre-form manufacture is injection moulding (Figure 9.5). PET granules are fed into the barrel of the injection moulder where they are plasticised into a molten mass at about 270–280 °C. This material is then injected at very high pressures into a multi-cavity mould where it is formed into the required pre-form shape. The mould is supplied with chilled water which circulates around the mould cavities and core pins to rapidly cool the molten PET. The mould then opens and the clear amorphous PET pre-forms are ejected.

Far less widely used is the tube process for pre-form manufacture. Here, a continuous tube of PET is extruded under pressure through a circular orifice. The tube is cut into predetermined lengths which are then further processed to seal off one end and to form the screw-threaded neck (finish) on the other end.

There are a number of important considerations in pre-form manufacture which strongly influence the quality of the completed bottle. Some of these are reviewed in the following sections.

(a) *Material drying.* In any pre-form manufacturing process, it is extremely important to ensure that the PET granules are thoroughly dried before the material is processed. PET is a hygroscopic material which can absorb up to 0.5% by weight of moisture during transportation and storage. If this moisture

CONTAINERS AND CLOSURES

(a)

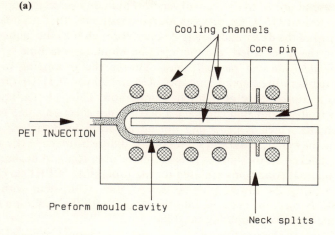

(b)

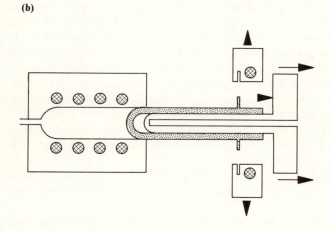

Figure 9.5 (a) Pre-form moulding. (b) Pre-form mould opening.

is not removed, to a level below 30 ppm (0.003%), hydrolysis of the polymer takes place at the elevated temperatures used in injection moulding and the material experiences a reduction in polymer chain length with an associated reduction in the physical properties of the finished bottle.

In practice, injection moulders used for PET pre-form manufacture are fitted with a dryer mounted above the material feed port. In the dryer the PET granules are subjected to a stream of very low humidity gas at a temperature of between 150 and 175 °C for between four and six hours. PET granules are fed into the top of the dryer and slowly descend through it as the material at the bottom of the dryer is fed into the injection-moulding machine.

(b) *Acetaldehyde and thermal abuse.* During the processing of PET at elevated temperatures, small amounts of acetaldehyde (CH_3CHO) are generated as a consequence of thermal degradation. This volatile liquid has a distinctly fruity odour that can have a detrimental effect on some beverages. Acetaldehyde, which is generated during pre-form manufacture, becomes trapped in the material and will subsequently diffuse from the bottle walls into beverage packed in the bottle. It is therefore important to ensure that the gentlest thermal conditions are used during processing of PET; in effect, this means ensuring the lowest possible temperature (consistent with actually being able to inject the material into the mould) with the minimum residence time in the injection-moulding machine.

In addition to the generation of acetaldehyde, thermal abuse of PET will also cause a reduction in polymer chain length – though to a lesser extent than hydrolytic degradation caused by inadequate drying.

(c) *Crystallinity.* The final consideration in relation to the temperatures used in PET pre-form manufacture concern the degree of crystallinity in the finished pre-form. One of the very desirable properties of PET is its glass-like clarity and sparkle. However, this property is associated with the amorphous state of the material and, when crystallised, PET becomes opaque. As supplied to the converter, PET is in the form of crystalline chips; and during melting in the injection-moulding machine, the crystalline structure must be destroyed. The most favourable temperatures for crystallisation of PET are between 85 and 250 °C – in effect, between its glass-transition temperature and a point about 10 °C below the completely molten state. Maximum crystallisation rate occurs around 160–165 °C.

It is important, therefore, to process PET into pre-forms in a highly controlled way to ensure that the properties of the pre-form are optimised.

1. Temperatures must be sufficient to ensure maximum destruction of crystallinity present in the raw material chips, while minimising acetaldehyde formation.
2. Once injected into the mould, the material must be cooled as rapidly as possible through the temperature range from 250 to 85 °C to avoid crystalline haze. Excessive cooling and/or excessive atmospheric humidity can, however, lead to condensation on the moulding surfaces, giving rise to hydrolysis of the molten material as it is injected into the mould.
3. Stronger bottles result from longer polymer chains and the rate of crystallisation falls as polymer chain length rises, but greater chain length requires higher moulding temperatures with the attendant risk of excessive acetaldehyde formation.

The best PET resins will allow the moulder as large a 'window' within these conflicting requirements as possible, because errors made at the pre-form manufacturing stage cannot be corrected in subsequent processing. Bad pre-forms will always make a bad bottle!

Bottle manufacture. Pre-forms are converted into bottles in the second stage of the manufacturing process in an operation known as 'stretch – blow moulding'. The steps involved in this process are as follows:

1. The pre-form body is reheated to a temperature slightly above the material's glass-transition temperature by moving it past a bank of radiant heaters (Figure 9.6).
2. The reheated pre-form, held by the neck, is placed inside a mould made to the dimensions and shape of the desired bottle.
3. The pre-form is stretched axially by a stretch rod which is forced down the inside of the preform (Figure 9.7).
4. The pre-form is blown out to conform to the inside of the mould by air pressure and held in contact with the mould until the temperature of the material cools sufficiently below the glass-transition temperature to maintain the shape as defined by the mould.
5. The completed bottle is removed from the mould and either placed on a conveyor to the palletisation operation or is fed to an intermediate operation where a base-cup is applied.

During these operations, the PET is converted from a relatively thick (3.5–

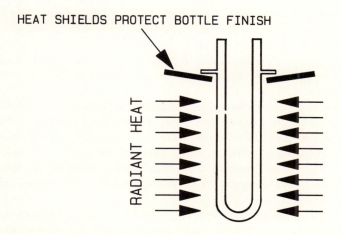

Figure 9.6 Pre-form heating.

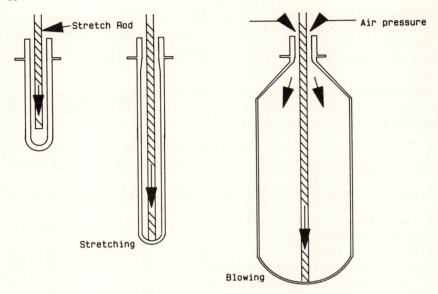

Figure 9.7 Stretch-blow moulding.

4.0 mm) section to a much thinner (0.3 mm) cross-section, and becomes biaxially oriented. The stretching of the material at around 85–90 °C, just above the glass-transition temperature, in the longitudinal and circumferential directions, induces a degree of alignment of the polymer chains within the material. In changing from an amorphous to the oriented state, small areas of strain-induced crystallinity are developed in the material; however, unlike thermally induced crystallisation, the regions of strain-induced crystallinity are small enough not to cloud the material. The process of orientation also has a marked effect on other physical properties of PET. The tensile strength and stiffness of PET is increased by approximately half in the axial direction and by between a factor of 2 and 3 in the hoop direction. Water vapour and oxygen permeability are cut by 50% and carbon dioxide permeability is reduced by about 25%.

The performance of the finished bottle in terms of strength, clarity and barrier properties is heavily dependent upon the degree and uniformity of orientation in the bottle walls and upon the consistency of material distribution throughout the oriented parts of the bottle. It is self-evident, therefore, that the relationship between material properties, pre-form design and finished bottle design will ultimately determine the overall balance of bottle properties.

The ideal container shape for carbonated beverages, in terms of maximised physical properties for the minimum amount of material, is a sphere. Such a container is obviously impractical and a cylinder with a hemispherical base is the next best thing. For this reason, all early PET bottles were of a two-piece construction, with a separately manufactured base-cup glued onto the hemispherical base to provide the means for the bottle to stand upright. Later developments brought about the possibility of free-standing single-piece bottles which had a number of feet moulded into the base. This obviously removed the need for separate base-cup moulding and gluing and assembly operations.

9.3.3 *PET bottle properties*

PET bottles have provided the carbonated beverage industry with a package which is able to deliver to the consumer volumes of product above the glass bottle's practical limit of 1 litre in a single container and have therefore contributed strongly to the overall growth in the industry's volume in the 1980s. It is rare for a PET bottle to break (unless it has been exposed to chemicals which cause the development of environmental stress cracks) and even when breakage does occur, no sharp fragments are produced. The light weight of PET bottles is also an obvious attraction.

In common with most plastics, PET is not a perfect gas barrier and a certain amount of carbon dioxide is lost from the packaged beverage during the life of the package. This has not, however, been a significant problem in the sizes where PET bottles are most widely used and of greatest value to the industry – over 1 litre. Smaller PET bottles do suffer to a greater extent from excessive carbonation loss owing to the larger surface area to volume ratio inherent in smaller packages. In recent years, however, advances in PET polymer and process technology have made 0.5-litre bottles viable in terms of cost, versus competitive packages and in terms of providing a reasonable shelf-life without unacceptable carbonation loss. A limit of 15% carbonation loss over twelve weeks of storage is specified by many beverage fillers.

External coatings of high gas barrier polymers such as PVDC (polyvinylidene chloride) and multi-layer packages consisting of co-injected or co-extruded layers of PET and other high-barrier plastics, such as EVOH and Nylon, do reduce the carbonation losses from PET bottles, although such packages do have certain disadvantages. They generally cost more due to the extra complication of their manufacturing process and they are less easily recycled due to the intimate intermixing of different polymers.

During the manufacture of PET bottles, the polymer chains in the material are 'frozen' in an unnaturally high energy state by the rapid cooling of the material below its glass-transition temperature after stretch–blow moulding. There is thus a natural tendency for the material to try to achieve a lower energy level, accompanied by a change in the properties of the container. This ageing tendency is manifested by an increase in density and yield stress (making the polymer more brittle) and by a shrinking of the bottle. The rate at which these changes take place is heavily dependent on the temperature at which they are stored and the closer this temperature is to the glass-transition temperature, the more rapid are the changes.

The thermal sensitivity of PET has restricted the use of the material to one-trip containers for most of the ten years since they were first introduced into the carbonated beverage industry. Advances in resin and processing technology have now advanced to the point where a refillable PET bottle is technically feasible. Refillable bottles must be tolerant of elevated temperatures inherent in the sanitising process through which they must pass before refilling; this requires careful selection of material and process variables. At least two separately developed refillable PET bottles are in limited use.

Another interesting development, which has so far enjoyed little commercial success, is the PET can. Several different manufacturing techniques have been developed to produce a can body that is filled and sealed in the same way as a metal can. The plastic can illustrates the versatility of PET but unfortunately suffers from the inherent property of PET – carbonation loss through permeation.

9.4 Metal cans

Metal cans have been in use as a package form for many centuries; it is known that cans were produced as far back as the fourteenth century in Europe. It was not until the early 1960s, however, that the metal can started to make any impact upon beverage packaging. Early beverage cans were little different from the tin-plated steel cans that had been used for packaging of various foods for many years. They had cylindrical walls formed by curling individual pre-cut rectangular blanks of tin-plate around a forming mandrel and engaging the two edges parallel to the axis of the can by hooks formed into the edges of the blank. The double-hook seam was then hammered flat and soldered with pure tin solder to form a hermetic seal. The cylinder of metal was then flanged on each open end and a separately manufactured tinplate end was seamed onto one end of the can. End manufacturing and seaming are described in more detail later.

Finished cans, with one end attached, were shipped to the filling plant where they were filled and sealed by the application of a second tin-plate end, this being sealed to the can with a seam identical to the one formed when the first end was applied by the can maker. This type of can had one important drawback which was that it required the consumer to use a piercing tool to create a hole in the end through which the contents could be dispensed and this severely limited the popularity of the can as a beverage container.

In the early 1960s, the easy-open aluminium can end was developed and the beverage-can business started to grow. The easy-open end incorporated the now familiar ring-pull opening feature and immediately provided the can with the necessary convenience feature – which the major beverage companies rapidly exploited.

The three-piece can described above is still in use in many parts of the world, along with similarly produced cans that have a welded side seam in place of the hooked and soldered seam.

In the biggest can markets, however, the three-piece can has now largely disappeared from use for beverages and has been replaced by the two-piece 'drawn and wall ironed' (DWI) can. This type of can has its cylindrical walls and bottom end formed from a single piece of metal, eliminating the side seam and base-attachment seam, with an aluminium end attached after filling in exactly the same as on any other can.

Two-piece DWI cans have several advantages over three-piece cans as containers for carbonated beverages:

1. They can be manufactured in aluminium as well as tin-plated steel. Aluminium is difficult to weld or solder into a three-piece construction.
2. They have no side seams, which are always potential sites for attack by acidic beverages.
3. External decoration is continuous around the walls of the can; there is no necessity to leave a vertical band clear of print at the side seam.
4. They use significantly less metal.
5. Seaming (by the filler) at the top end is considerably easier owing to the absence of a junction of the side seam and the end seam.

9.4.1 Raw materials

Two-piece beverage cans are manufactured either from tin-coated low-carbon steel or from a special aluminium alloy (3004-H19). A description of the production processes for the raw materials is beyond the scope of this review. However, it is important to understand that these materials are used in such vast quantities that specialised plants are operated by the metal producers solely to produce can stock.

The can-manufacturing process itself requires that the metal used is of the highest quality and major investments have been made by the metal producers to ensure that the exacting requirements of the can makers are met.

9.4.2 Can production

Virtually all of the metal for two-piece can manufacture is supplied in coil form in widths determined by the width of the metal rolling mills used to produce the stock. The production of cans from the coil takes place in the following steps:

1. The coil of metal is placed on a horizontal uncoiler and fed through an

inspection and lubricating unit which checks the material for pinholes and applies a controlled amount of lubricant.
2. The metal strip passes into a heavy-duty, double-action press which stamps out a number of discs across the strip and draws the discs through a cupping die to form shallow cups in one operation. The scrap material from between the discs falls away and is removed to a scrap baler for recycling into raw material production.
3. From the cupping press, the individual cups are conveyed to the 'bodymakers'. Usually, a production line will have one or two cupping presses feeding up to ten bodymakers. The bodymaker performs three operations. First, it redraws the cup down to the final diameter of the can and increases its height, but makes no change to the thickness of the metal. Second, it carries the cup on a punch through a series of carbide ironing rings which effect a greater than 50% reduction in the thickness of the metal in the walls of the can and increases its height. Third, at the end of the stroke, the metal at the bottom of the can is reformed to create the pressure-resistant dome.

In the bodymaker, a great deal of work is done to the metal and substantial quantities of lubricant are used both to assist the passage of the metal through the tooling and to cool the operation. The metal in the can walls is cold-worked substantially and becomes work-hardened. The metal in the can base is unchanged in thickness from the original coil and is not work-hardened; for this reason the pressure-resistant dome is formed to ensure that the can base will not blow out under the pressure generated by carbonated beverage.

By shaping the punch at the point where the top edge of the finished can will be at the end of the ironing process, the metal at the top of the can becomes slightly thicker than further down the walls. This is done to ensure adequate metal for the formation of the seam by which the end is attached.
4. The wall ironing (Figure 9.8) in the bodymaker produces a can with a wavy top edge. This is trimmed away by a rotary trimmer to produce cans with a constant height with reference to the base.
5. After formation in the bodymakers, cans are mass conveyed through a multi-stage washer which removes all traces of lubricants and any metallic swarf from the metal, ensuring the very high degree of cleanliness essential for satisfactory adhesion of external decoration and internal protective lacquer. Normally, adhesion-enhancing surface treatments are included in the final stages of washing and the cleaned cans are hot-air dried and ready for subsequent processing.

The effectiveness of the can-washing process is highly determinative of several aspects of finished can quality and a significant part of the investment in a can-making plant goes into the washer and its associated support systems.

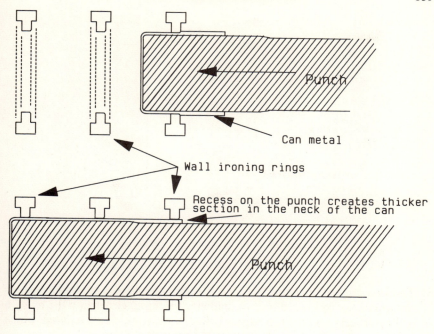

Figure 9.8 Wall ironing.

6. The can bodies pass next to the external decoration processes. If required, an all-over basecoat is applied to the can walls and oven-dried. Multi-colour decoration up to six colours is then printed onto the walls of the can followed, in some cases, by a clear varnish. The whole decoration system is then stoved by passing the cans on a peg-chain through a hot-air curing oven.
7. After external decoration, the cans are fed to one of a bank of internal lacquer spray application machines. The can is spun about its vertical axis, inclined to horizontal and a fixed pattern spray-gun applies a film of solvent or water-based lacquer to all of the interior surfaces. The lacquer is dried by again passing the cans through a curing oven. Depending upon the metal (steel or aluminium) and type of product to be packed, one or two coats of lacquer may be applied.

Since the lacquers prevent direct contact between the products, which are often corrosive, and the metal, the importance of achieving a continuous pinhole-free internal coating is paramount. Not only must the metal be totally clean and the lacquer spray precisely controlled but the handling of the can must be carefully controlled to avoid runs and drips forming in the lacquer.

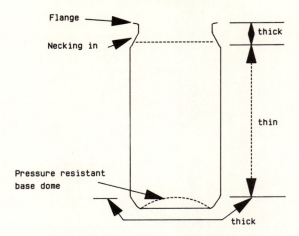

Figure 9.9 Two-piece can nomenclature.

8. The final stage of can manufacture is a further metal-working stage in which the top of the can is reduced in diameter (necking) either in a die or by a rolling process and the top edge is rolled outward to create a flange which will be used to form the double seam to attach the top end.
   The necking of the top of the can is done for two main reasons: it reduces the amount of metal required for the can end and prevents seam-to-seam contact of filled cans, thus avoiding damage.
9. Finally, the finished cans are inspected to ensure that no pinholes are present in the metal and are packed in layers on pallets for shipment to the filler.

Two-piece can nomenclature is illustrated in Figure 9.9.

9.4.3 *Double seaming*

The method of attachment for ends on all types of cans is identical; Figure 9.10 illustrates the formation of a double seam.
   Double seaming is a two-stage process; in stage one the metal of the can-flange and the end-curl are interlocked in a double-hook arrangement and in stage two the hooks are flattened together and, in conjunction with the small amount of sealing compound in the end curl, they form a hermetic seal.

9.4.4 *Can dimensions and production rates*

Whereas many designs exist for beverage bottles (according to fillers preferred image, size, etc.) few variations occur in can design and size. In fact, the fillers' and can-makers' industry associations liaise with each other to determine

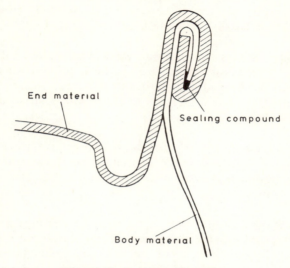

Figure 9.10 Double-seal formation.

overall industry specifications for cans and often the agreed standards transcend international boundaries.

The reasons for this are largely economic but have their roots in the can-manufacturing process. Can-production lines do not lend themselves readily to frequent changes in the size and shape of cans. It is difficult to describe in words the intricacies of can-making operations but one visit to a can-making plant will make it very clear that non-standard cans would cause extreme difficulty for the can maker. Any downtime associated with size changes is extremely expensive for an operation capable of making up to 1400 packages per minute.

Fillers of beverage cans have taken advantage of the standardised dimensions of cans to create extremely high-speed filling operations. Because allowance is not necessary for variations in size and design, can-filling lines are built to handle the standard can at higher rates (packages per minute) than any other beverage package. Bottles are seldom filled at faster rates than 1000 packages per minute but can lines capable of filling at over 2000 cans per minute are in operation in several countries.

9.4.5 Which metal?

It is a much debated question as to which is the best metal for beverage cans and there is certainly no definitive answer. Indeed, several attempts have been made to develop plastic cans, though none of these has yet achieved wide-scale commercial use.

The factors affecting the choice of metal for cans are many and complex and it is beyond the scope of this review to debate all of them fully. However, some important technical factors are worth brief consideration:

Abuse resistance. Steel is a tougher material than aluminium and is therefore more resistant to physical abuse. This can be important where shipping distances are extended or subject to rough conditions, because even minor dents can cause the internal lacquers to crack with a subsequent risk of corrosion and leakage.

Taste. Most products will slowly dissolve areas of exposed metal inside cans through microscopic pinholes in the internal lacquer. In general, the human ability to taste the dissolved metal in the product is much greater in the case of steel than aluminium.

Runnability. It is often said that steel cans, being heavier, are easier to run on filling lines compared with aluminium cans. This is usually an opinion based on personal bias because both types have been run at the highest speeds used by the industry. The runnability of a particular can type will depend upon the knowledge and skill of the operators in understanding the relationship between the filling equipment and the can's characteristics.

Recycling. It is often claimed that one material has certain benefits over the other in terms of the ease with which the material may ultimately be recycled. Again, this is largely a debate based upon personal or commercial bias because both materials are effectively recycled into raw material production operations.

9.4.6 *Can lacquering and corrosion*

There are various types of corrosive reactions that may take place between the metal of a can and its contents. All corrosion is undesirable since it can lead to deterioration of both the properties of the product and the performance of the can. In addition, external corrosion (due to high levels of atmospheric moisture or product spillage) can have a detrimental effect on the can's appearance and, ultimately, lead to perforation.

For carbonated beverages, internal corrosion is linked closely with the formulation of the product and residual oxygen content in the filled can. Several product constituents and water impurities have been found to have particularly corrosive effects on exposed metal and, while much of this experience is now taken into account during product formulation, great reliance is placed by the industry on pre-production test packing of new product formulations, new lacquer materials and new or modified can-making technology. Ultimately, however, the most effective control over can integrity is derived from the formulation and application of the internal lacquer.

In general, three-piece cans are made from pre-lacquered plate where the lacquering is performed before the can is formed. Post-repair of the soldered or welded seam area is obviously required and this is frequently the area where most corrosive activity initiates. Two-piece cans must be lacquered after body manufacture because coatings are incapable of withstanding the stresses and deformation involved in the DWI operation but, in general, higher levels of overall lacquer integrity are achievable in two-piece cans.

Coatings are supplied in the form of solutions or dispersions in solvents or water. The precise formulation of a lacquer will depend on the product type, the can metal and the can-maker's application and curing equipment. Modern can lacquers use synthetic polymer resins as the film-forming component; epoxy, vinyl, acrylic and phenolic resins are common and a constant research and development effort is undertaken by the lacquer suppliers and can makers to find more resistant and economic resins. In common with all fields of technology where synthetic polymers are used, the potential for development of improved systems is endless. Water-based coatings are becoming increasingly widely used because of the expense of operating to the extremely stringent solvent emission control regulations in can-making plants.

Internal lacquer systems are constantly evolving as the industry strives to achieve new levels of quality and economy.

One important new lacquering system, projected to be of great future importance, is a method employing electrolytic deposition of the film-forming component from a solution. This promises to offer unparalleled levels of film integrity but at much lower film weights because more consistent control and continuity of lacquer film are possible.

Others are examining methods by which the metal for DWI cans may be pre-coated. Such systems would have a significant impact on the design of can-making plants as the need for post-forming application of lacquers and their curing would be eliminated.

9.4.7 *Easy-open ends*

No review of modern beverage can technology would be complete without a review of the easy-open ends which have, to a large extent, been responsible for the can's convenience and popularity.

The earliest type of easy-open end was the ring-pull type and, in many parts of the world, this type remains the industry standard. In other countries, concern over the litter of thoughtlessly discarded ring-pull tabs has prompted the industry to adopt a lever-assisted 'Stay-on-Tab' end type where no component is detached from the end during opening (Figure 9.11).

Can ends are manufactured in a multi-stage synchronised operation which stamps out discs of metal from coil or pre-cut sheets, reforms the metal with the desired cross-sectional profile, partially scores the metal around the opening, forms an upwardly protruding rivet onto which is attached the ring or opening

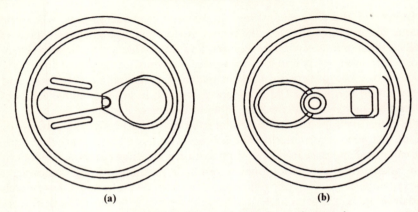

Figure 9.11 (a) Ring-pull can end. (b) Stay-on tab can end.

lever and secures the ring or lever to the rivet. Finally, the cut edge of the end disc is partially rolled over to form the curl, which becomes the end's part of the double seam that attaches the end to the body. A small amount of sealing compound is placed into the inside of the curl to help form a hermetic seal in the double seam. The ring or lever is produced from a separate coil of metal in a multi-stage stamping operation synchronised to the main end-forming operation (cf. Figure 9.12).

This description is a massive oversimplification of the actual process. The tooling for stamping and forming the end is extremely complicated and is designed to perform most of the sequence of forming operations in a series of steps as the metal is indexed forward one step at a time within the forming tool.

The vast majority of aluminium ends are made from a special aluminium alloy (5182-H19) around 0.3 mm thick. The scoring around the opening of a can end penetrates to roughly half of the thickness of the material and is a very precisely engineered feature: variations in the cross-section of the score are built-in to regulate the way in which the tearing of the metal progresses during opening.

The material used in end manufacture is pre-lacquered in the sheet or coil form; the lacquers which are intended to prevent direct metal contact with

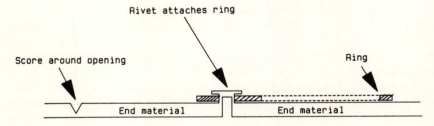

Figure 9.12 Section through can end.

beverage must, therefore, be resistant to the metal deformations which take place during end manufacture. An entire classification of lacquers, different from can body lacquers, has been developed for beverage can ends.

9.4.8 Steel can ends

Steel ends are used extensively for the bottom end of three-piece cans but are seldom used for easy-open applications. This is largely because of the difficulty in producing a score in the material along which the end can be opened. Steel is a far tougher material than aluminium and producing a score that will open with a consistent tearing force is extremely difficult. Rusting of the cut edge of the score is also a problem.

The steel industry is obviously keen to remove the exclusivity that aluminium has in beverage end production. In Europe where steel can bodies are extensively used (unlike the USA and Australia where virtually all beverage cans are made of aluminium), the steel producers have expended a good deal of effort to develop end designs that remove the inherent disadvantage of steel.

One development heavily promoted by the steelmakers is the push-button end (Figure 9.13) in which two circular tabs are formed into the end by cutting almost completely around the opening aperture, leaving only a small tab to act as a hinge. The cut edge of the tab is 'spread' slightly so that it tucks under the edge of the aperture to form an interference which holds the tab in place. The internal cut edge is sealed with a sealing compound and internal pressure in the can ensures that the two openings remain sealed. The consumer opens the can by pushing the smaller of the two circular tabs (to vent the headspace gas pressure), followed by the larger of the tabs to dispense the product. This type of end can be manufactured in steel or aluminium and does not require a ring or lever to be attached.

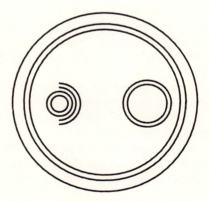

Figure 9.13 Push-button can end.

Another steel end under development is the composite metal/plastic end in which the end is formed conventionally from pre-coated stock, but instead of a score or partial cut around the opening aperture, a piece of material is completely removed. The end blank is then placed in an injection-moulding tool and a polymer resin is injected to seal the aperture and form a ring for opening it. The resin completely encapsulates the cut edge of the metal, avoiding any corrosion problems, and adheres to the specially formulated coating which is pre-applied to the inside surface.

9.4.9 *Future can developments*

It is almost certain that, for the foreseeable future, can developments will focus (as they have in the past) on gradually evolving improvements to the same basic technology. All will be driven by a desire to reduce the cost of cans and improve their physical properties. Further lightweighting of cans is almost certain as improved metallurgy allows the material to be drawn down to ever-thinner gauges while maintaining the ability of the package to hold pressure without deformation. Metal-forming techniques are also being constantly refined in order to gain maximum performance out of minimal amounts of metal.

In particular, the design of the can base, the easy-open end and, most importantly, how they are both formed from flat metal, will have an important impact on can lightweighting.

Further into the future, the radical developments which will allow the use of pre-coated stock for two-piece can production could revolutionise the canning industry.

9.5 Closure systems

All packages require some form of closure system through which the contents can be filled and ultimately dispensed by the consumer. Beverage package closures, unlike many others, have to be capable of holding an internal pressure inside the package and of venting that pressure safely as the package is opened.

All closure systems consist of at least two parts – the 'finish', which is an integral part of the container, and the closure itself. Neither can be considered in isolation because they have to work together to provide an effective seal and both must be manufactured within agreed, predetermined tolerances. Figure 9.14 depicts the two basic types of finish that are used for carbonated drink bottles. The carbonated beverage industry uses three main types of closure; can ends are a specialised form of closure applicable only to cans and are discussed elsewhere.

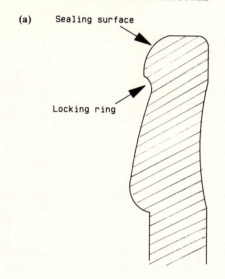

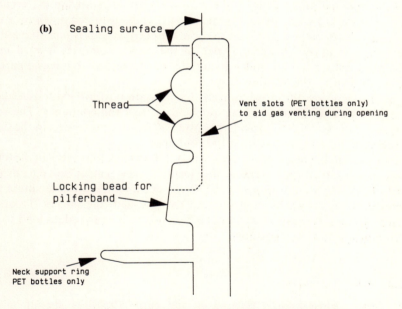

Figure 9.14 (a) Crown-finish nomenclature. (b) Screw-finish nomenclature.

9.5.1 Crowns

Crowns or crown-corks are the oldest form of closure still in widespread use in the beverage industry. The crown was invented in the USA in 1892 by William Painter and rapidly became a widely used industry standard in place of several, sometimes complicated, devices used to seal beverage bottles. They derive their name from their appearance and the fact that the earliest ones used discs of cork inside the crown to seal against the bottle finish.

Crowns are manufactured from pre-coated and printed tinplate or 'tin-free' steel sheets in a stamping operation which cuts the metal discs and forms the cut edge into a special corrugated shape. A sealing medium is then inserted into the crown either as a pre-cut disc of compressible material or as a liquid polymer 'compound' which is then cured by the subsequent application of heat in a tunnel oven through which the crowns are passed.

The application of a crown to a filled package takes place immediately after filling, in a machine specially designed for the purpose and usually driven in close synchronisation with the filler. The crowner dispenses a crown from a feed chute onto each bottle in turn and applies a vertical load to the crown to ensure that the sealing medium is compressed between the metal of the crown and the material (usually glass) of the finish. Without removing the vertical load from the crown, a specially profiled circular die is forced down over the corrugated edge of the crown. The shape of the corrugations is such that as the crowning die moves down, the edge of the crown collapses inward in a controlled fashion to lock under an annular bead which is an integral part of the bottle finish – the 'locking ring'.

In this operation, the crown is firmly locked onto the finish with its sealant liner in a permanent state of compression, to seal the container. The sole purpose of the metal is to provide a constant state of compression in the liner to maintain an adequate seal on the container.

Conventional crowns suffer from the major drawback that a tool in the form of an opener is required to remove the closure. This is remedied in part by the twist-off crown. Modifications to the design of the bottle finish facilitate the removal of the crown by grasping and twisting with the fingers. Twist-off crowns are not, however, very popular because their sharp edges tend to be uncomfortable for the consumer.

9.5.2 Roll-on closures

The roll-on closure (first devised in the early 1920s) is similar in its fundamental principle of operation to the crown, i.e. a partially finished metal cup containing a compressible sealing medium on the inside of its upper surface; the closure is placed over the finish of the bottle, compressed by a vertical load and the metal in the sides of the closure is mechanically deformed to lock onto appropriate parts of the bottle finish (Figure 9.15).

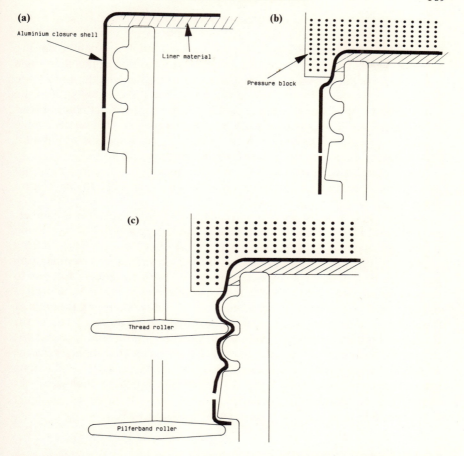

Figure 9.15 (a) Closure in position on finish. (b) Closure reformed over sealing surface by pressure block. (c) Closure fully formed by rollers.

The major innovation made by the roll-on closure, however, was that instead of a simple annular locking ring used in the crown finish, a single-start continuous helical thread is moulded in the container finish. The roll-on closure is usually made in aluminium by a stamping and drawing process followed by liner insertion, again either in the form of a disc cut from a sheet of material or as a curable liquid compound.

Application of the closures after filling is effected by a multi-head machine located next to the bottle filler and driven in synchronisation with the filler. Each bottle picks up a closure from a dispensing chute located on the closing machine infeed and is then located in one of several sealing stations mounted on a rotating turret. The sealing head is lowered onto the bottle and a vertical load applied through a shaped pressure block; this reforms the metal at the top corner of the closure, around the top outside corner of the finish. With the reform completed, free-running rollers mounted on lightly sprung spindles – offset from the vertical by an angle equal to that of the helix angle of the thread in the finish and driven so that they rotate around the finish – are pushed into the metal of the closure skirt. The metal is thus forced into the finish thread as the thread-forming rollers rotate to form a matching thread in the closure. This thread engagement then provides the vertical force required to maintain the liner in a compressed state to seal the bottle until the consumer unscrews the closure by hand to open the bottle.

The roll-on closure provides two major advantages for the consumer – no need for an opening tool and resealability, allowing partial consumption of the contents on several occasions.

Another feature incorporated into many roll-on closures is a tamper-evident band. The tamper-evident band is formed by a series of scores and bridges around the circumference at the base of the closure shell; during application of the closure, the band is tucked underneath an annular ring around the base of the finish by rollers similar to the thread rollers. In order to remove the closure from the bottle, it becomes necessary to break the bridges or score lines in the tamper-evident band. This is actually done by the angular force of unscrewing the closure being converted into a vertical force by the thread engagement between the closure and the finish, and because the tamper-evident band is located underneath an annular ring in the finish the vertical force causes the band to break away from the closure or to split along pre-made score lines in the band. It is therefore apparent that when the closure has been removed and replaced it serves as a warning to the consumer that the bottle may previously have been interfered with.

9.5.3 *Plastic closures*

Plastic closures with preformed threads have made significant inroads during the 1980s into the previously dominant position of the aluminium roll-on closure.

Plastic closures are injection or compression moulded from either polyethylene or polypropylene and have pre-moulded threads which take the responsibility for thread formation out of the hands of the bottler. In most cases, the moulded closure shell has a liner of an appropriate sealing medium inserted in a similar way to roll-on closure shells. However, because it is not possible to perform a reforming operation as part of the closure application cycle, the internal design of the top corner of the plastic closure shell (combined with the formulation and distribution of the liner material) becomes critical to the effective creation of the all-important side seal which is needed to ensure a durable and efficient seal.

The application of plastic closures is a comparatively simple, though precise, operation. Closures are dispensed from a chute onto the necks of the bottles and then screwed into place under a light, compressive load by a torque-regulated rotating chuck with inside surfaces that are made to engage with the external knurling of the closure skirt. Often, the equipment used for plastic closure operation is modified roll-on closure application equipment.

Various means are used to provide plastic closures with tamper evidence: either a heat-shrinking band or some form of one-way mechanical interference band is incorporated into the bottom of the closure skirt, engaging with the annular ring in the bottle finish.

In general, plastic closures have been designed for application to existing finish designs in order to avoid the need to replace both refillable bottle populations and bottle finish production tooling.

9.5.4 *Seal formation*

The most effective type of seal between a closure and bottle finish is not a simple vertical compression of the sealing medium against the top surface of the finish because the carbonation pressure inside the bottle is acting against the inside top surface of the closure. A much more effective seal, requiring less force to maintain it, is a top-and-side seal, where the seal is formed around the top of the outside of the finish. During the reforming operation at the beginning of the closure application cycle, the vertical load applied through the pressure block reshapes the metal and squeezes the sealing compound around the top outside corner of the finish, thus forming a side seal which is maintained by the hoop strength of the metal rather than by vertical forces acting on the thread engagement.

Another advantage of this type of seal formulation is that each closure is tailor-made to the finish on which it is formed and in this way otherwise unacceptable variations in finish quality can effectively be sealed.

9.5.5 *Proper application*

The primary function of a closure is to contain effectively its pressurised contents. Considerable pressures are built up inside carbonated beverage bottles (up to 4 bar) and if, for any reason, this force manages to overcome the forces holding the closure onto the bottle (i.e. the thread engagement), the closure may be ejected from the finish with considerable velocity and present an obvious hazard to anything that gets in its way. Closures should be capable of withstanding internal pressures of at least 10 bar without detaching from the finish, and the pressure holding capability is an important measure of closure application quality.

It is essential, therefore, to ensure that roll-on closure threads are fully formed into the bottle thread and that the closure metal is not cut through by a sticking thread roller. It is equally important to ensure that the finish conforms fully to specification, especially in regard to thread formation (length and depth of thread).

Although plastic closures remove the responsibility for formation of the closure thread from the bottler, the pre-formed threads may still be damaged by incorrect application procedures with such damage leading to less-than-satisfactory holding power. Of course, the holding power of a closure may be perfectly adequate in a particular instance, but if the force required to remove a closure is too high, a consumer may resort to the use of implements that may damage the closure and thus reduce its holding power. The torque required to remove a closure is, therefore, another important measure of application quality.

Another closure ejection hazard may result from inadequate venting of headspace gas during removal of the closure. It is obvious that if all of the headspace gas is not vented before the point at which the finish and closure threads disengage, the remaining pressurised gas can project the closure violently from the finish. Since a person is always in close proximity to a bottle being opened, the potential for personal injury is evident.

Clearly, larger bottles have the potential for greater build-up of headspace gases, so large PET bottles are equipped with vertical vent grooves cut into the outside of the finish, through the threaded area, to assist the outward flow of gas during opening.

In addition, aluminium closures often have venting ports cut into the metal between the threaded area and the sealing area to assist the rapid depressurisation of a bottle once the seal is broken during opening. Plastic closures normally have vertical vent paths cut through the threaded portion of the closure skirt to speed gas flow.

The responsibility for ensuring a safe as well as an effective seal is a joint one between the supplier of the bottles, the supplier of the closures and the bottler. All closure systems are exhaustively tested prior to use in the market, but the need for vigilance in ensuring correct application of properly made closures to properly made finishes is ever present at all levels. In times when consumers are increasingly aware of their rights to compensation in the event of a malfunction in something they buy, any lack of attention of good closure application principles and practice can have serious consequences for a bottler.

All closure suppliers have experts on their permanent staff who are trained in the application of closures. The services of these experts are available (usually as a back-up to the supply of closures) and should be fully utilised whenever any doubt exists at the bottling plant regarding closure application standards.

9.6 Future trends

The factors that have influenced beverage packaging in the past – cost, quality, safety and convenience – will clearly have an important role in the future; but they will be joined by another increasingly important factor – environmental impact. Generally increasing public concern over the impact on the environment of man's activities is resulting in growing public and governmental criticism of 'one-trip' packaging. Because returnable multi-trip packages are used for carbonated beverages, and because many consumers dispose of beverage containers thoughtlessly as litter, beverage packaging in particular is coming under a great deal of critical scrutiny. In the future, beverage packaging will not only have to be low cost, safe, convenient and capable of preserving beverage quality, but it will also be required to have some form of further use – either as a refillable package or via a recycling system where the material can be converted into another useful article.

All of the materials used in beverage packaging are capable of being recycled – provided, of course, that the used packages are collected in sufficient quantity to make their recycling a viable economic operation.

A balanced package mix, which can provide beverages to consumers for a variety of consumption situations, will continue to be one of the important cornerstones for the industry. Plastics will almost certainly play an increasingly important role as the existing trend continues towards plastic closures, plastic bottles, plastic labels and plastics in secondary packaging. If the international glass strength research programme succeeds in developing significantly lighter, stronger bottles, the advance of plastics may be slowed.

The one thing that can be said with certainty about the future, is that the rate of change in beverage-packaging technology will not slow down. Packaging represents a major part of the cost in producing carbonated beverages, and such is the level of competition between beverage producers that innovation in packaging, ahead of the competition, is essential.

10 Handling empty containers

C. PETTITT

10.1 Introduction

The handling of empty containers has now become a specialised field in bottling and canning technology. No longer is it feasible to link machines together in a production line and hope that the empty container handling will be a simple matter that will look after itself.

It can be seen today that an ever-increasing variety of containers is available on the carbonated soft drinks market which may be a marketing man's dream, but the widely varying handling problems must be resolved before any container can be filled, packed and distributed.

The empty containers are normally stored in environmentally sound conditions in an area from where they can be delivered to the bottling line. This usually involves bulk goods on pallets which are either 'returnables' or 'non-returnables' in all shapes and sizes. It is critical to ensure that sufficient quantities of the desired containers are always available in a good condition to keep the bottling line fully primed. The efficiency of a modern line depends absolutely on this important function.

The empty container now encompasses a wide variety of bottles and cans in different carry containers such as wooden boxes, plastic crates and cardboard cases. The empty containers will be made from various materials such as steel, aluminium, PET, PVC and other plastics-based materials.

This chapter explains in general terms the handling techniques involved in modern lines to prepare the containers for filling. Each particular area of subject matter is examined briefly but could demand a detailed study of its own.

The machinery involved will be of particular interest to plant engineers who design systems of machines with modern conveying technology and line speeds of up to 120 000 containers per hour, e.g. 2000 cans/min.

The empty container journey starts on the carbonated soft drink line at the depalletiser and continues until the bottle or can arrives at the filling machine in a suitable condition ready to accept the beverage.

10.2 Depalletisers

This operation describes the method of unloading pallets and when pallets were first introduced into the soft drinks industry they were used with wooden boxes and unloading was carried out by hand, a rather labour-intensive and strenuous operation. Plastic crates gradually replaced wooden boxes and standardisation resulted in increased line speeds and the use of semi-automatic machines; these were introduced to assist the operator and improve the line performance. Semi-automatic machines allow the operators to unload from a staging at high level using a pallet hoist in the machine which indexes the load upwards as each layer is removed. From this concept was developed the fully automatic machines, described in this section.

When non-returnable bottles were initially introduced, it was convenient for these to be supplied to the production line packed in cardboard cases which could be hand unloaded and the cases re-used for packing the finished goods. The unloading of the empty bottles was slow and laborious using this method and it led to the development of a bulk system of bottle supply; bottles were close-packed on pallets in layers separated by cardboard pads or trays. Machines of increasing complexity were designed to unload or depalletise these bulk supplies at high outputs at the feed end of the production line.

10.2.1 Wooden boxes and plastic crates

For returnable bottles in wooden boxes or plastic crates a 'lift off' type depalletiser is used (Figure 10.1). This is a system having a 'gripper head' or frame which lifts off a layer of crates or boxes on to a table from which they are combined into a single file on a conveyor. The boxes or crates are then conveyed to a decrater or unpacker so that the bottles can be unloaded on to the bottling line.

In order to lift the plastic crates one type of machine arranges for a lightweight hook-type gripper head to be lowered over the layer so that the hooks interlock into the hand holes of the crates. The hooks are normally adjustable on shafts so that they can be spaced at various pitches to accept alternative layers on the pallet stack as well as a single patterned configuration. These lifting heads can also be equipped with outer guide plates to ensure that they locate over the layers and centre onto the stacks without any manual help. The mechanism usually incorporates a system of clamp levers, quick-acting cylinders and spring-loaded outer clamping plates. The crates or cases are thereby lifted together during their transfer to the waiting table by the combined system of clamps and hooks.

Clamp heads alone are used on patterned layers where the crates are interlocked to form a solid pattern without gaps and the hand-holes are too small or at a low level on the crate making it difficult to locate any lifting hooks. Clamping type gripper heads are often used when depalletising bonded

Figure 10.1 Depalletiser unloading plastic crates. (Courtesy of Kettner GmbH.)

layers of crates (where the pattern of each layer varies from layer to layer) and the pattern allows outer pressure plates to retain all the crates during the lifting operation.

The clamping type of head is often used when handling small crates which may be supplied misaligned to the machine when they are returned to the factory from distribution: the clamps help to realign any tilted crates.

10.2.2 Bulk bottle supplies

Sweep-off machines. These machines operate at either 'high' or 'low' level. This reference describes the level at which the bottles are swept off the pallet in the machine. In order to achieve a high-level sweep, the pallet of loaded containers is usually indexed upwards in a hoist inside the depalletiser, the layers being located in a box sweep frame which transfers the layers sideways on to a waiting conveyor table or platform. With a low-level sweep-type machine, an intermediate conveyor table elevates or descends between the top of the pallet and the fixed level of the discharge conveyor; a 'double sweep' operation is thus required which reduces the overall output of this type of unit.

The transfer on sweep-type machines must be very smooth with flush-fitting deadplates to allow the bottles to transfer on to the conveyor table without

Figure 10.2 Sweep-off bulk glass depalletiser with automatic inverted tray removal. (Courtesy of Kettner GmbH.)

losing stability. Various devices on deadplate transfer are used such as finger-type interlock plates to form a continuous surface over which the bottle can travel. Vacuum or air-pressure holes in the deadplates are also used as a method of physical bottle control to stop the container falling over during the transfer. In most cases the bottles can be crowded together with sufficient back pressure to maintain good stability.

The pallets of bottles are often supplied in trays or on layer pads which are made of cardboard or plastic sheet material. Trays are sometimes fitted inverted or 'lips down' to ensure stability of the layers and in this case the whole layer may be swept off complete with tray which may be removed at a later stage; alternatively, the tray may be removed automatically (before the sweep operation) and deposited in a magazine for re-use (Figure 10.2). When bottles are supplied *in* the trays ('lips up') a knife will cut the corners to allow the sides to be folded down and allow the sweep to take place; alternatively a 'lift-off' unit may be used (see next section).

When sweeping bottles off flat layer pads, gripping clamps are used to hold the pad in position during the sweep, after which suction clamps will transfer the pad into a magazine.

Lift-off machines. This type of depalletiser is an alternative to the sweep unit and handles non-returnable bottles which are delivered to site in trays with

Figure 10.3 Lift-off bulk glass depalletiser with automatic tray removal. (Courtesy of Kettner GmbH.)

'lips up' tray walls; in this case depalletising may be effected by a 'lift-off' type gripper head. If layer pads are used on the complete pallet/bottle assembly, then either a lift-off or a sweep-off type machine may be employed (Figure 10.3).

Lift-off machines are usually manufactured in versatile assemblies to be built up to suit a variety of layouts. The position of the operator and the infeed of raw material and discharge of bottles, pads, trays and pallets must always be borne in mind. The machines are operated normally in a cyclical motion with a slow start and stop but quicker operation during the transfer of the containers from the pallet to the reception table.

The use of a gripper-head assembly suspended on a chain hoist and guided in running tracks or by cams allows positive positioning of the heads over the containers for good location before lifting. Counterweights are often used to balance the load and to reduce power requirement of the machine.

A choice of discharge for the bottles at high level or low level is possible. Usually the infeed of pallets is at low level, thereby the pallet can remain on conveyors throughout the process until unloading is completed and the empty pallet is transferred to a magazine for storage or re-use. Discharge tables at low level, allowing for direct connection onto a conveyor at operator working level, are most common.

Some 'lift-off' depalletisers are built on an in-line construction whereby the infeed of pallets is at the front of the machine and the lift-off mechanism is in a straight line motion through the machine frame. The containers are then lifted by a scissor lift or hoist motion and lowered to the discharge table at the rear of the machine with the empty pallets discharged either to the rear (underneath the bottle conveyor) or at right angles to the movement of the bottles.

In some constructions the transfer of bottles is through 90° whereby the gripper head is suspended from a post that turns through an arc allowing the discharge to be set at an angle to the infeed direction. The rotation of the post is controlled by a crank arm on this type of machine to enable the position of the gripper head to be positively located during each cycle.

The sequential movements are controlled by proximity switches, etc., which ensure that operations are carried out in a logical pattern and any disturbance to the correct function (e.g. misalignment of bottle layers, etc.) is monitored and corrective action taken.

Some machines are equipped with electronic control systems (PLC) and these allow the operator to programme the operation of the machine. A PLC system is also used to provide the automatic control of operation obviating the need for an operator except to rectify a malfunction. Fault indicators on a mimic panel or diagram board show up the reason for any interruption, such as fallen bottles, damaged pallets, etc. Most depalletisers are equipped with outer guarding systems whereby the operating parts of the machine are totally enclosed in a cage to prevent access during the operation of the machine.

10.2.3 Cans

The depalletising of cans has involved the manufacturers of the machines in the design of high-level depalletising systems as most cans are delivered in pallet loads up to 3 metres high. The empty can pallets are usually made of wood and are lightweight in construction as the loads are minimal; they cannot be used for full goods and are therefore returned to the can makers for re-use.

Both lift-off and sweep-off types of depalletisers are available, but the use of lift-off machines requires different types of lifting heads since the delicate cans must not be gripped by physical means. This has led to the development of both vacuum and magnetic plate-type heads for use on different can materials. In the case of magnetic heads, steel cans are magnetised for attraction to the lifting plate; aluminium or non-magnetic cans are lifted by vacuum suction-type heads.

The popular sweep-off machines operate with a sweeping device based on a box or rail unit which moves the cans on to a reception table from which they are conveyed forward to a single-laning combiner (Figure 10.4). Conventional

Figure 10.4 Automatic can depalletiser and single-laner. (Courtesy of N.S.M. GmbH.)

cans have very good stability and can be conveyed on normal slat band or belt type conveyors which are usually used in the combiner conveyor to single-file the cans. Single-laning is a delicate operation and side pressures on the cans must be avoided as the thin material of the open-top can is very susceptible to damage. Layer boards are normally re-used by the can makers and the depalletiser system has a magazine for collecting these.

The pallets are often supplied with a metal top frame which gives protection to the top layer of the cans while in transit. These metal frames are usually lifted off by the depalletiser using a magnetic lift device or a clamp-holding system and the frames are either released onto the empty pallet after can unloading or stored in a special magazine; the programming of the machine operation enables the frames to be stacked together for easy removal and return to the can makers.

All common can sizes are catered for on sweep-off machines and speeds in excess of 2000 cans per minute on one machine are possible. Future trends in can depalletising are towards even higher speed systems, but to obtain high efficiencies many canning companies are using twin systems so that 60–75% of the operation can continue in the event of a stoppage on one of the units.

Combinations of de/re-palletisers on canning lines are also used to save space and allow common high-level walkways so that one operator may control both machines simultaneously.

10.2.4 *Pallet magazines and conveyors*

Pallet magazines are used to stack and store empty pallets after the unloading operation; sometimes the empty pallets are redispensed to feed to the full goods palletiser in the line although a working reserve must always be accommodated to cater for inevitable fluctuations in throughput.

In most cases a pallet magazine will hold 10–20 empty pallets and when a maximum stack has accumulated, the magazine will dispense it on to an adjoining conveyor where it can be fork-lifted away for permanent storage. When being discharged a stack of pallets may need re-alignment and in some systems a pallet centring device is used. This device will re-align the pallet stack to ensure good stability when lifted by the fork-lift truck.

In most modern soft drinks factories palletised goods are conveyed on either roller or chain conveyors. The layout of conveyors will be designed to meet the different formats of the line design or building confines. With the aid of a programming function the conveyor motors can be automatically controlled to route the pallets sequentially into the depalletiser using proximity sensors, electric eyes or microswitch controls.

When selecting conveyors, care needs to be taken to use either chain or roller types which do not interfere with the pallet boards on the underside of the pallet. Normally a fully boarded pallet will travel over rollers without jamming. Pallets constructed with runner rails are usually designed to run on chain conveyors, spaced appropriately; if used on roller conveyors the boards can drop between the rollers interrupting the normal, smooth movement.

10.3 Unscrambling machines

These machines are often used instead of depalletisers or where the bottles are contained in boxes or silos. When bottles are not previously arranged on pallets, such as pre-produced PET bottles that are manufactured on site and stored in silos, it is usual to unscramble on the infeed end of the bottling lines and single lane to the infeed of a rinsing machine or filler.

There are several types of unscramblers but basically two types are used on soft drinks lines. One is a vertical-lift and gravity-drop type machine and the other is a rotary-dish unscrambler.

10.3.1 *Vertical unscrambler*

The vertical-lift or 'up and over' unscrambler is basically an elevator that picks up bottles from a low-level hopper and takes them vertically on a belt over a head stock and dispenses each bottle into a gravity chute ensuring that every bottle feeds down the chute base foremost. The bottles are then deposited on an intermittent belt ready to be conveyed away. Several rows of bottles are lifted and inverted at the same time, thus determining the output of the unscrambler.

This type of machine tends to have a limited output of approximately 30 000 bottles per hour, and for higher speed lines a number of machines grouped together to give a combined output will be required. The surfaces of the machine where containers are in contact with the inclined chutes are usually

Figure 10.5 Rotary unscrambler. (Courtesy of Tolke GmbH.)

lined with wearing materials or non-abrasive finishes to avoid scratching or damaging the surface of bottles.

10.3.2 *Rotary unscrambler*

Unlike the comparatively tall vertical unscramblers, this type of machine (Figure 10.5) is normally available for operation within a single-storey building at the feed end of the bottling line. It employs a rotary selector dish running in a continuous manner to unscramble the containers and dispense them onto a discharge conveyor. Wherever possible these machines have a large storage capacity within the infeed hopper or unscrambling dish to maintain a continuous flow of bottles.

The units are usually variable in speed and currently units of up to 36 000 bottles per hour are commonplace. The bottle enters a dished cylinder or drum where it is rotated under centrifugal force and then dispensed on to a conveyor, using a scroll or worm to select it from the dish wall and place it upright on the conveyor track.

10.4 Decraters

The decrater is used to unload crates (or cartons) of bottles that have been returned for refilling; new glass can also be introduced by this machine to top up the line requirements, provided that supplies are suitably packaged.

There are basically three varieties of automatic decrating machines: the rotary continuous movement, the in-line continuous movement and in-line intermittent lift. The output of decraters or unpackers is normally a balanced requirement to suit the needs of the filling machine. The actual output is controlled by using pack-, back- or line-control switches by which the decrater may be slowed or stopped if it produces too many containers for the next machine in the line.

10.4.1 *Gripper heads*

These are individual holders which pick up the bottles from inside the crates; suitable gripper heads are manufactured to suit most bottles and containers. They are usually air-pressure or mechanically operated and arranged in a formation to match the corresponding pockets within the crates; some degree of flexibility is designed into the location of gripper heads in the holding frame so that the inevitable variations in crates and bottle centres may be accommodated (Figure 10.6)

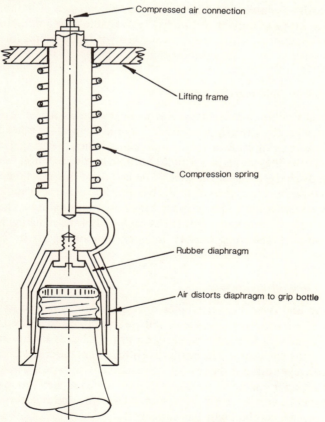

Figure 10.6 Typical decrater gripper head (air-operated.)

10.4.2 *Rotary decrater*

This type of decrater is fitted with a series of gripper heads which rotate on a carousel guide chain in a horizontal plane; a cam or guide rail lowers the gripper heads into the passing crates. Flight bars on the crate conveyor ensure that the crates are offered to the machine at the same pitch as the gripper-head frames on the carousel.

The continuous motion of this machine enables the movements of the gripper head and crate to be timed to coincide. The gripper heads then descend into the crate and grip the bottles, lifting them clear of the crates; the formations of bottles are carried in the gripper-head frames around the carousel and lowered on to the bottle reception table where they are released and conveyed to the next machine on the line.

The mode of operation of these types of machines makes them very suitable for variable-speed operation which can be modulated to meet the flow of the bottling line. They are particularly good for high-speed operations with outputs in excess of 100 000 bottles per hour, depending on the number of gripper-heads assemblies used in the original design size of the machine.

The rotary type of unpacker is normally available with a minimum of four and a maximum of twelve stations; twin conveyor lane systems for the crates are also used on these machines for higher outputs.

10.4.3 *In-line continuous decrater*

This type is similar to the rotary unit in operation as it has a continuous movement whereby the gripper head is lowered from an overhead chain or track into the moving crate in a horizontal plane. When the gripper heads have lifted the bottles they travel in a straight line to a bottle reception table and the crates are discharged below the table. The layout of this machine allows the bottles to travel at a higher level on the bottling line with the crates running at a lower level underneath. The gripper heads are mounted on a chain or belt which returns via sprockets mounted at each end of the machine frame. The unit is compact and very smooth in operation owing to its continuous movement.

10.4.4 *In-line lift-up decrater*

The 'lift-up and over' type machines operate on an in-line basis where the crates enter on a conveyor at low level and are separated and pitched by a flight-bar arrangement; a lifting frame carrying gripper heads locates inside each crate and grips the bottles. The bottles are lifted out of the crates and placed on a bottle conveyor table at the rear of the machine. This type is available in size from one lifting frame up to six lifting frames in a row, each frame having the same number of gripper heads as bottles in the crate and registering with the crate positioned by the flight-bar conveyor.

Figure 10.7 In-line lift-up decrater. (Courtesy of Kettner GmbH.)

The up-and-over operation of the lifting heads needs careful control of the crates to enable the operation to proceed without difficulties. Before the gripper heads are lowered over the bottle mouths, a centring frame is often used to square up the crate position to the gripper. Additionally, it is sometimes necessary to restrain the crates to prevent any upward movement during unloading. The bottles are lifted out of the crate and carried vertically up and over a sprocket or pulley system and then lowered on to the bottle reception table, from where they are transferred to the next machine. Meanwhile the gripper-head assembly returns to a position over the crate conveyor ready to extract the next load of bottles (Figure 10.7).

10.5 Crate washers

Most soft drinks producers now operate with plastic crates when handling returnable bottles, although there are many companies still using wooden crates. After the removal of the empty bottles, wooden crates are normally turned over to eject any rubbish, and then re-varnished and repaired as necessary. The advent of smart plastic crates in bright colours has created a requirement to install automatic washing machines to maintain the crates in a clean condition, free from traffic grime, etc.

After decrating, when the bottles have been removed, the crate is conveyed to the crate washer, where it enters the machine and is inverted to remove any debris. It is then washed out, cleaned with detergent, rinsed, dried and re-inverted ready to be used again at the packaging end.

10.5.1 In-line crate washers

Crate washers have developed over the years from simple bath-soaking systems to more sophisticated machines having a multiple number of baths through which the crates are conveyed continuously from tank to tank with intermediate jetting sections where they are subjected to more aggressive spray and jetting techniques. These machines are usually of the tunnel-type with a chain conveyor to transport the crates. A crate-tipping system at the infeed to the tunnel is often used to invert the crate to eject extraneous matter, the inverted crate then passing through the machine to allow effective washing and draining. (Some machines are equipped with a twisting conveyor to rotate the crates axially through 180° at infeed and discharge.)

In each tank section a recirculation pump recovers the washing water from below the chain level and recirculates it to overhead sprays via a filter where any rubbish or debris in the water can be removed. Included in the washing zones are areas where chemical treatments and anti-static solutions can be applied. Both hot and cold cleaning agents are used, and more recently a system of ultrasonic vibration that disturbs the surface grime on the plastic crates has been developed and used in many machines (Figure 10.8).

The output of in-line tunnel machines is normally from 1000 to 6000 crates per hour and the machines vary in length up to 12 metres. After treatment, a crate tipper similar to that at the infeed is fitted at the discharge end to re-invert the crates to the normal open top position onto the crate conveyor leading to the packaging machine.

Tunnel-type machines take up a considerable amount of room owing to the spacing of the crates in order to wash the ends and to the prolonged treatment required; they also use a large amount of energy. Waste hot water from boiler systems, bottle washers, etc., is often recirculated and the crate washers are insulated to avoid drastic loss of heat.

10.5.2 'S' type and twist crate washer

The 'S' type machine has been developed from the tunnel variety. The crates enter the machine at high level and follow the path of a letter 'S', travelling on a conveyor or in guide rails through the machine to emerge at low level. The purpose of this development is to produce a more compact machine. Obviously the higher level of infeed increases the machine height to a level where the crates fall through the machine either by gravity or with conveyor drive assistance.

Figure 10.8 In-line crate washer. (Courtesy of MCG Techno Pack Ltd.)

Similar compact models of crate washer are created using a spiral-type guide-rail system through the machine, with pressure being applied to the infeed crates to push them through the unit.

Such compact styles of machine can be used for high-speed systems provided the surfaces of the crates are fully accessible in each section for washing treatment. In spiral type machines the guides are used to twist the crates to turn them for jetting.

10.6 Decapping machines

Screw-cap bottles that are returned to the bottling line for washing and refilling will invariably have had the caps replaced by the consumer. Sugar deposits, etc., usually attach these caps very securely and they must be removed from the necks (without damaging the bottles) before the washing process. Decappers are therefore only needed where returnable screw-cap containers are handled and many bottlers will design the line with a bypass round the decapper to allow non-returnables or crowned-bottles to be conveyed direct to the rinser/washer.

Several types of decapping machines are available to carry out this difficult operation, and 'in-line' and 'rotary' versions are equipped with heads that remove the caps from the bottles. Some machines use a belt system and others have discs that unscrew the caps. Normally the caps are deposited in a bin for collection.

Most decappers are used 'on stream' within the actual bottling line layout, and therefore need to be sized to operate at a high enough output to ensure that sufficient bottles are available to feed the next machine. In some cases the decapper is linked to a bottle inspector and the incidence of rejected bottles will affect the volume of empty containers available at the washing machine. In this instance the output of the decapper/inspector will be substantially in excess of that of the washer. Decappers are normally available with very high outputs of up to 100 000 bottles per hour.

When bottles are sorted and decapped in a separate area as an 'off-line' operation, the required empty containers are usually replaced in the crates for later use on the bottling line. In this case the efficiency of the bottling line is not directly affected by the lack of bottles at the washer or filler due to rejects or removal of foreign bottles; this 'off-line' sorting therefore increases the overall efficiency of the line although the added cost of the separate sorting/decapping operation is considerable.

10.6.1 Decapping heads

Decapping heads are the working part of this machine and are normally fitted with a cap-detection device which senses the presence of a cap before the jaws or collet close onto the bottle. In most cases this is achieved using a central spindle in the head which will enter the open bottle mouth if no cap is present; a mechanism then stops the jaws closing on the fragile screw thread. Some decapping heads have special facilities, such as knives or spikes, which pierce the metal caps to obtain a positive grip during the turning action of the removal head.

10.6.2 In-line decappers

In-line decappers are so called because the bottles pass underneath the decapping heads, continuing through in a straight line. On some machines the decapping head revolves to unscrew the cap; alternatively the decapping head is static and the bottle is revolved between belts or rotating discs as it passes under the decapping heads.

Care must be taken within the machine to ensure good bottle centring and to avoid damage to the mouths and screw threads on the bottles. Good bottle handling is needed and infeed worms are used to pitch the bottles at the correct distance between the decapping heads.

Where revolving discs are used these are usually contrarotating and rub against the side of the screw-cap to unscrew the cap. This type of decapper has the advantage of allowing for pitch variation and reasonable differences in height of bottles.

Figure 10.9 Rotary decapper. (Courtesy of Alcoa GmbH.)

10.6.3 *Rotary decappers*

Rotary decappers (Figure 10.9) possess the advantage that the bottles are under positive control during the cap-removal operation; the decapping heads are similar in design and function to those on in-line machines with the exception that the bottles are always held stationary and the heads are rotated by a sun-and-planet gearing system. Some designs have adjustability of the cap-gripping forces and a decapping efficiency of 99.9% is claimed.

10.7 Bottle washing

This is a very important area in the handling of empty returnable bottles; these machines are built with many variable features to suit both the required treatment of the bottles and the line layout in the factory.

Heat, detergents, jetting and soaking are combined to effect a cleaning treatment, a perfectly clean and sterile bottle being the key to the success of the whole bottling operation. Returnable soft drink bottles usually need from 6 to 12 minutes of caustic soaking time to loosen the soil and residues followed by jetting and rinsing periods to scrub clean the bottles. Although returnable soft drink bottles are generally easy to clean and sterilise, it should be noted that, owing to the current increase in the fruit juice content of carbonated drinks, the residues may be more difficult to remove and the treatment should be modified accordingly.

10.7.1 *Types of bottle washer*

Basic standard designs comprise single-ended and double-ended machines with high and low tanks in addition to hydro and hydro-soaker treatment systems.

Double-ended units transport the bottles in one direction through the machine (Figure 10.10). In single-ended machines the bottles are introduced into, and discharged from, the same end with the bottles making a 'double pass' along the length of the machine; hence, this type is usually taller than the equivalent double-ended machine but occupies much less floor space, which may be an important factor. On the other hand, a single-ended machine has dirty and clean bottles in close proximity and, for some bottlers, this is unacceptable.

Variations of the machines include hydro units (with jetting only) and

Figure 10.10 Double-ended, multi-tank bottle washer.

Figure 10.11 Feed and discharge of single-ended bottle washer. (Courtesy of Ortmann and Herbst GmbH.)

hydro-soaking with both jetting and soaking tanks; in recent times the hydro has diminished in popularity in favour of hydro-soaking machines which provide a greater treatment potential.

In most washers the bottles are carried through the machine in banks of cups, holders or clip attachments which secure the bottles to carrier bars in the machine. The carrier bars are fixed on an endless chain running on sprockets positioned at each end of the machine housing. The chains dip in and out of tanks on the soaker type of unit, but on the hydro version the chains are horizontal during their journey from one end to other. On double-ended machines the chains and carriers return from the discharge to the feed end underneath the frame of the machine; in single-ended units it is usual for over 95% of the bottle holders to be occupied since most of the complete chain travel is used for treatment (Figure 10.11).

10.7.2 *Typical treatment sequence*

To eliminate load variations on the drive mechanism, the chains and bottle carriers operate in a continuous movement; in this circumstance outside jetting of the bottle presents no problem since the bottles pass through zones in

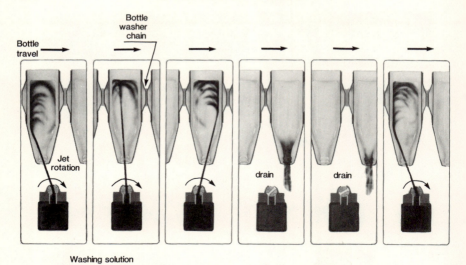

Figure 10.12 Rotating jet pipes giving increase in internal treatment times. (Courtesy of Ortmann and Herbst GmbH.)

which they are sprayed from all angles. However, the internal jetting of a bottle is more difficult due to the relatively small area of the bottle mouth through which the liquor must enter and discharge; the period of time that a moving bottle is over a stationery jet is so brief as to be non-effective. Therefore, jets are designed to reciprocate, swivel or rotate in sympathy with the movement of the bottle and this development greatly increases the actual internal jetting time (Figure 10.12).

On soaker-hydro machines the bottles are immersed in a series of tanks holding caustic solutions at progressively higher temperatures (to prevent thermal shock) followed by further tanks of caustic or water at lower temperatures; between the various tanks, jetting zones may be provided to remove the soil loosened by soaking and the final section usually involves jetting with warm and then cold water to remove all traces of caustic and ensure the bottle is discharged at an acceptable temperature ready for filling.

A typical treatment would be:

1. Pre-warming by rinse water 30 °C.
2. Empty the residue.
3. Pre-rinsing by warm water 50–55 °C.
4. Empty out.
5. Immersion in soak tank 1 at 60–65 °C detergent solution.
6. Rinsing at 60–65 °C and empty out detergent.
7. Immersion in soak tank 2 at 80–85 °C detergent solution.
8. Rinsing with water 80 °C and empty out detergent.

9. Rinsing with water 60 °C and empty out detergent.
10. Rinsing with warm water at 50 °C for removal of detergent and empty out.
11. Rinsing with water at 30 °C and empty out.
12. Final rinse with clean fresh water and empty out.

These are not necessarily the treatments and temperatures for all bottles as special cleaning facilities are provided on different types of machines. There are many cubic metres of water and cleaning agents within these machines and since operation is expensive, heat-recovery systems and insulation of the outer tank walls are often advantageous. The effective use of pre-rinsing liquor has now gained importance and in some cases re-use of pre-heated water can be effective in rinsing away large particles, raising the container temperature, reducing the load on the main washing area and prolonging the life of the detergents. Often this pre-rinse liquor can be recovered from the final rinse sections in which the effluent is normally discharged to waste. The prime objective of a bottle washer is to produce clean and sterile bottles and, to achieve this, full consideration must be given to the temperatures, caustic strengths and contact times in order to effect a balanced treatment appropriate to the condition of the incoming soiled bottles.

Consistent with the maximum diameter of bottle to be handled, the bottle holders will be positioned as closely together as possible on the carrier bars across the width of the machine; at the infeed a magazine marshals the bottles into lanes from which they are lifted and placed (usually horizontally) into the holders. At the discharge point, the bottles are allowed to descend from the holders and are placed gently in an upright position on the discharge conveyor. Adequate clearance of the bottle within the holder is required to allow the circulation of the washing liquors and permit the label residue to escape. On some washing machines neck holders or clamps are preferred to afford the bottles free space in the tanks.

The infeed and discharge mechanisms operate at an optimum speed of approximately 15 cycles per minute; therefore, for a required output of, say, 27 000 bottles per hour the width of the machine should be

$$\frac{27\,000}{15 \times 60} = 30 \text{ bottles wide at the necessary pitch.}$$

The actual operating speed of a washer should always be between 5 and 10% in excess of the filler speed in order to compensate for reject bottles removed at the inspection point prior to the filler and to ensure an 'over-supply' of bottles at the filler to maintain continuity of operation of that unit.

10.7.3 Label removal

When labelled bottles are being washed it is vital that the soaking and/or jetting system is designed to remove labels in one piece and allow the label debris to be efficiently extracted from the machine. In order to accomplish this, the circulation of washing liquors (carrying the removed labels) is directed onto sieve belts or drums outside the machine; brushes, air-blowers or similar flushing devices then transfer the label residues into a suitable receptacle for removal from site. Additionally, filters are fitted into pump discharge pipework to entrap label pulp, etc., which may block jet orifices.

It is also necessary to empty the tanks from time to time to clean out any broken glass fragments which tend to accumulate in the bottom of the washer. Any excessive breakage of bottles within the machine will be avoided if the heating steps and temperature changes are designed to avoid thermal shock. Most washers have a series of low-level manholes from which the rubbish can be withdrawn and which may be used to check the internal condition of the tankwork.

10.8 Container rinsing

Bottle rinsing has become more predominant following the introduction of non-returnable containers including glass, PVC and PET. The purpose here is to remove any residual dirt, dust or particles before filling, often still necessary even with bottles straight from the container manufacturing plant. For simple rinsing and draining, a range of in-line and rotary machines has been developed. The rinsing medium is often treated water containing chlorine of 9–12 ppm to act as a germicide; this water can be obtained from the water-treatment plant prior to the carbon filter.

It is obviously important to remove as much residual rinsing water from within a bottle and any residues containing chlorine to avoid contamination of the finished product. Container-rinsing machines for soft drinks, therefore, tend to have a rinsing section followed by a generous drainage time.

10.8.1 *In-line bottle rinsers*

In-line machines for this operation are usually simple rail-guiding units which twist the bottle into the inverted position and carry it over a series of jets or sprays that rinse the bottle internally. As an example, rinsing times of 10 seconds followed by draining times of 4 seconds are usual for these in-line rinsers, depending upon bottle sizes. The bottles are pushed through the machine by an infeed worm or starwheel.

In-line rinsers that have bottle-neck holders or clamps mounted on a chain or carrier between vertical sprockets are also available. This type of rinser allows for longer jetting and draining times, depending upon the length of the chain; the jetting section tends to be fixed to the machine frame with the carriers passing over the top of the jets. In many cases this type of rinser is directly

HANDLING EMPTY CONTAINERS

coupled to the infeed of the filling machine, thus providing full control of the empty bottle on its path to the filler.

10.8.2 *Rotary bottle rinsers*

Rotary rinsers are basically machines having jetting tubes or vents arranged around a revolving carousel or ring vessel. The machines normally have an infeed worm which loads the bottles into a starwheel. This transfers the bottles to their correct position to be gripped in a neck holder or body clamp which inverts the bottles. During the passage of the bottles around the carousel the jet or cleaning nozzle enters each bottle mouth, positively rinsing the bottles internally (Figure 10.13); finally, the bottles are allowed to drain before they are re-inverted to the upright position and passed into a discharge starwheel for placing back on to the conveyor. These rinsers are now available from small units up to machines with 80 jetting heads, thus catering for line speeds of 60 000 bottles per hour.

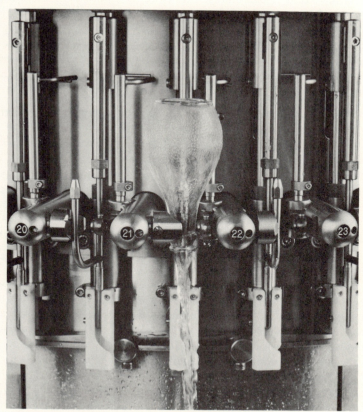

Figure 10.13 Internal jetting on a bottle rinser. (Courtesy of Perrier.)

10.8.3 *Can rinsers*

Can rinsers are used in canning lines to remove any dust or particles in the cans before filling (Figure 10.14). The rinsers are usually in-line twist units, built on an incline so that the rinse water will drain away. The cans are arranged to fall under gravity feed between guide rails passing over the jetting sprays. This ensures that the cans do not get crushed inside the rinser as a result of any back pressure.

10.9 Container conveyors

Much of the equipment described in this section also relates to the handling of filled containers. The design of conveyors for various types of container is an important technology that has developed considerably over the last fifteen years: conveyors are now considered to be machines within their own right. The layout of bottling lines is finally established by the design of the conveying system and it is recognised that the conveyors can contribute to the overall efficiency of the line.

At one time the use of single-lane conveyors linking each machine was considered to be the only method of transporting the containers between the machines; now it is necessary to engineer a system that will meet many criteria, including the following:

- Systems should be commercially reliable in operation, avoiding damage to containers.
- They should operate at acceptable noise levels and avoid any violent collision of containers. Noise can reach unacceptable levels and most safety regulations require operation up to a maximum of 85 dB.
- The conveyor chosen for any particular operation must also be easily maintained to avoid expensive maintenance work and overhauls.
- Owing to hygienic standards being necessary in soft drinks factories, all conveyors should be easily cleanable and the design of the chassis and table areas should be such that they can be washed down effectively.
- In most soft drink bottling lines the speeds of operation require the use of accumulation to allow the intermittent stoppage of a machine and to enable an adjoining machine to continue running. As an example, if a washing machine operating at 30 000 bottles per hour were to stop running for a period of one minute, then the accumulation between washer and filler should be capable of providing 500 bottles before the washer restarts; this allows the filler to continue in operation while the washer is stopped and promotes continuity of running at the filler with consequent uniform filling quality. Subsequent to the washer restart, this machine would operate at a slightly faster rate than the filler and

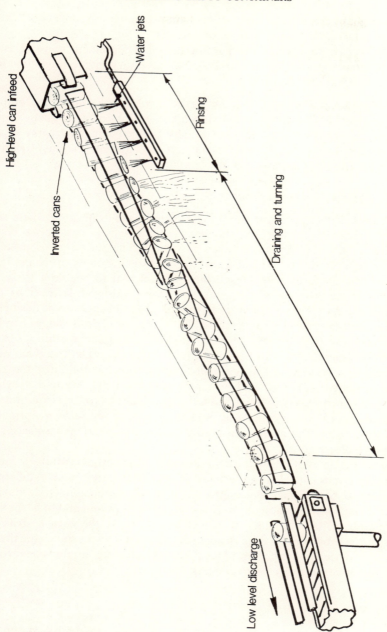

Figure 10.14 Principles of can rinsing. (Courtesy of H. Erben Ltd.)

eventually create a stock of bottles on the accumulation conveyor to cater for a further brief stoppage of the washer.
- The system should allow any accumulation of containers amassed, as in the previous item, to be marshalled as necessary into a single line for presentation to the next machine.

It is widely acknowledged that the high-speed conveying of returned, empty soft drink bottles is one of the most difficult operations on a production line. The bottles are invariably dry, are usually labelled and may carry sugar deposits on their exterior – all factors that can prevent the easy relative movement that is so necessary when handling bottles 'en masse'.

10.9.1 *Slat band chain conveyors*

These are still the most common in carbonated soft drink lines where the conveyor section is made from a stainless steel box structure with a slat band running on the top surface between sprockets. The slat band is usually stainless or a plastic and is sometimes referred to as the conveyor chain. It is purpose manufactured with hinged lugs on each slat to enable it to bend readily over the drive and idler sprockets. Various widths of slat sections are available (e.g. nominally $3\frac{1}{4}$ in. wide or 7 in. wide) and can be mounted in single-lane or multi-lane table arrangements. Slat chains of various shapes can be used to bend in both planes – often referred to as a biplanar chain.

Considerable attention to the problem of friction between the chain and container is necessary, and to minimise pressure on the container the slat chains are often lubricated with water or soap solutions. Many conveying designs have drip trays mounted under the slat chains to collect the lubricant and recirculate it via a pump and holding tank. The lubricant is normally sprayed on to the slat chain or a timing device is fitted to give a measured amount to atone for any losses.

Guide rails on each side of the containers keep the bottles centrally located on the slat bands, and these are usually adjustable to accommodate different diameters of bottle; various designs of guide-rail supports include quick-release clamps which enable the rails to be raised or lowered and to be moved horizontally. Often the guide rails will have gaps at the corner areas of conveyor systems through which a fallen bottle can roll off the slat chain; this is purposely designed to prevent fallen bottles jamming in the conveyors, especially at corners or crossover areas.

10.9.2 *Roller conveyors*

These special conveyor types are often used where a pressureless system is needed to avoid crushing the bottles and are ideal when handling empty PVC or PET bottles.

The conveyors are manufactured using small-diameter driven rollers

HANDLING EMPTY CONTAINERS

mounted in a frame, so that the bottle bases can glide from roller to roller without toppling. The drive to the rollers is normally carried out with some degree of slip, so that any resistance to the flow of the bottles will stop the rollers driving until the bottles are free to move again.

A similar design of slat chain conveyor is also available with rollers mounted in the chain slats; these rollers are normally free-running (without drive) and allow the bottles to pack back, minimising the friction between the bottle base and the moving slat chain.

When using roller systems care must be taken to ensure stability of the bottles, especially at corners and crossovers where change in speed and direction can cause bottles to fall over. With this type of conveyor suitable guide rails are still required to keep the bottles centred on the tracks, and in some cases belts and side-wall propulsion techniques are used to assist the container travel.

10.9.3 *Air-driven conveyors*

Conveying lightweight plastic bottles can be effected using neck support conveying rails with air-blowing techniques (Figure 10.15). This is possible when there is a neck ring, which will allow suspension of the bottle from support rails; the upper half of the bottle runs in an inverted 'U' section assembly through which air is blown via fan ducts or louvres. Owing to the

Figure 10.15 Air-driven conveyor for PET bottles. (Courtesy of Miker Enterprises Ltd.)

small clearance between bottles and side walls, the bottles are pushed along the support guide rails by the air movement and a virtually pressureless condition is obtained when the flow of bottles is stopped by the next machine downstream.

10.9.4 Conveyor accumulators

Within conveying systems there can be found accumulating tables or areas where containers can assemble while one machine in the system is temporarily stopped. These tables can accept the empty containers and enable the flow of the line to be continued despite short interruptions. Obviously, when the table becomes full, the containers need to be blended back into the main conveying system as the line restarts. The slats on the accumulating table are sometimes speed-controlled to achieve an empty-out operation.

The accumulating tables can be designed for in-line operation having a parallel feed and discharge to the direction of the main flow. This enables the containers to enter at one end and discharge at the other. As an alternative the accumulating table can be mounted at right angles to the conveyor, but when required to empty out the drive is reversed and the slat chain table pushes the containers back into the main stream. This type is usually controlled with photocell switches or proximity controls (Figure 10.16).

Figure 10.16 Bi-directional conveyor accumulator for filled cans. (Courtesy of Ortmann and Herbst Ltd.)

10.9.5 Conveyor combiners

These units are built into a conveying system to combine bottles or cans from multi-lane movement to single lane, usually at a machine when single-row entry into an infeed worm is required.

There are basically two types of combiner, i.e. converging and pressureless. In the case of a converging design the area over the slat bands is reduced to a single lane using guide rails and the containers push against each other until they are in a single row. This method can be very noisy and cause extensive damage to the containers. In some designs the guide rails have vibrators or reciprocating rail sections to jostle the bottles so that they do not lock together in the entrance throat of the single lane. This system is comparatively old-fashioned compared to the more modern alternative of pressureless combining (Figure 10.17).

Figure 10.17 Pressureless bottle combiner. (Courtesy of Gebo (UK) Ltd.)

As indicated in the name, a pressureless combiner offers an alternative method using slat chains in a multi-lane construction but each slat is moving at a slightly faster speed. As each chain is increased in speed the containers gradually accelerate away from each other and a guide rail is used to move the containers across the combiner table so that they are guided onto the appropriate tracks. The container finishes on the fastest outlet track at the exit from the pressureless combiner and is slightly spaced apart from its neighbour, which eliminates a noise problem at this point. This type of combiner can be used at the infeed to a machine or at the discharge of an accumulation section.

10.9.6 *Can conveying*

When conveying empty cans, care has to be exercised to ensure that the fragile empty cans are not squeezed out of shape. Therefore, most conveying is pressureless and of multi-lane design to enable slow movement of the cans in mass with a higher speed single-lane section feeding the filler. In many systems the conveyor between depalletiser and filler comprises steel ropes between pulleys which exercise a slight side-wall friction on the can thus conveying it along between two or more guides. Cable conveyors with plastic covering are also used when handling empty cans as these avoid any damage to the can surface.

In some can lines the conveyors are all slat-chain type from the depalletiser up to the filling machine, and in this case use of accumulation tables and combiners is similar to the handling of bottles.

10.9.7 *Container flow monitoring and conveyor speed regulation*

When operating an empty bottle conveying system it becomes apparent that when certain machines stop or slow down the conveyors must also follow the speed changes to avoid containers colliding or jamming on the conveyor tracks. In order to achieve this a control system has to be introduced to monitor and to regulate the conveyor speeds and start and stop each section in balance with the flow of empty bottles, etc. The use of photocells, pressure switches, and proximity sensors enables conveyor motors to be regulated. Many systems now have frequency converters fitted to the motors to allow speed variations from 0 to 100% of the maximum speed, thus giving a complete regulation to the empty container flow rate. Centralised monitoring systems are now commonplace to enable full microprocessor linking to bottling line drive units, allowing totally automatic regulation and control of the flow rates.

The design of production lines must also take into consideration the sizes of containers and the speeds at which they are to be handled. Successful handling of empty containers – either glass, PET, PVC or cans for carbonated soft drinks production – is a prime part of the bottling line effectiveness, and thus the basis for high line-operating efficiencies.

11 Carbonation and filling

A.J. MITCHELL

11.1 Introduction

In any bottling or canning operation, the filling machine is the interface between the container and the product. For the majority of carbonated soft drink installations, the final product does not exist until a point immediately prior to the filler. The processing operation involves combining finished syrup, treated water and carbon dioxide gas in the correct proportions, and normally this is carried out on a continuous basis, with the beverage being produced at a slightly faster rate than the requirement of the filler. Although there are many plants still operating on the old method of pre-dosing the empty bottle with the requisite amount of syrup and then topping-up with carbonated water, the modern process of pre-mix filling, where the carbonated, finished product is transferred to the container in one filling operation, is now predominant. The accuracy of the syrup/water proportioning and the control of the degree of carbonation are vital to the commercial success of carbonated soft drink production. This chapter details some of the latest advances in the processing equipment. The filling machine is unable to improve the standard of the beverage and the correct function of the processor plays a major role in predetermining the ultimate quality of the product and the overall success of the filling operation.

In the last ten years filling units have continued to increase in complexity – and price. Filling valves have been developed in three main areas: to encourage a streamline pattern of liquid flow into the container (vital to successful ambient filling); to improve the control of fill height in the container; to adapt to changes in container design. The introduction of the large PET bottles (2-, 3- and, recently, 5-litre sizes) has necessitated the production of fillers with appropriately wide filling valve centres to accept these containers; hence, although filling machines have increased in the multiplicity of filling valves, they have also increased in physical size owing to wider valve centres. At the time of writing, Krones AG of Neutrabling, West Germany, claim the *largest* filler of 176 valves and in excess of 6 metres in diameter; the *fastest* filler would appear to be the 120-head can filler from Holstein and Kappert GmbH of Dortmund, West Germany, capable of speeds over 2000 cans per minute.

The advances in filler design have coincided with the development of ambient filling facilities and the almost total replacement of glass by PET plastic in the large bottle sector. The inexorable progress of the latter material casts a shadow over the future of glass for carbonated soft drinks as PET bottles become larger and, more important, are breaking into the less-than-1-litre range. The possible success of a *returnable* PET bottle is a fascinating scenario with its effect on bottling line composition and design, together with the added complications in the areas of distribution and return.

The filling machine still holds pride of place in the production line; the individual handling of containers at current outputs cannot fail to be impressive – although it is recognised that the modern, high-speed labelling machine is a tribute to engineering excellence in dealing with bottles, labels and adhesives at speeds never contemplated twenty years ago. At the filler, the formulated product is introduced into the prepared container, and the management of this operation is crucial; high outputs of large bottles demand huge quantities of product, and flow rates of 50 000 litres per hour through fillers are commonplace. Inaccuracy, malfunction or human error can have disastrous effects on yields and operating efficiencies, and the production technologist must be aware of how the equipment is intended to function so that, in the event of failure, corrective actions may be speedily taken before substantial losses are incurred.

11.2 Carbonation

The artificial inclusion of a dissolved gas in a soft drink beverage was developed from the popularity of natural occurring mineral waters, which are discharged in a slightly carbonated form from rock formations in many of the well-known spa resorts around the world. The medicinal advantages of spa products have been panegyrised to the point of exaggeration and the ingestion of the dissolved carbon dioxide has always been considered an important part of the therapeutic process. What cannot be denied is that the addition of carbon dioxide makes any soft drink more palatable and visually attractive.

11.2.1 *The nature and effects of carbonation*

Carbonation may be defined as the impregnation of a liquid with carbon dioxide gas (CO_2). When applied to a soft drink product, the result is a beverage which sparkles and foams as it is dispensed and consumed. This escape of CO_2 during consumption of the drink should complement and enhance the flavour, and will add an exciting tingle that stimulates the palate. The amount of CO_2 gas producing the carbonation effect is usually specified as 'volumes' – meaning, in broad terms, the number of times the total volume of dissolved gas may be divided by the volume of the liquid. For example, a 3.0

volume drink will contain CO_2 to the extent of three times the volume of the beverage. A more accurate definition involving the parameters of pressure and temperature will be explained later in this chapter.

The organoleptic effects are not the only benefits of the CO_2 content; at carbonations above 3.0 volumes the CO_2 has a preserving property, the extent of which is dependent upon pH, sugar, initial microbial load and the nature of the micro-organisms.[1] This desirable attribute of carbonation should be considered as an added 'bonus' and must not be a substitute for other precautions taken to ensure the safety and extended shelf-life of a product.

Each soft drink formulation requires a particular degree of carbonation so that the effervescence is appropriate to the flavour and nature of the beverage. Fruit drinks – such as orange, bitter lemon, etc. – should contain low levels of carbonation whereas juice-based drinks – colas, ginger beer and cream soda – should be in the medium-to-high range of CO_2 content. Mixer drinks, so-called as they are used mainly to mix with spirits, require high carbonations since the addition to other still (non-carbonated) liquors dilutes the original carbonation level. Drinks in this category would be tonic water, ginger ale and soda water. Soda water filled into syphons contains the maximum degree of carbonation usually encountered in the industry. These particular containers rely upon internal pressure to dispense the contents and, as the syphon empties, this pressure is replenished from the carbon dioxide dissolved in the product.

In practical terms, carbonation levels vary between 1 volume of CO_2 in fruit drinks to 4.7 volumes in mixers and up to 6.0 volumes in soda syphons. Figure 11.1 shows typical carbonation values for a range of well-known drinks; a degree of latitude is indicated since individual recipes require their particular carbonations. The increasingly popular, large PET bottles constitute a special case; not only does the CO_2 gradually escape through the permeable polymer material producing a marked reduction in the carbonation of the contents over a period of time, but the repeated opening and closing of the container for occasional consumption can result in the final 25/30% of the contents having an unacceptably low carbonation. This latter problem arises because each time the bottle is resealed, CO_2 gas escapes from the residual product to pressurise the headspace volume and most of this gas then escapes to atmosphere on the next occasion the bottle is opened. These large containers are not really intended for intermittent consumption and to compensate for the future loss of carbonation, the product is carbonated to a slightly higher level than would be appropriate for the particular drink.

11.2.2 *Properties of carbon dioxide*

Carbon dioxide* is a colourless gas with a slightly pungent odour; when dissolved in water the resultant carbonic acid mixture has an acidic and biting

---

* Technical information kindly supplied by Distillers MG Ltd, Reigate, UK.

CARBONATED SOFT DRINKS

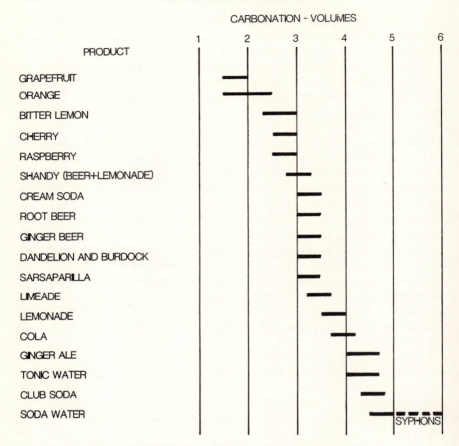

Figure 11.1 Typical carbonation levels for well-known products.

taste which is not unpleasant. CO_2 does not support combustion and is used extensively in fire extinguishers. High concentrations in the atmosphere will quickly suffocate respiratory animals and since the gas is 1.53 times heavier than air at 70 °F, great care must be taken when entering vessels that have contained CO_2 and may not have been sufficiently vented and purged; in these circumstances, residual CO_2 will lie in the base of the tank to trap the unwary entrant. Carbon dioxide is usually present in atmospheric air at a level of approximately 300 ppm by volume and it is dangerous to breathe atmospheres containing more than 5% by volume; it has been postulated that workers may be safely exposed to a maximum concentration of 5000 ppm CO_2 by volume for 8 hours per day.

CO_2 is one of the very few gases suitable for providing the effervescence in soft drinks. It is non-toxic, inert, virtually tasteless, readily available at moderate cost and may be liquefied at reasonable temperatures and pressures, allowing convenient bulk transportation and storage. The solubility of CO_2 in both water and soft drink product allows an acceptable retention of the gas in solution at atmospheric pressure and room temperature, although slight agitation will promote an evolution of gas bubbles from the body of the drink which creates the attractive sparkling effect.

The bulk storage of liquid CO_2 in soft drinks plants is now commonplace all over the world; pressurised and insulated tanks holding CO_2 at 20 bar, and maintained at a temperature of $-17\,°C$ by a small refrigeration unit, are available in various sizes from 5 to 50 tonnes in both horizontal and vertical modes. In order to obtain a sufficient supply of dry gas to feed the carbonation equipment, etc., it is necessary to utilise a suitable vaporiser unit which may be heated by water, steam or electricity.

Small-scale production plants still use thick-walled cylinders containing approximately 25 kg of liquid CO_2 at 60 bar and these are usually connected in banks to allow a reasonable off-take rate without recourse to additional heating. In the more remote areas of the world, carbon dioxide is still generated on site by the chemical action between an acid and a carbonate – e.g. sulphuric acid and sodium bicarbonate; alternatives would be the combustion of fuel oil or the extraction of CO_2 from the flue gas of a boiler operation or similar heating facility.

11.2.3 *Equilibrium pressure*

In common with other gases, carbon dioxide increases in solubility as the liquid temperature decreases, and for every combination of (i) the amount of CO_2 in solution and (ii) the temperature of the liquid there is a finite *minimum* pressure that is necessary to retain the gas in solution. This is a condition known as 'equilibrium' where, owing to molecular movement, the gas leaving solution is equalled and balanced by the gas entering solution. At equilibrium pressure the gas/liquid mixture is *just* stable but any decrease in pressure or increase in temperature will render the mixture metastable (or supersaturated) in that the pressure/temperature combination is insufficient to keep the CO_2 in solution. In this circumstance, gas will be spontaneously released (particularly if there is some agitation of, or irritant applied to, the solution) – a condition known as 'fobbing' or 'foaming' and usually apparent when a bottle of carbonated product is opened to atmospheric pressure. The inability of a carbonated beverage to retain its full CO_2 content in solution at atmospheric pressure gives rise to the attractive ebullience observed during the act of pouring the drink into a glass and the liberation of further CO_2 during the actual consumption.

Carbonated product held in a container that is open to the atmosphere will gradually lose carbonation as the gas is liberated and escapes from the liquid. In a closed container, this evolution of gas proceeds to fill the headspace volume and gradually increases the pressure, quickly at first and then more slowly as the equilibrium condition is approached. The actual rate of the transfer of gas from product to headspace depends not only on the proximity of the headspace pressure to equilibrium pressure but also on the temperature of the liquid, the nature of the beverage and the extent of any agitation or irritation imposed on the liquid. A quiescent, stable product not subjected to vibrations or movement may take many hours to reach equilibrium – whereas the same product, roughly shaken, will take only seconds to attain the equilibrium state. The CO_2 gas leaves the beverage and collects in the headspace volume to provide the necessary equilibrium pressure to keep the remaining gas in solution – at a slightly lower carbonation than the original value. This condition applies to all bottles and cans that have been filled with carbonated beverage and then sealed with the appropriate closure.

11.2.4 *Measurement of carbonation*

Since the degree of carbonation is such an important factor in the formulation of a soft drink, it is imperative that a standard form of measurement of carbonation should be available; this would allow the production of particular products at different times and in different locations and yet ensure that the carbonation of these products meets the required, agreed standard. Previously, it has been mentioned that carbonation may be quantified in 'volumes'; a volume of gas is indeterminate unless the parameters of pressure and temperature are specified and two scales are in current use. In the UK the term 'volumes Bunsen' is popular where the gas volume is measured at atmospheric pressure (760 mm of mercury) and the freezing point of water (32 °F or 0 °C); an alternative scale used in the USA (and therefore followed by many franchise bottlers throughout the world) is 'volumes Ostwald', where measurement is also carried out at atmospheric pressure but any temperature adjustment is ignored. A third method, used on the European continent, measures carbonation in grams per litre and since one volume (Bunsen) is equivalent to 1.96 grams CO_2 per litre, doubling the volumes of carbonation will give an acceptable approximation of the grams of CO_2 per litre of product.

11.2.5 *Carbonation determination*

An obvious method of gauging the degree of carbonation would be to extract the total CO_2 content from a known volume of product, adjust the gas volume to atmospheric pressure and, where desired, mathematically convert this volume to 0 °C; the ratio of gas volume to original beverage volume will give the figure for carbonation. This procedure is used in some QC laboratories but is usually restricted to low carbonation products as the routine is somewhat cumbersome.

A superior procedure (and still used extensively in the industry despite the

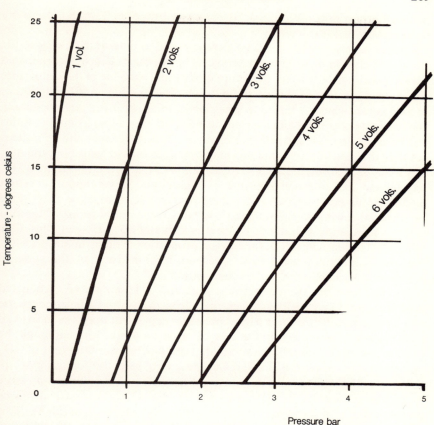

Figure 11.2 Gas volume chart. Volumes of CO_2 (reduced to 0° Celsius and 1.013 bar) dissolved in a unit volume of water at various temperatures and pressures assuming 100% purity of CO_2.

introduction of later, sophisticated techniques) makes use of the equilibrium phenomenon described in Section 11.2.3. If the temperature and equilibrium pressure of a product are known, then there must be a fixed carbonation level based on these two factors. From laboratory measurements of the maximum amounts of CO_2 dissolved in water at various pressures and temperatures and by the application of Henry's Law, a graph may be produced of the three-way relationship between dissolved volumes, temperatures and equilibrium pressures. Figure 11.2 shows the maximum volumes of CO_2 (adjusted to 760 mm Hg and 0 °C) which may be dissolved at various temperatures and pressures. Fitting a pressure gauge to a container of carbonated beverage and shaking vigorously until the pressure stabilises will give the equilibrium reading; having also noted the temperature of the product, these two readings may be applied to the graph and the degree of carbonation determined. For example, an equilibrium pressure of 2 bar at a temperature of 6 °C would indicate a carbonation of 4.0 volumes.

Unfortunately, this simple procedure is prone to inaccuracy owing to the possible inclusion of air in the beverage. (The presence of air in a carbonated product will radically affect the future quality and shelf-life of the drink since the oxygen element of the air promotes aerobic spoilage and oxidation of certain constituents.) A further complication is that air is roughly one-fiftieth the solubility of carbon dioxide,[2] and although it may be considered that this small fraction renders the presence of any air inconsequential, the opposite is actually the case; any air contained in the product will *exclude* approximately fifty times its own volume of CO_2. In fact, air in its normal composition of 21% oxygen and 79% nitrogen (ignoring trace gases and water vapour) does not dissolve in liquids to produce dissolved oxygen and nitrogen in these same proportions; owing to the differing solubilities and proportions of the two main constituents, the dissolved 'air' is actually 35% oxygen plus 65% nitrogen. This enrichment of the oxygen proportion is unfortunate since it is this particular component that is responsible for many of the spoilage problems associated with 'air' contamination.

When measuring the equilibrium pressure of a sample of carbonated product, other gases (such as oxygen and nitrogen dissolved from atmospheric air), if present, will also exert partial pressures dependent upon the individual presences and solubilities. In the case of oxygen and nitrogen, although the proportions may be small, the solubilities are much lower than that of CO_2 and therefore the partial pressures necessary to keep the foreign gases in solution will be higher. In this eventuality, the total equilibrium pressure (being the summation of the partial pressures of CO_2, O_2 and N_2) will be greater than that produced by the carbon dioxide content alone. This enhanced pressure gauge reading during the carbonation test will indicate a higher carbonation level than is actually present. A high 'air' content not only produces a false reading of the carbonation level but also imparts a 'flatness' or lack of sparkle to the beverage and results in a stale flavour.

In order to avoid a distortion in the carbonation reading produced by dissolved oxygen and nitrogen, a simple modification applied to the pressure/temperature procedure previously described will allow a more accurate determination of the CO_2 content: when the container is shaken to equilibrium, the *first* pressure is allowed to escape slowly to atmosphere and the container then agitated again to the equilibrium condition, which is the pressure used to compute the carbonation level from the graph in Figure 11.2. This adaptation is often referred to as the 'second shake' method and eliminates the misleading effect of any dissolved 'air' since the latter, being less soluble than CO_2, will leave solution during the first shake and will be vented-off as the container is de-pressurised.

The equipment used to determine the equilibrium pressure in a container will vary according to the supplier, but basically consists of a hollow piercer (or

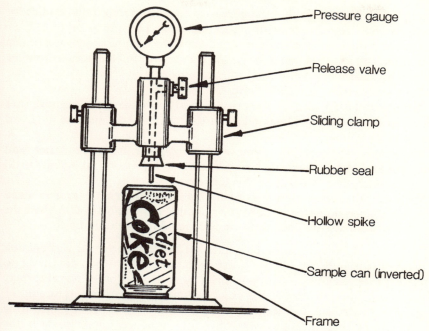

Figure 11.3 Pressure tester for cans and plastic bottles.

lance) which enters the bottle cap or can lid and is connected to a pressure gauge and a release valve. Glass bottles should always be enclosed in a metal cylinder during the agitating operation but plastic bottles and cans may be held in an open frame as shown in Figure 11.3.

When a filled bottle or can is selected for a carbonation test it may have been obtained directly from the production line, or it may have been extracted from warehouse stock, or it may have been delivered to a central laboratory from a satellite factory; in the last two instances, the containers will have attained equilibrium pressure but bottles and cans taken from the production line will not have reached that condition and require special attention before being tested. The container should be wrapped in a cloth dampened with cold water and shaken vigorously for *at least* 30 seconds and then allowed to rest undisturbed for another 5 minutes; this procedure will ensure that equilibrium conditions have been obtained for all gases in the package, i.e. the gases will have been distributed between the headspace and the product according to their solubilities and presences.

The test may now proceed as follows:

(1) Check that the release valve is tightly closed.
(2) Place the bottle or can in the apparatus and puncture the crown, cap or can base with a firm movement to ensure that a seal is made.

(3) Check that the pressure gauge has registered and record this reading for possible comparison with the second-shake pressure as in (6).
(4) Open the release valve slightly and allow the top pressure to escape in a controlled manner until the gauge reads zero or bubbles escape from the beverage.
(5) Firmly close the release valve.
(6) Shake the container and tester vigorously until the pressure gauge rises to a maximum reading and no further shaking will increase it; record this pressure.
(7) Release the pressure, remove the container from the tester and immediately measure the temperature of the product – taking care to do nothing that would raise this temperature.
(8) Using the pressure reading obtained in (6) and the temperature obtained in (7), determine the carbonation from the chart in Figure 11.2 where the vertical pressure line meets the horizontal temperature line.

If the point of intersection of these lines lies between particular carbonation curves, it is usually satisfactory to estimate the actual carbonation level; to obtain greater accuracy, the proportion of the *horizontal* (pressure) values may be used for interpolation of the exact carbonation. Instruments should be carefully maintained and regularly checked; any inaccuracies in measuring temperatures and pressures will obviously produce imprecision in the carbonation reading.

The difference between the 'first shake' and the 'second shake' pressures is significant. If the beverage is completely free of all gases except carbon dioxide and the headspace volume is of the order of 5% of the total container capacity (as is usual in most bottles and cans), then the percentage fall in equilibrium (gauge) pressure between first and second shakes will be approximately 5%; if the process is repeated several times, the pressure will reduce by 5% each time until all the carbonation is lost. Taking into account the CO_2 lost between shakes and the internal volume of the apparatus plus pressure gauge, etc., a reduction of 7–8% between first and second shakes could be considered acceptable; results in excess of this figure will indicate that dissolved 'air' is probably present and the cause should be investigated and rectified if reasonably air-free (and *oxygen*-free) products are to be produced.

11.3 Carbonators

In its basic form, a carbonator allows close contact between carbon dioxide gas and the liquid to be carbonated. The factors determining the degree of carbonation are:

- the pressure in the system
- the temperature of the liquid; as the temperature is reduced, the solubility of the CO_2 in the liquid increases

CARBONATION AND FILLING

- the time during which the liquid is in contact with the CO_2
- the area of the contact interface between the liquid and CO_2
- the receptivity (or affinity) of the liquid to carbon dioxide; water is more receptive to CO_2 than solutions of sugar or salt
- the presence of other gases mixed with the carbon dioxide; dependent upon their quantities and solubilities, proportions of these extraneous gases will be dissolved instead of CO_2.

Of these six factors, the pressure, contact time and contact area are utilised in the operation of all carbonating systems; additionally, carbonators with integral coolers use the variable of liquid temperature. The methods by which close association of the liquid and the carbon dioxide is obtained are rather limited; the fundamental principles are shown in Figure 11.4 and all carbonators are based on one or more of these basic designs.

11.3.1 Designs of carbonators

Current carbonators may be divided into two main groupings: those that carbonate the finished product mixture of syrup and water, and those that carbonate water only, which is then added to the syrup component. Finished product carbonators may also be equipped with a cooling facility which allows the product to be reduced in temperature as it is carbonated, and these units are often referred to as 'carbo-coolers'. Possibly the most well-known carbo-cooler is the Mojonnier unit which is of stainless-steel construction and comprises an outer shell forming the main pressure vessel capable of internal pressures up to approximately 6 bar. Within this pressurised CO_2 vessel are a number of vertical heat exchanger modules (actual number dependent on the

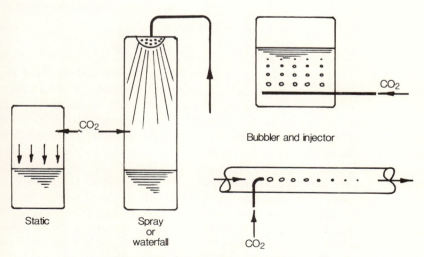

Figure 11.4 Basic carbonating methods.

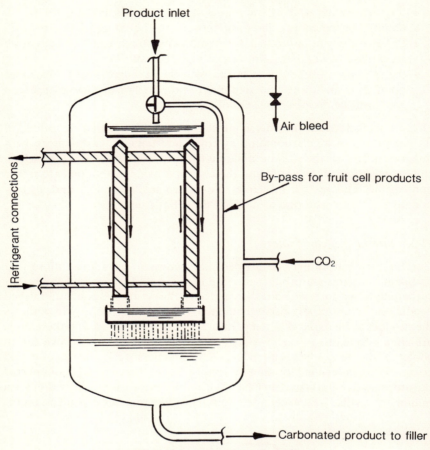

Figure 11.5 Mojonnier carbo-cooler.

total liquid throughput) through which is passed a refrigerant, usually ammonia. As shown in Figure 11.5, the incoming product is distributed into troughs from which it flows in controlled films over the corrugated surfaces of the cooling modules; the product takes a finite time to flow down the surfaces and also offers a large (and changing) surface area of contact to the gas, thus complying with two of the CO_2 dissolution factors mentioned previously. The relatively agitated flow pattern, coupled with the thin film characteristic, also promote efficient heat transfer from the product to the refrigerant and this ensures that the beverage is cooled during its passage down the plates.

An alternative unit operating on similar lines is the 'Predosix' model (depicted in Figure 11.6), manufactured by Simonazzi Srl of Parma, Italy. A vertical shell-and-tube heat exchanger is located within the main vessel pressurised with CO_2; product flows in thin films down the bores of the tubes

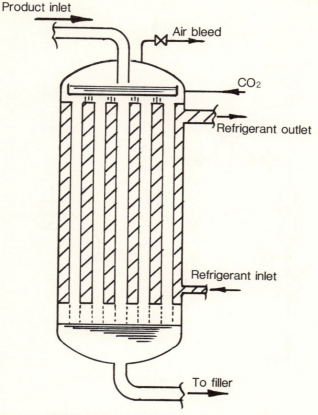

Figure 11.6 Simonazzi 'Predosix' carbo-cooler.

and refrigerant is passed through the shell of the heat exchanger to effect cooling of the beverage.

On both these designs of carbo-cooler, it may not be possible, under certain conditions, to achieve the desired level of carbonation within the confines of the pressure capability of the unit; a pre-carbonating device is often used, taking the form of controlled injection of CO_2 into the pipeline conveying product to the carbonating tower, as the typical arrangement shown in Figure 11.7. Carbon dioxide is fed (at a pressure approximately 4 bar above that of the tower) through a flow indicator, a flow-adjustment valve and a solenoid valve into the product stream immediately after the pump which conveys product to the carbonator tower. The solenoid valve is linked to the pump electrical circuit so that gas only flows when the pump is operating. This device allows a balance to be obtained between the pressure at the filler (normally equal to the pressure in the carbonator) and the desired level of

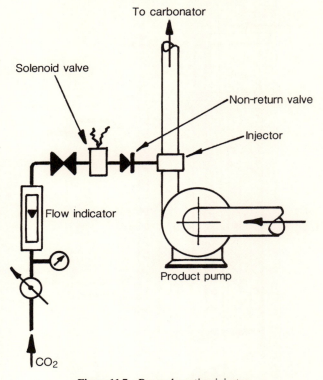

Figure 11.7 Pre-carbonating injector.

carbonation, the latter depending upon the equilibrium pressure and temperature of the product. For the best standard of filling, the pressure at the filler should be between 0.5 and 1.0 bar above the equilibrium pressure of the product. This extra pressure is referred to as 'over-pressure' and promotes the stability of the carbonated beverage as it is passed from the carbonator to the filler and through the filling valves into the containers. As an example, if a carbonation of 3.75 volumes is required in a product at a temperature of 10 °C, then the equilibrium pressure should be 2.15 bar (from Figure 11.2); the carbonator and filler pressures should therefore be $2.15 + 0.75 = 2.9$ bar. Using the data supplied by the manufacturer, the injection system should be regulated so that CO_2 at the correct rate is forced into the product. The desired result is:

Carbonation at injector + Carbonation in tower
= Required carbonation in product.

Alternatively, the same degree of carbonation could be obtained by operating the carbonator (injector + tower) at a higher saturation efficiency and then adding a further 1.0 bar of pressure to the product by means of an 'overpressure pump' fitted into the pipeline between the carbonator and filler. This pump has no effect on the carbonation level but, by administering the extra pressure, it quietens the product and encourages a high standard of filling. Operating the filler at a higher pressure than that at the carbonator gives a measure of control over the condition of the carbonated beverage and allows the best performance to be obtained from the filling machine.

The CO_2 injector device used to supplement the carbonation in the main tower need not necessarily be of sophisticated design; it is usual to insert a spigot (drilled with a number of small holes) into the product stream, the result being an array of small gas bubbles which should enter solution before the product arrives at the carbonator tower. At least one carbonator design (the Crown-Century unit by Crown Cork and Seal Co Inc., Baltimore, USA) relies on an injector to provide virtually all the carbonating effect. Figure 11.8 shows the CO_2 infusion unit, consisting of a porous, micronic diffuser tube inside a

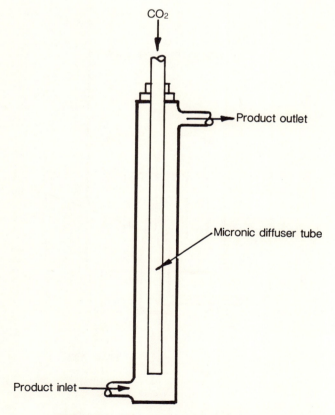

Figure 11.8 Crown-Century CO_2 infusion unit.

stainless-steel outer shell; the large-diameter pipe produces a low linear velocity of product flow which, coupled with micron-size CO_2 bubbles, ensures speedy and efficient dissolution of the gas. The carbonated beverage is then fed into a stabilising tank via a downheader below the liquid level and a baffle plate prevents splashing and turbulence. An additional feature is a built-in, 100 mm diameter serpentine stabilising coil through which the beverage travels before entering the pipeline feeding the filler; any entrained gas bubbles are therefore given every opportunity of fully dissolving *before* the product undergoes the filling operation, where the undissolved CO_2 could create problems.

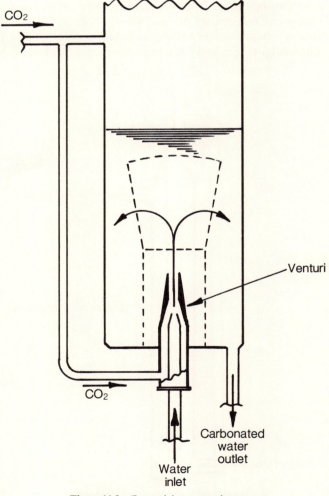

Figure 11.9 'Intermix' water carbonator.

Some carbonation towers are manufactured too small in diameter for the volumetric throughput, with the result that entrained bubbles of CO_2 (which will invariably occur when a falling liquid meets a free surface) cannot regain the liquid/gas interface because the upward buoyancy velocity is less than the downward velocity of the liquid as the filler requirement is supplied. Owing to the approach to equilibrium conditions, these loosely bound bubbles of CO_2 will require *time* to dissolve, and the most satisfactory course of action is to encourage them to rise and break free at the beverage surface in the tower before they can cause complications at the filling machine.

The type of processing system that carbonates water only and then combines carbonated water and syrup, may use any of the usual methods of impregnating the water with CO_2; widely used for over twenty years is the Intermix unit from Ortmann and Herbst GmbH of Hamburg, West Germany, which injects the water vertically into the base of the carbonator, entraining CO_2 through a venturi arrangement as shown in Figure 11.9. High turbulence and agitation are encouraged and it is claimed that 98% of full saturation may be obtained.

11.3.2 Air exclusion

The detrimental effect of air (or, more accurately, oxygen) in a soft drink has been highlighted in Section 11.2.5; it is vital that great care is taken to avoid air contamination at every stage of beverage production, particularly during the following operations:

- filling and transferring tanks in the syrup room
- agitation of syrup batches during blending and mixing
- pumping the syrup to the processing plant.

Allowing bottling syrup to settle after compounding is very important and is always beneficial as even one hour's dwell time will permit much of the entrained air to escape from the product. Using mechanically de-aerated water in the production of syrup batches has become more popular in recent years but the efficacy will depend upon subsequent handling as splashing and agitation in the presence of air will soon aerate the liquors.

The treated water, which is combined with the syrup to form product, should be fully de-aerated and the majority of carbonation systems include a de-aeration module. This may be a vacuum tower into which the water is introduced in a finely divided form; at least 95% vacuum is required to remove the undesirable air content of the water. An alternative method is known as 'reflux' and involves spraying the water into a chamber in which there is a pressure of approximately 0.25 bar CO_2; the effect of the pure CO_2 is to 'flush-out' the air content from the water and a vent at the top of the tower allows the air (and, inevitably, some CO_2) to escape to atmosphere, being replaced from the CO_2 supply to maintain the purity within the tower. The vacuum system is energy-intensive owing to the large pumps (vacuum and water) that are required, whereas the reflux de-aeration unit is wasteful in CO_2 and produces slightly carbonated water that may cause problems with cavitation during subsequent handling. These disadvantages may be overcome by the use of nitrogen as the flushing gas since this is sometimes cheaper than CO_2 and has a lower degree of solubility.

11.4 Proportioners

The commercial success of filling pre-mix or finished product into bottles and cans may depend a great deal on the efficiency and reliability of the proportioning system; in the last thirty years many types of proportioner have been available, some enjoying greater success than others. Current models have high degrees of accuracy, are capable of handling syrups of widely varying characteristics (including fruit cell products), can adjust the syrup/water ratio to fine limits and may be effectively sanitised without major dismantling. A high-quality proportioning unit will possess all these virtues and will deliver product that varies in strength within a narrow band of control.

Present-day models may be grouped under four main headings:

- Flow control valves regulating instantaneous pumped supplies of syrup and water
- Gravity-flow system with regulation of syrup and water rates
- Positive pumping of syrup and water
- Filling/emptying of syrup and water chambers of known (and adjustable) volumes.

The examination of all available proportioners is beyond the scope of this chapter but some of the more popular and well-known types are described hereunder.

Bran and Luebbe pump unit. Two or more reciprocating, positive-displacement pumps are driven from a common motor, each pump being infinitely adjustable in stroke; syrup and water volumes are handled by the different pump heads according to the desired ratio. By phasing the cycles of the pumps and using double-acting pistons, the flow pulsations can be minimised producing a reasonably smooth operation; however, some high-gallonage models are fitted with pressurised pulse dampers to avoid liquid hammer shockwaves although, in many applications, the syrup/water mixing chamber fulfils this function. At pump speeds of approximately 100 strokes per minute the volumetric efficiency approaches 100%, allowing precise stroke settings to produce the desired throughputs; an in-line mixing arrangement ensures homogeneity of the finished product. The syrup pump gland assembly

Figure 11.10 B and L proportioning unit with twin pumps. (Courtesy of Bran and Luebbe (GB) Ltd, Brisworth, England.)

is fitted with a water-flushed lantern ring to prevent the accumulation of sugar deposits, which could cause excessive wear.

The B and L pumps are well-established, very reliable, of sanitary design and are incorporated into many of the processing systems offered by leading manufacturers of soft drinks plant in the past twenty-five years; Figure 11.10 shows an example of this popular unit.

Mojonnier 'Flo-Mix' proportioner. The principle of operation of this unit is that a constant head of liquid over a fixed orifice produces a constant rate of flow. As shown in Figure 11.11 (Mark I), the syrup and water chambers discharge by gravity into a lower vessel from which the product is pumped to the carbonator; the syrup orifice is fixed (but is interchangeable to allow selected rates of flow) and the water orifice is variable by means of a micrometer screw device. Constant liquid heads are maintained by overflowing stand-pipes on older models or by accurate level controllers on the current units; the whole assembly is pressurised to about 0.2 bar CO_2 to exclude air contamination.

The pronounced rise in beverage throughputs in recent years has produced a Mark II variant in which the syrup and water components are passed

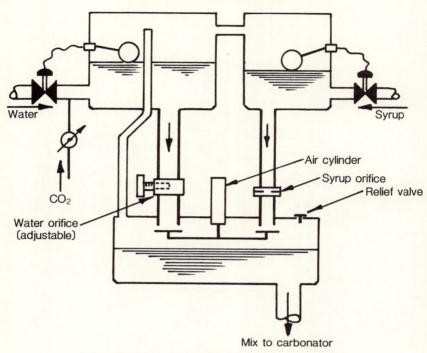

Figure 11.11 Mojonnier Mark I 'Flo-Mix' proportioner.

through their orifices by means of CO_2 pressure, thereby substantially increasing the flow rates compared with the gravity model; the arrangement is shown diagrammatically in Figure 11.12 and the pressure control of flow is claimed to maintain a higher accuracy of proportioning than the Mark I 'Flo-Mix'.

The combination of de-aerator, proportioner and carbonator which form the Mojonnier packaged unit is shown in Figure 11.13.

Crown-Century beverage preparation unit. Syrup and water centrifugal pumps (driven by a common motor) are fed from constant-head reservoirs and deliver the two components through control valves into an in-line static mixing chamber from which the product is transferred to the carbonator. The manually adjustable micrometer flow control valves are of special design with parabolic plugs to give uniform alterations in flow for equal movements of the handwheel; accurate dial indicators are provided on both valves to allow repeatability of settings. The principles of this simple but accurate unit are depicted in Figure 11.14; the complete carbo-cooler system is shown in Figure 11.15.

CARBONATION AND FILLING

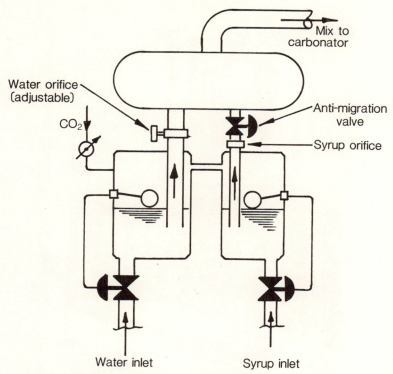

Figure 11.12 Mojonnier Mark II 'Flo-Mix' proportioner.

Ortmann and Herbst volumetric proportioner. This is known as the Series 'C' Intermix and operates on the principle of filling and emptying syrup and carbonated water reservoirs of known volumes as shown in the diagrammatic arrangement in Figure 11.16. The syrup-dispensing tank is completely filled and the water-dispensing tank is filled to a level determined by the required syrup/water ratio; the supplies are then shut-off and the system drains into the lower mixing/storage tank. The action of the carbonated water 'chasing' the syrup batch out of the dispensing tank aids accuracy and mixing; the homogeneity of the final product is ensured by a centrifugal pump circulating the beverage in the lower tank. Level controls, check probes and sensitive adjustment of the carbonated water volume (by slow addition of the last few litres) make this a reliable unit with a claimed accuracy of $\pm 0.025\,°$Brix in the finished product; microprocessor control allows pre-programmed ratio selection.

All types of proportioner coupled with a de-aerator and a carbonator to form a processing system must be capable of being effectively cleaned in place;

Figure 11.13 Mojonnier unit with de-aerator, Mark II 'Flo-Mix' proportioner and carbonator. (Courtesy of Meyer Mojonnier Ltd, London.)

the CIP facility usually includes the filler and would follow the principles specified in Chapter 8 on syrup room operation. Systems of varying sophistication are available, ranging from those relying on manual operation and supervision to microprocessor-controlled, fully automatic units.

11.5 Fillers and filling valves

Enormous strides have been made in the last fifteen years in the size, speed, quality of performance and complexity of filling equipment for carbonated soft drinks. Mention has been made in the introduction to this chapter of the current *large* machines, but a separate development has produced a new generation of fillers where the 'link' or 'bloc' system has been substantially expanded. Close-coupled fillers and crowners (or cappers) have been in fashion for over thirty years and this design allowed absolute synchronisation of the

CARBONATION AND FILLING

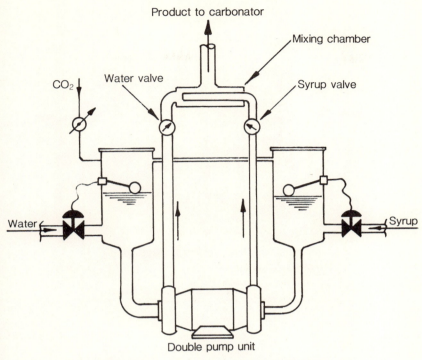

Figure 11.14 Crown-Century proportioning unit.

machines with positive handling of the bottles from inlet to discharge of the duplex arrangement. The development of the high-speed, rotary labelling machine promoted the inclusion of this unit into the filler grouping; the main advantages were reductions in floor area, bottle conveyors and the number of operators. This process was continued with the addition of filled/labelled bottle inspectors, bottle rinsers, empty bottle inspectors, etc.; these latest 'superbloc' systems are extensive and complicated and operate more efficiently when only one size of bottle is handled since change-over procedures are lengthy and involved. Among other factors to be considered is the loss of overall production when one element of the group has a defect and needs to be stopped for corrective action to be taken; in this circumstance, the *whole group* must stop and the total lost production in a day's operation can be appreciable.

Apart from the first crude units, can fillers have always been designed 'monobloc' with seaming machines; owing to the high fill level in the can, serious product loss would result if the open, filled containers were allowed to travel freely and impact together during normal conveying. For the same reason, the filled cans must not be handled by a rotating starwheel at the filler

Figure 11.15 Crown-Century proportioner/carbonator. (Courtesy of The Crown Cork and Seal Co Inc., Baltimore, USA.)

outfeed and a tangential discharge is always supplied resulting in a smooth, positive transfer of the can to a perfectly synchronised seaming unit. This refinement has recently been made available on high-speed bottle fillers to prevent loss of product from the open bottles between filler and closing unit.

The requirements of a filling machine are:

(1) The containers must be filled as quickly as possible with a commercially accurate quantity of product.
(2) The quality of the product must be maintained during the filling operation.
(3) There must be no mutually inflicted damage between the container and the filling mechanism.
(4) The product contact surfaces must be capable of being effectively cleaned in place and sanitised.

The exception to the third objective is bursting glass bottles; there will always be an incidence of burst bottles, the extent of which will depend upon a combination of several circumstances – glass quality, temperature stresses in the glass, operating pressures, etc. Fillers handling frangible bottles must be well guarded externally to protect operators and should be fitted with guards between the filling valves to deter flying fragments entering other bottles;

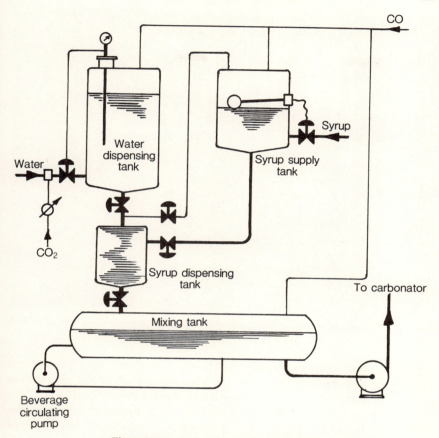

Figure 11.16 Series 'C' intermix proportioner.

ideally, each bottle should be filled in an enclosed compartment from which broken glass cannot escape except vertically downwards for easy removal. Automatic flushing devices are now available to deluge filling valves and associated items with cleaning water after a burst; the vast majority of bursts occur in the counter-pressure area and the auto-flush action usually takes place later in the filling cycle. A bottle filled subsequently on the same filling valve may be removed (sometimes automatically, as explained in Section 13.4 of Chapter 13) for further examination.

High-speed filling machines demand reliability of operation and perfect container handling. The latter starts at the infeed worm where the containers are separated to suit the pitch of the filling valves and continues through the starwheel-handling system to the discharge point where the filled and closed containers are deposited on the outfeed conveyor. Although cans and PET bottles are commendably uniform, glass bottles (particularly the returnable

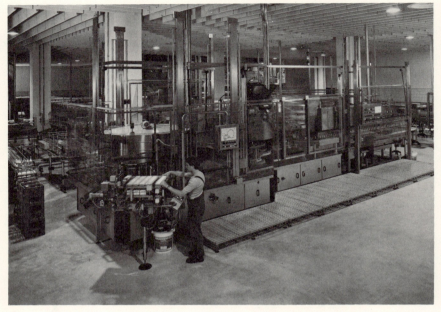

Figure 11.17 Example of a large filler/capper/labeller unit. (Courtesy of Krones (UK) Ltd, Manchester, England.)

types) vary considerably – but must still be accommodated. Lining starwheels and guides with resilient material not only reduces noise but also allows positive handling and precise positioning of these irregular bottles.

A well-designed, easily maintained filler of robust mechanical construction is still only as good as the filling valve it carries. Although the other qualities of the machinery are important, the prime purpose of the filler must be to satisfy the four performance conditions mentioned previously. From the simple filling valves of thirty years ago has been developed the wide range of modern valves which are fitted to present-day machines; over the years, modifications have been carried out to improve the operation and performance of filling valves and to accommodate the new containers that have been developed for soft drinks. Following a description of the function of the original type of valve, the modern variants will be detailed in the subsequent sections of this chapter.

11.5.1 *Basic filling valve operation*

The counter-pressure, gravity-flow filling valve is the origin of all the sophisticated, current valves; it was referred to as the 'automatic' valve since, on attainment of full counter-pressure in the container, the liquid valve opened *automatically* to allow product to flow into the container.

CARBONATION AND FILLING   229

The elements of this type of valve are shown in Figure 11.18 and the method of operation to fill a bottle is as follows:

(1) Filler receiver pressure maintains the counter-pressure and product valves in the closed position (against the action of their respective springs).

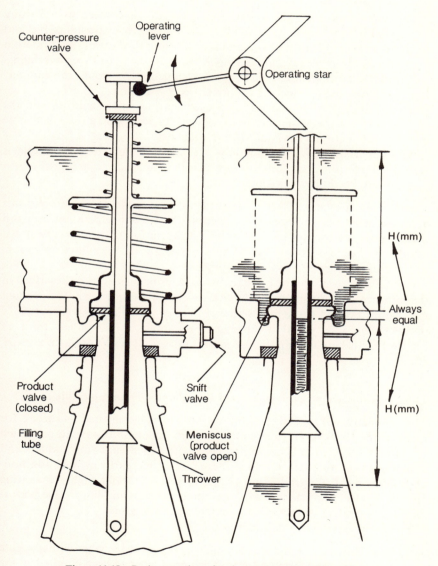

Figure 11.18 Basic operation of carbonated product-filling valve.

(2) The bottle is lifted to seal against the filling valve.
(3) The external operating star is turned to open the counter-pressure valve and allow gas into the bottle.
(4) When bottle pressure equals receiver pressure, the product valve opens under the action of the spring; product flows over the thrower into the bottle (under gravity) and the gas in the bottle returns to the receiver headspace through the filling tube hole.
(5) Flow ceases when the hole in the filling tube is covered by the rising product; product rises in the filling tube to find its hydrostatic level and the product below the product valve falls into the bottle. The formation of a stable meniscus and a 'gas-lock' prevents the bottle from overfilling.
(6) After a short settling time to allow turbulence to subside, the counter-pressure and product valves are closed by the star.
(7) The snift valve is opened (by a cam bearing on the button) and the pressure in the bottle is allowed to escape to atmosphere via a small hole usually between 0.5 and 0.75 mm diameter.
(8) After a snift of approximately one second's duration, the bottle is lowered from the filling valve which is then ready to accept the next empty bottle in the system.

Immediately after stage (4) the star is turned back a small amount and this action, coupled with the intentional clearance between the operating lever and the flanges of the counter-pressure valve, allows the filling valve to close down should the bottle burst (or has already burst) during the filling operation; wastage of product and gas is thus prevented.

Can filling valves operate on the same basic principles as those described in the foregoing, but owing to the large diameter of the open can, the product may be introduced through a series of nozzles which allow a high rate of product flow. A typical arrangement is shown in simplified form in Figure 11.19 where the nozzles are angled in two planes to produce a 'swirl' effect.

11.5.2 *Filling valve development and the influence of ambient filling*

The simplicity and reliability of the 'automatic' valve ensured its success for many years. However, it suffered from several inherent disadvantages which rendered it less and less suitable to cope with the changes in products and containers that were introduced in the seventies and early eighties; additionally, following the introduction of the larger fillers and higher operating speeds, bottlers and canners *expected* improvements in filling valve performance, particularly regarding air contents and fill levels.

The early valves were deficient in the following areas:

(1) The filler receiver and valves could not be cleaned in place with absolute certainty, particularly after fruit cell products had been handled; in the

CARBONATION AND FILLING

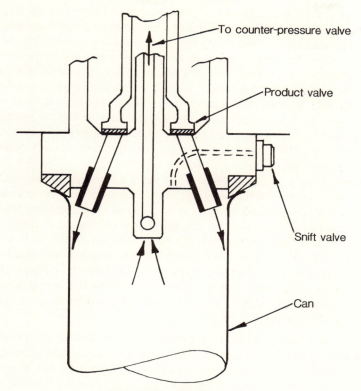

Figure 11.19 Can-filling valve.

latter case, the more discerning bottlers and canners would remove and dismantle all the filling valves to effect a reliable cleaning routine.
(2) The air in the container was returned to the filler receiver head-space. Even if the latter were initially charged with pure CO_2, after a short period of operation the headspace would be almost 100% air; product in the filler receiver and product flowing into the containers (particularly 'down-the-wall' filling) would pick up undesirable amounts of air. Agitation of the product in the filler receiver exacerbated this condition and the inertial swirl of beverage past the filling valves during the starting and stopping actions of the filler was a contributory factor.
(3) During the snift operation, the product in the bore of the filling tube was pushed into the container by the expanding volume of gas trapped below the (closed) counter-pressure valve; forcible transfer of both product and gas into the container during pressure reduction disturbed the product, causing foaming and possible loss of liquid.
(4) The final fill level in the bottle was produced by the position of the hole in the filling tube (or the end of the tube) plus the product which was in

transit at the moment the hole (or tube end) was covered – see stage (5) in the previous section – plus the product in the bore of the filling tube. This 'addition' of product above a fixed point measured below the top of the bottle gave rise to variations in final fill levels due to the inevitable differing neck shapes. The extra product above the filling tube hole (or tube end) also had a detrimental effect on the condition described in (3).

(5) Different fill levels (for example, in different bottles) required various lengths of filling tubes – positioning the hole or end to produce the desired level; this entailed an interchange of filling tubes which screwed into the body of the product valve, and frequent changes resulted in worn threads, loose tubes, etc.

(6) The 'automatic' action of the filling valve meant that the product valve should open when the pressures in the container and the receiver head space were equalised. In practice – owing to manufacturing tolerances in springs and the need for the initial flow of product to move cleanly from the filling tube thrower onto the inside surface of the bottle – the product valve was designed to open when the bottle pressure was still 0.1–0.2 bar below receiver pressure. This had the desired effect of making the product nappe spring clear of the thrower but, unfortunately, also caused agitation as a result of the sudden surge of the initial product entering the bottle.

The above disadvantages and faults, although serious by present-day standards, still allowed acceptable filling qualities and outputs to be obtained from the fillers manufactured up to fifteen years after the Second World War. The major palliative factor was the almost universal employment of efficient chilling systems, resulting in a product temperature in the filler receiver of 3.0 to 4.0 °C; this low temperature made even the high-carbonation 'difficult' beverages relatively quiescent and easy to fill. Influenced by the policy of some manufacturers to design fillers with a maximum working pressure of about 3.0 bar (making product cooling essential for carbonations above approximately 3.75 volumes), the inclusion of refrigeration units in the beverage-processing systems became the norm. The capital and running costs of the cooling plants were certainly affordable compared with the benefits to be obtained, i.e.

- a high standard of filling
- maximum possible output from a given size of filler
- the minimisation of the detrimental effects of bad syrup-room practices (refer to Section 11.3.2) and less-than-perfect filling valves.

In the appropriate conditions of humidity, filling a glass or metal container with cold product results in condensation on the exterior surface. In the era when returnable bottles were packed in wooden crates, the timber element absorbed the moisture and there was no real difficulty; wet cans, however, presented a twofold problem – possible rusting of the can material and deterioration of the cardboard packaging. It was necessary to warm the cans (above the dew-point level) by spraying them with hot water in a tunnel-type machine; a final operation of turning the cans and subjecting them to blasts of compressed air ensured that all traces of moisture were removed.

The introduction of plastic, non-absorbent crates for returnable bottles caused problems with label mould and label 'shift' that were mostly overcome by modifications to label materials and adhesives. The greatest challenge, however, was the advent of non-returnable bottles in the early 1960s which produced the same difficulties as had been experienced with cans, i.e. damage to cardboard packaging by moisture impregnation. One remedy would have been to follow the technique of the canners and warm the bottles above the dew-point but this would have necessitated the installation of extremely large warming tunnels – heating steps would have to be small (to avoid thermal shock on the rather weak non-returnable glass) and heating zones needed to be long, owing to the poor thermal conductivity of the bottles. The only viable alternative was to dispense with the cooling process and reduce the speed of the filling unit – introducing the product more slowly into the bottle, giving more settling time and a longer snift time. A suitable combination of these three factors would quieten the beverage, allowing a reasonable standard of filling – albeit at a higher pressure, which inevitably would produce a higher incidence of burst bottles. Dependent upon the design features of the particular filling valve in use, the reduction in filler speed was between 20 and 35%. Financially, the extra cost of a larger filling unit (even 50% greater than would be considered when filling cooled product) was less than the cost of purchasing and running the necessary refrigeration and heat-exchanger system.

Thus the concept of 'ambient' filling was explored and developed in those areas of the world where climatic conditions were generally suitable to allow this non-cooling operation to function in an acceptable manner. Filling valve designs were modified to improve and streamline the flow pattern of product and to eliminate the intrinsic disadvantages to which reference was made at the commencement of this section. The massive increase in energy costs in the early 1970s made the operation of refrigeration equipment even more expensive and gave tremendous urgency to the development of improved valves. Modern valves can now allow fillers to operate satisfactorily at product temperatures of 15–20 °C subject to a reduction of only 15% on the speed attainable with full cooling to 2.0 °C – and the situation is still improving. Even can-filling valves, notorious in the past for necessitating full product cooling, may now fill soft drink beverages up to 3.7 volumes carbonation at 20 °C.

In parts of the world that are subject to yearly cyclic temperature changes, it may be possible to pursue ambient filling for most of the year; however, there could be a period of several weeks when beverage temperatures will be too high to allow the usual speed of production to be maintained. Under this 'hot weather' condition, demand will probably be at a high level and cutting production rates would be deplored; apart from large and expensive stocks to compensate for temporary reductions in outputs, the answer is to use a modest-size refrigeration unit to cool the product sufficiently to allow normal output speed to be sustained. Hence, 'partial, occasional cooling' is now adopted by many discerning bottlers and canners to permit them to take full advantage of the 'ambient filling' concept and yet accommodate those abnormally high temperatures that are encountered from time to time.

The alterations in design, which improved the general operation of filling

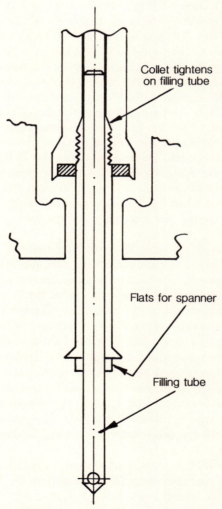

Figure 11.20 Adjustable length filling tube.

valves, made them more suitable for the changes in containers and enhanced the ambient filling capability, are many and varied. The following review gives brief details of the major advances, but is by no means exhaustive.

1. Adjustable-length filling tubes (Figure 11.20) allowed precise settings when fill levels were altered; an alternative was the 'snap in, snap out' tube retained by a rubber 'O' ring located in twin semi-circular grooves.
2. To prevent product rising up the bore of the filling tube, a simple ball valve was incorporated in the tip of the tube (Figure 11.21) which allowed the counter-pressure and gas return operations to function in the normal way but rising product lifted the ball to seal the tube. This was a delicate mechanism and was not always successful – partly because of the interference of fruit-cell particles.
3. Designed in a similar way to the snift button, a second button was incorporated into the lower body of the filling valve; when depressed, this button connected the bottle to a vacuum source which extracted

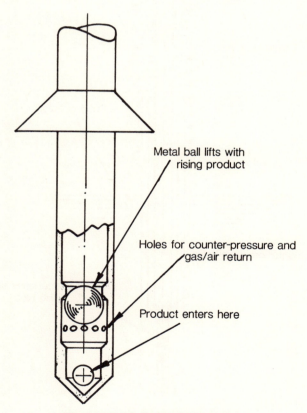

Figure 11.21 Ball valve in filling tube.

approximately 90% of the air as an initial operation before normal counter-pressurisation. Coupled with CO_2 purging of the receiver headspace (addition of gas at one point and release of gas and air at a point diametrically opposite), this resulted in a marked reduction in air (and oxygen) pick-up by the product. For those products considered to be supersensitive to the effects of oxygen, this procedure could be extended by arranging that the bottle, after the first pre-evacuation, was filled with pure CO_2 to approximately atmospheric pressure (by momentarily opening the counter-pressure valve) and then pre-evacuated again; this had the effect of further reducing the 10% air content by 90%, i.e. to as low as 1.0%. Owing to the cost of the CO_2

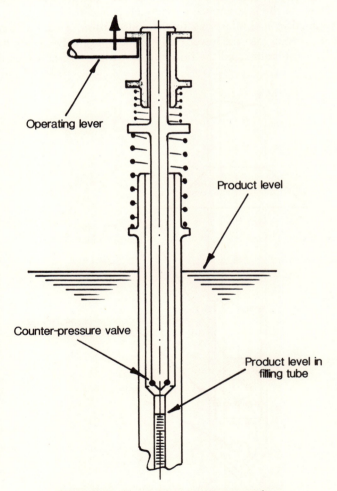

Figure 11.22 Low-level counter-pressure valve.

wasted and the sacrifice of valuable filling time in the operating cycle of the filler, it is considered that measures to improve this low air content cannot be justified.
4. The manifold which supplied the filling valves with vacuum – as in (3) above – was used to convey sanitising solution; by adding dummy bottles to the valves (clipping or screwing in position) the solution was circulated through the filling tube and up the product valve in a reverse direction to the normal flow, giving an effective cleaning and sterilising effect.
5. The counter-pressure valve was positioned below the liquid level in the filler receiver in a cup arrangement to exclude the product, as shown in Figure 11.22; this had the effect of reducing (or even eliminating) the column of trapped gas which, during snift, drives product into the container causing agitation and wildness.
6. The agitation caused by filling valves protruding into the product in the filler receiver may be avoided by fitting the valves into small chambers on the outside of the receiver as depicted in Figure 11.23; this

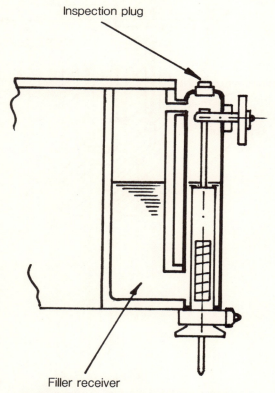

Figure 11.23 Filling valves in separate chambers.

modification also allows the receiver to be smaller and permits the valve and operating mechanism to be removed as a composite assembly.
7. The incorporation of the 'long filling tube' into valve design allowed a smoother entry of product into the bottle and greater control over the fill level; more importantly, since the product flowed through the bore of the filling tube (contrary to the operation of the 'thrower' type of filling, or vent, tube shown in Figure 11.18) it was possible to separate the product and gas passageways within the valve itself and feed the two constituents from a centre receiver with separate stars controlling the

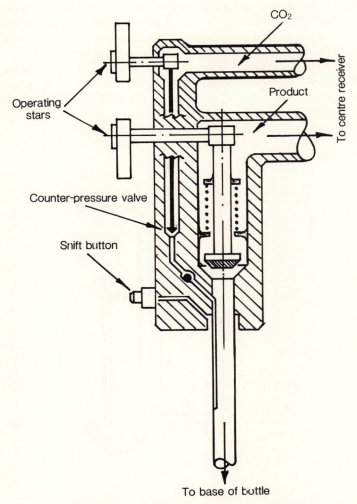

Figure 11.24 Long filling tube with variable flow rate.

counter-pressure and product valves. This was beneficial in itself but was further developed by adding a facility to restrict the rate of gas/air return – which, in turn, restricted the product flow into the bottle. Hence, the initial flow could be slow, then fast for the bulk of the product volume and then slow again for the final 20–30 mm of the neck, producing a very quiet fill and exceptional level control. The valve, shown diagrammatically in Figure 11.24, also includes a ball valve in the gas-return passage, to prevent product rising to find its own level, and a drain vent to allow product in the filling tube to fall into the bottle after snift. An alternative version of this 'long tube' valve used the tube as an electronic probe to control restrictor valves in the gas/air return and produce a slow–fast–slow sequence of product flow as described above; additionally, the probe could be used to close the product valve and produce exactly the desired level in the bottle.

8. The majority of soft drink bottles are either crown cork or 28 mm screw finish, both of which have bores of between 16 and 20 mm – quite suitable for the thrower type of filling tube. The introduction of wide-mouth bottles with 38 mm screw finishes posed a problem with this design of tube and a 'swirl' modification was conceived to spin the product radially and circumferentially outwards to cling to the inside surface of the bottle.

9. The vertical load holding a can or bottle against a filling valve must be able to withstand the separating force when the container is pressurised; before the latter event, the full upward force is applied and, in the case of a thin-gauge can, the sealing surface may be distorted resulting in difficulties in seaming. To overcome this problem on can fillers, the 'differential' filling valve was conceived where the valve was divided into two sections, one sliding within the other on a piston-and-cylinder principle. The moving half of the valve was lowered gently onto the can (which remained at the same level as the conveying system, i.e. no can-lifting cylinders) and the counter-pressure gas built-up in the can and also behind the piston section of the filling valve. The piston diameter was slightly in excess of that of the can and this differential ensured that the end force applied to the can was always just greater than the separating force produced by the internal pressure. Thus, any damage to the can sealing surface through an excessive end force was prevented.

Some types of PET bottle also cannot withstand the end forces required for sealing against the filling valve, and several modifications have been designed to overcome this problem. It is possible to restrict the travel of the pneumatic cylinders which lift the bottles to the filling valves and, by careful adjustment of the position of the filler receiver, the bottle can be *just* trapped to allow the filling operation to proceed; this method had varying success owing to slight differences in bottle heights and the vertical variations due to receiver 'wobble'. The more popular modifications took advantage of the large neck rings positioned immediately under the screw finish of the large PET bottles. One version relied upon a post, secured to the lifting cylinder, carrying a claw at the upper end which located under the neck-ring flange thus transferring the vertical load of the cylinder to the neck ring. Alternatively, later designs incorporated the claw into the filling-head mechanism so that the bottle was pulled to the head to effect the necessary sealing action. The latest development utilised the principle of the 'differential' can-filling valve in which the head descended onto the bottle and counter-pressure gas produced the required force. In the instance of bottle filling, it was necessary to make the filling tube retractable to avoid excessive movement of the lower part of the valve and this feature rendered the filling valve mechanism even more complicated.

10. The latest lightweight cans and fragile PET bottles will not withstand the application of vacuum to extract the undesirable air. With these containers the procedure of pre-purging with CO_2 (or another inert gas such as nitrogen) has been adopted. The can or bottle makes a seal against the filling valve and either the counter-pressure valve or a special plunger-operated valve supplied with CO_2 is opened to introduce the gas into the container. A suitable exit from the container is provided for the gas/air mixture to escape to atmosphere after which the outlet valve is closed and the container counter-pressurised and filled in the normal way. The extent of the purging action will be optimised by evaluations of gas consumption compared with the resultant oxygen levels in the product. A method of pre-evacuating cans by vacuum is currently under development. The can is sealed to the filling valve and enclosed in a pressure-tight compartment; the complete compartment and can are then subjected to the vacuum, thus eliminating any external forces on the flimsy wall of the can. The can is then counter-pressurised, the vacuum in the outer annulus dissipated and filling proceeds.

11. Undercover gassing on canning lines has been followed for many years. This process involves jetting the can headspace with carbon dioxide or nitrogen as the lid is offered to the open can, thereby greatly reducing the air (and oxygen) content of the package. For particularly oxygen-sensitive beverages in bottles, a similar arrangement is available at the crowning or capping machine; the inert gas is sprayed under the crown or cap as it is positioned on the bottle. As with any system of this type, a balance must be drawn between the cost of the gas used and the benefits obtained regarding reduced oxygen contents in the product.

12. The variations in fill levels in varying bottle necks produced by the 'thrower' type of filling tube have troubled the industry for many years but a method of overcoming this problem has now been developed. At the end of filling (but before the snift operation) the product valve is closed, the counter-pressure valve retained open and CO_2 at approxi-

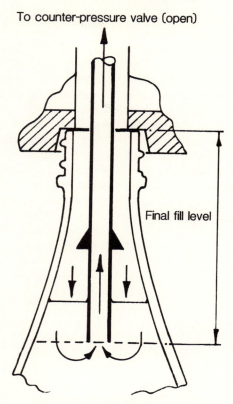

Figure 11.25 Fill level correction.

mately 0.2 bar above receiver pressure (equals bottle pressure) is introduced into the bottle. This extra pressure pushes product in the neck of the bottle up the filling tube until the lower end of the tube is uncovered, as shown in Figure 11.25; the product flows out of the counter-pressure valve to rejoin the beverage in the filler receiver. Thus an exact fill level is obtained with no wastage of product and the filling tube is cleared of product which, when blown into the next bottle at counter-pressurisation, may cause a problem with the filling quality. After closure of the counter-pressure valve, the bottle is snifted in the usual way.

The majority of the above refinements are aimed at improving the *quality* of the packaged product; invariably, their incorporation will result in a reduction in the operating speed of a given-sized filler. The instantaneous rate of flow of product into a bottle or can has increased only marginally in the last twenty years since compliance with the gravity-flow principle allows only two variables – head of liquid over the product valve and the area of the bottle mouth (less filling tube space). Meticulous attention to flow patterns through the filling head will further reduce the adverse differential between 'ambient' and 'full cooling' outputs. This process may well allow a departure from the gravity flow concept and the adoption of a 'twin-tank' system where the product and counter-pressure gas are separated with the former at a slightly higher pressure than the latter; higher rates of flow (varied at will) could then be entertained, resulting in enhanced outputs.

Whatever form future filling-valve development takes, there can be little doubt that ever-increasing standards will demand valves capable of producing the necessary results. These valves will be more complicated and sophisticated and, as with the processing and other production equipment for the 1990s and into the next century, the high levels of technology will require equally high standards of maintenance and supervisory expertise. As in the past, the fullest possible co-operation and exchange of technical information between manufacturers and bottlers/canners will be of paramount importance to the continued success of the carbonated soft drinks industry.

References

1. V.J. Batchelor, in *Developments in Soft Drinks Technology*, Vol. 3, Elsevier Applied Science Publishers, Barking (1984), Chapter 5.
2. A.J. Mitchell, in *Developments in Soft Drinks Technology*, Vol. 1, Elsevier Applied Science Publishers, Barking (1978), Chapter 6.

12 Container decoration

D. KAYE

12.1 Introduction

Soft drinks containers have, until recently, been 'decorated' simply by the application of conventional paper labels. This is no longer the case and, resulting from innovative technology with regard to materials and application systems, there is a trend towards a diversity in container decoration, with the newly available technology being used in addition to, or as an alternative to, conventional paper-label application. Hence, it is felt that the term 'decoration' underlines the fact that containers can now be prepared for point-of-sale in a variety of ways, including paper labels, various types of plastic sleeves, metallic neck foils/capsules, medallions, neck-strip labels and shrinkable closure bands.

The way in which a container is dressed is influenced by sales and marketing considerations, although obligations to the consumer (in respect of legislative and security aspects) influence the nature of the dressing and the appearance of the container in the market-place.

Considering the complexity of the modern consumer market and the need to meet the requirements of so many industries, it follows that many different machines/systems are available for decoration of a diverse range of products/containers. For example, there are machines for applying wet glue labels (pre-cut and reel-fed), pre-gummed labels, pressure-sensitive (self-adhesive) labels, thermoplastic (heat-seal) labels, metallic neck foils and plastic shrink sleeves/stretch sleeves. Given the strictures of space, this chapter can only attempt to deal with a relatively narrow segment of 'container decoration', namely that relating to the machinery, systems and materials used traditionally, currently or potentially in the future by a modern, automated soft drinks production plant (Figure 12.1).

Container decoration can be regarded as a 'high-speed assembly' job in which the application machine (e.g. labeller) brings together several components – in the case of conventional labelling, the label, the container and the adhesive. The interdependence of machinery and materials cannot be over-emphasised and so we shall review not only the types of application machinery available, but also the important factors influencing the quality and efficiency of container decoration. Paramount among these is the quality of the materials to be handled.

12.2 The aims of container decoration

Before reviewing the application machinery and decorating materials, it is worth examining the main reasons for decorating a container in the first place. While the following is by no means a totally comprehensive list, it does include the major functions of container decoration.

(a)
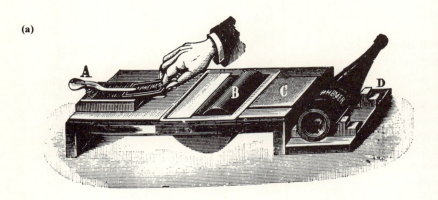

(b)
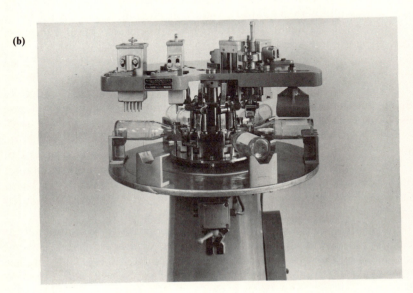

Figure 12.1

(c)

(d)

Figure 12.1 Labelling systems over the years: (a) a simple manual operation used 100 years ago; (b) a labelling machine (c. 1950) used for the semi-automatic application of body and shoulder labels; (c) a 1960s vintage automatic labelling machine using the vacuum principle for label transfer; (d) a modern, high-speed, rotary labeller used in an automated soft drinks bottling plant.

To sell the product. Without doubt the prime purpose of container decoration is to assist sales of the product by rendering the total package visually attractive at the point of sale. The most successful types of decoration are able to generate 'impulse buying', and a constant objective must be to try to ensure that any individual product is chosen ahead of its competitors as a result of its 'eye appeal'. Many factors combine to produce a container dressing which is aesthetically pleasing and sales-effective, and much depends upon individual circumstances. There is no magic formula for success, but a combination of experience, careful market research and design flair can produce highly effective results.

To inform. The label conveys information to the consumer, some of which is associated with sales/marketing considerations, while other information is presented in the direct interest of the consumer. For example, information relating to a description of the product, the container contents (i.e. volume) and price all have a bearing on the decision to purchase, whereas an analysis of the product's ingredients may be very important for health or dietary considerations.

To comply with legislation. Increasingly, international legislation is directly affecting the nature of container decoration. In most countries there is a legal requirement to indicate the shelf-life (e.g. 'best by' or 'sell by' dates) for foodstuffs, as well as the need for a detailed analysis of ingredients of certain products (a notable example of this is the law applicable to mineral waters).

To instruct. Instructions are often necessary for the safe and correct use/consumption of the product.

To protect. In recent years, the subject of 'tamper-evidence' has assumed tremendous importance. While, in the drinks industry, the most common method currently employed is the use of tamper-evident closures, these are not suitable for all applications and container decoration (in the form of specially extended neck or shoulder labels, U- or L-shaped neck seals, or shrinkable PVC neck bands) is now used for this purpose.

12.3 Main types of decoration application machinery

Soft drinks containers are still predominantly 'dressed' by the application of labels by means of wet (i.e. liquid) glue systems, although the methods of application are constantly diversifying. (NB. For the purpose of this chapter, wet glue labelling systems can involve the use of either cold adhesives or hot-melts.)

Automatic labelling machines can be broadly split into two different categories: indirect-transfer machines and direct-transfer machines.

12.3.1 Indirect-transfer labelling machines

In this context, the term 'transfer' refers to the travel of a pre-cut label from the stacked condition in the magazine to the point of application to the container being labelled. Indirect-transfer machines incorporate some form of mechanical device interposed between the label magazine and the container, and this device, by means of a sequence of operations, removes individual labels from the stack and transfers them, under controlled conditions, to the point of application.

Indirect-transfer machines have been extensively used in the soft drinks industry for many years, primarily because of the flexibility they offer, coupled with the capability for high outputs and high efficiency levels. Indirect-transfer labelling machines can be further subdivided into two types: in-line and rotary. These terms basically describe the paths of the containers through the machines.

In-line machines. The containers are fed in single file to the machine where they are pitch-spaced by means of an infeed worm (feed-screw), following which they continue to travel in a straight line through the machine. They remain on a slat conveyor which conveys them past the point of label application and subsequently through an 'after-roll' station which fully applies the labels to the bottles by means of a rolling action.

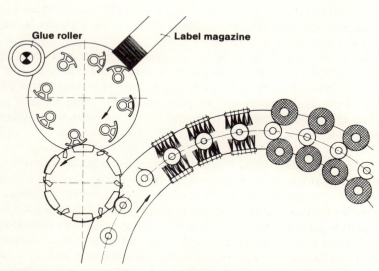

Figure 12.2 Principle of the modern rotary, indirect-transfer labelling machine. Pre-cut labels are removed from the stationary magazine by means of a controlled glue film applied to the front surfaces of the eight glue segments. The labels are subsequently removed from the segments by a six-station mechanical gripper cylinder that transfers them to the containers on the main carousel after which the labels are fully applied to the container surface by means of brushes and sponge rollers.

248 CARBONATED SOFT DRINKS

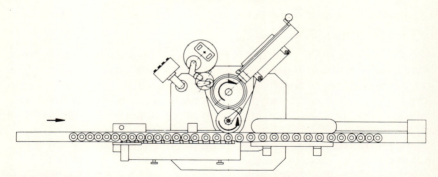

Figure 12.3 Plan view of an in-line, indirect-transfer labelling machine. Machines of this type are limited in their flexibility and are mainly used for body label application only.

There are several disadvantages with this method of bottle handling which greatly restrict the use of machines of this type. Firstly, since the bottles must be free to rotate they cannot be clamped overhead in any way, and accordingly are relatively unstable. Bottle stability is directly affected by the diameter/height ratio and, of course, linear speed (i.e. line output), and accordingly these factors must be carefully assessed before using an in-line machine for any given application. Because of their relative instability, it is important that the forces applied to the containers during label transfer and final application are close to the centre of gravity, otherwise adequate control will be impossible. In most cases, this precludes the application of labels to the neck or shoulder areas of a bottle.

While, in principle, it is possible to apply front and back body labels simultaneously (by positioning label stations on each side of the conveyor), the lack of total control of the bottle at the application point means that precise register of the two labels cannot be guaranteed. For these reasons, in-line machines are normally only recommended for the application of single body labels on to cylindrical containers.

Rotary type machines. These offer many advantages when compared with in-line machines, primarily as a result of the positive control of the containers throughout the labelling cycle. As with in-line machines, containers are conveyed to the machine in single file and separated by an infeed worm, but they are then subsequently transferred from the conveyor to a central rotary carousel via an infeed transfer starwheel. Once on the carousel, containers are positively and accurately located between a base platform and an overhead centring device which is mechanically lowered to the containers. As they travel around the carousel, the containers are rotated axially as necessary (by means of a gearing mechanism which controls the movements of the base platform) dependent upon the application. Accordingly, at all points throughout the cycle, the containers are totally under control. This system allows precise and efficient labelling to be carried out at the highest production line speeds. Equally important, however, are the flexibility and versatility afforded by rotary-type machines. Multiple-label application can be achieved using one, two or three separate labelling stations positioned around the periphery of the container carousel. Most rotary machines can handle many combinations of front body, back body, shoulder, medallion, neck-around, deep-cone and aluminium neck foil labels.

While emphasising the versatility of rotary machines, one important limitation should be caefully noted. *Rarely is it practicable to apply full wrap-around body labels on rotary type machines with indirect transfer.* The reason for this is that the size of the developed label (i.e. in its flat form in the magazine) will, in the majority of cases, exceed the maximum label width capability of the machine due to the gluing/transfer mechanism design. It is sometimes possible, with small-diameter bottles, to handle a full wrap-around label but this will normally greatly reduce the machine's overall flexibility and it is normally more practical (and cost-effective) to purchase a machine that is designed for the application of wrap-around body labels. (See Section 12.3.2.)

Labelling stations. Without doubt, the labelling station can be considered the 'heart' of any indirect-transfer labelling machine. While in-line and rotary labellers differ with regard to the methods of container handling, both types often incorporate labelling stations of similar design.

The labelling station is a very important sub-assembly which is mounted on the machine base plate and which incorporates two key operations, namely the *label transfer system* and the *glue handling system*. While individual designs vary, the labelling stations on most modern machines use very similar concepts which comprise, briefly, a stationary label magazine holding pre-cut labels in stack form; a vertical glue roller/scraper assembly; a glue segment carousel; and a mechanical label transfer device.

The design and configuration of the labelling station is vitally important, since the number and pitch-spacing of the individual glue segments on the carousel directly determine the maximum width of labels that can be handled. When selecting a new labelling machine, all current and potential future applications should be carefully assessed to ensure that the labelling station design does not prove to be a constricting factor.

(a) *Label transfer systems.* The overwhelming majority of applications in the soft drinks industry involve the use of pre-cut labels for which the labelling stations on indirect-transfer machines are designed. (It should, however, be noted that reel-fed labelling systems are likely to create an important niche in the market in the future, for reasons outlined later in this chapter.) Pre-cut labels are guillotined or die-cut and delivered by the label printer normally packaged in bundles of 500 or 1000. After removal of the transit/storage packaging the labels are loaded – normally manually – into a magazine which, on most modern labelling machines, is static and will have a capacity of between 3000 and 8000 labels dependent upon label thickness and size of magazine. One of the most critical aspects of the labelling operation is the removal of individual labels from the stack and the subsequent transfer to the container. It is essential that this operation is carried out extremely efficiently, even at the highest operating speeds, otherwise the operation of the entire production line is seriously jeopardised.

Over the years various methods of label removal from the magazine (often referred to as label 'picking') have been used, the three main ones being:

- by vacuum
- by mechanical means (using mechanisms similar to those used in high-speed bank note counters)
- by glue film.

The first two methods have several inherent drawbacks and limitations, not the least of which is their inability to operate satisfactorily above certain (quite moderate) speed thresholds. Accordingly, almost without exception, all modern indirect-transfer labelling machines incorporate methods of label removal using a carefully controlled glue film.

In most cases the system adopted is a rotary carousel comprising a number of vertical shafts (this can vary dependent upon label size range, operating speed, etc.,) to which are fixed glue segments (normally made of aluminium or brass and frequently referred to as 'glue pallets') which are individually profiled to the size/shape of the labels to be handled. (*NB*: These glue pallets are 'change parts' and hence should be designed for very quick and easy removal and replacement when changing label sizes.)

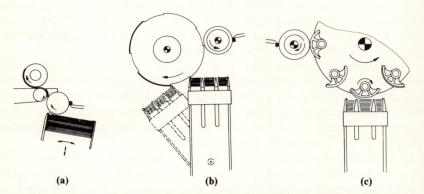

Figure 12.4 Different types of 'glue-picking' labelling stations: (a) an oscillating label magazine from which individual labels are removed by means of a rotating grooved glue roller. Labels are removed from the surface of the roller and guided directly onto the container to be labelled by means of a 'comb' arrangement; (b) an oscillating label magazine from which individual labels are removed by means of a rotating glued cylinder, on the periphery of which are four glued areas which correspond to the size and shape of the labels being handled; (c) a stationary label magazine from which labels are removed by means of oscillating glued segments housed within a rotary carousel.

Axial movements of the shafts/pallets are accurately controlled via a mechanical system housed within a gearbox located below the carousel. The front surfaces of the glue pallets are curved to allow precise and uniform contact with a vertical glue application roller as the carousel rotates. The individual shaft-drive mechanisms induce a planetary movement of the pallet when in contact with the glue roller, resulting in the application of a uniform glue film over the entire segment surface. The fully glued segments are then brought into contact with the rear surfaces of the labels in the magazine stack and, due to a combination of a 'rolling' movement of the pallet surface across the labels and the tack of the adhesive, individual labels are removed. (*NB*: The label stack is held in the magazine by a series of retaining 'prongs' around the periphery of the labels at the exit end of the magazine. Clearance slots are milled into the periphery of the glue segment at corresponding positions, to provide clearance when the segment contacts the end of the label stack.) The milled recesses on the leading edge of the segment perform a double function, since they also allow cam-operated mechanical gripper fingers (which comprise part of the transfer gripper cylinder assembly) to grip the leading edge of the label (against an 'anvil' on the gripper cylinder) and to 'peel' the glued label from the segment. As the gripper cylinder rotates, the glued labels are transferred to a point where they come into contact (tangentially) with the containers travelling around the main machine carousel. In most cases, the labels are applied to the containers on their vertical centre-lines and, assisted by slight pressure from a sponge incorporated into each station of the gripper cylinder, they are affixed by a narrow vertical strip of adhesive. For most applications an overall film of adhesive (i.e. across the entire rear surface of the label – with the exception of the small cut-out areas) gives the best results from both application and label presentation view points and this overall coverage is readily and efficiently achieved by the glue segment system described. However, sometimes it is desirable (or, indeed, necessary) to have only partial gluing and in such cases the surfaces of the pallets are machined in the desired pattern. One of the most common reasons for reducing the glued area is where returnable containers are being handled and difficulties in label removal in the washer are being experienced. By incorporating partial gluing (e.g. by using pallets with a 'grid' or 'honeycomb' pattern), penetration of the caustic solution can be accelerated, thus facilitating label removal. However, it should be pointed out that partial gluing patterns can have an adverse effect on the visual appearance of the label, especially in unsuitable storage conditions. Moisture from the air readily permeates the unglued areas of the label causing bubbling or rippling due to the localised expansion of the paper fibres. (*Note*: It is strongly recommended that partial gluing should *not* be introduced as a glue-saving measure. Long experience has shown that rarely are any nett reductions in glue usage achieved and since there are several other negative aspects partial gluing should be avoided except where absolutely essential for ease of label removal.) A second example of the need for partial gluing is in the case of special promotional labels where a section of the label is to be removed by the consumer as a 'proof of purchase' coupon or to gain access to information printed on the back of the label.

(b) *Glue-handling system.* Given the fact that most modern high-speed labelling machines rely upon glue film for label removal and transfer, it follows that the handling and control of the adhesive on the labelling station is of paramount importance. The properties of the adhesive are critical, as emphasised later in this chapter. However, it should be appreciated that many types of adhesive can lose their 'machinability' (i.e. the facility for satisfactory control and handling on the machine) if they are subjected to excessive mechanical stresses resulting from the machine operation. Such stresses can also result in temperature increases which can adversely affect the performance of the adhesives. It is therefore desirable to restrict the mechanical stresses as much as possible in the glue-handling system. A primary step that can be taken in this respect is the use of a recirculating system which limits the amount of adhesive 'in work' at the glue roller/scraper position at any given time. (Earlier types of labelling machines frequently incorporated a small glue bath within the machine into which relatively small amounts of glue had to be regularly loaded by hand. This small volume of adhesive was subjected to constant mechanical stressing as it 'churned' in the glue bath, with the subsequent fall-off in adhesive and cohesive properties.) Most modern high-speed automatic machines incorporate a recirculating system with a pump feed from a large-capacity (usually approx. 30 kg) container supplying the adhesive via a nozzle to the top of the vertical application roller. The adhesive travels down the surface of the roller (by gravity), but owing to the close proximity of the adjustable scraper blade and the rotation of the roller, the glue forms a relatively uniform 'sausage' within the 'vee' created by the scraper and the glue roller surface. Thus, the adhesive is only subjected to mechanical stresses for a relatively short period prior to returning (again by gravity) to the large container. This recirculation also ensures that there is no tendency for the adhesive to overheat.

Since the mid-1970s casein-based adhesives have gained in popularity and now have a very large share of the market for labelling returnable and non-returnable glass containers. Casein adhesives have a number of important advantages, but perhaps the main one is their ability to satisfactorily label very wet and cold glass containers. By their nature, casein adhesives give optimum efficiency and 'yield' (i.e. the number of labels applied per unit volume/weight of adhesive) when operating at a temperature in the range of 26–30 °C. Since this is often higher than the ambient conditions, some labelling machines incorporate electric heaters within the glue supply system which can be thermostatically controlled to ensure that the operating temperature is precisely regulated. (Conversely, ambient temperatures in tropical climates can be damaging to certain grades of glue and in these circumstances it is necessary to incorporate a glue cooler in the supply system.)

The importance of glue film thickness control on the glue application roller cannot be over-emphasised. The entire labelling cycle begins with the removal of the labels (i.e. by means of the glue film) and hence the efficiency of this determines the ultimate efficiency of the entire operation. The more precise and controlled the glue film thickness, the more efficient and economic is the labelling operation. In this respect, there has traditionally been an inherent weakness in the glue-handling system which has now been overcome by newly available technology.

On all vertical glue-roll machines, in order to obtain a thin and uniform film of adhesive on the glue pallet surface, a 'shearing' action is necessary between the glue roller and the glue segment. Because of the very close proximity of these two components, it is not practical to have a 'metal-to-metal' condition and it is important that one of the two components has a degree of resilience to facilitate an efficient operation. Historically, this resilience has resulted from the use of a rubber surface on the glue roller, operating in conjunction with metal (aluminium or brass) glue segments. However, being a natural product, rubber is prone to distortion, and no matter how precisely the rollers might be machined initially, after being stressed mechanically during operation and regularly cleaned in warm water, etc., some degree of ovality or eccentricity invariably results. To accommodate these conditions, a minimum clearance must be present between the scraper blade and the roller surface, and often this is greater than the label thickness, thus allowing any stray label from the magazine to become adhered to the roller. (This can be the result of incorrect setting of the label magazine, under-size labels, etc.) Extensive studies have shown that one of the most common causes of labelling machine stoppages is labels adhering to the glue roller and this, in turn, often exposes the biggest weakness of a rubber roller, namely its susceptibility to damage. Often, blunt instruments are irresponsibly used to scrape labels off the roller and this can easily result in 'grooving'. However, the sharp edges of a crumpled label trapped between the roller and scraper blade can also damage the rubber surface. It follows that any deformation or damage to the rubber roller immediately results in a variation of the glue-film thickness which, in turn, can directly and adversely affect the efficiency of label removal. This basic weakness in the glue roller design has been positively overcome by substituting the rubber roller by a precision-manufactured stainless steel roller which eliminates any risk of deformation. Only by gross negligence can the glue application surface of the steel roller be adversely affected. The required 'resilience' – as mentioned above – is achieved by the use of special glue segments which are rubber-faced. The rubber is fixed to an aluminium casting by an advanced vulcanising process involving extremely high temperatures and pressures to ensure a permanent bond.

This important technical advance in gluing technology allows the scraper blade to be set very closely to the surface of the roller and this has resulted in a number of significant advantages. Firstly, the labelling station can be adjusted

(a)

(b)

Figure 12.5 (a) Damaged rubber glue roller. Caused by loose labels trapped between the glue roller and scraper assembly, this damage results in loss of perfect film control, increases glue consumption and reduces overall efficiency. (b) Modern stainless steel glue roller with plastic scraper. The advantages of this type of assembly are a perfectly concentric roller (which can only be damaged by gross negligence), higher operational efficiency and reduced glue consumption.

to prevent individual loose labels from adhering to the glue-roller surface, since they cannot penetrate the narrow gap between scraper and roller; this eliminates many of the individual stoppages which, although quite short in themselves, contribute to an overall loss of efficiency. Secondly, the consistent glue-film thickness results in improved overall performance of the machine, with reduced risk of labels 'skidding' when placed in contact with the container, owing to an excess film thickness. Thirdly, a significant reduction in adhesive usage can be achieved and in many cases reductions of up to 25% have been reported.

12.3.2 *Direct-transfer labelling machines*

The term 'direct-transfer' indicates that labels are transferred directly from the label magazine to the container without the use of an intermediate transfer mechanism. As with indirect-transfer machines, a stationary magazine containing a stack of pre-cut labels is utilised but in this case there is no labelling station as such. The container to be labelled acts as its own label-removal mechanism, and this is done by directly applying the adhesive to the container surface. The containers are then presented to the label magazine in a controlled manner and the tack of the adhesive on the container removes the individual labels.

It must be emphasised that, compared with indirect-transfer machines, direct-transfer machines are very restricted in their versatility and are used, in the main, for application of full wrap-around body labels. They are not suitable for multi-label application, for profiled labels, or for application on anything other than vertical surfaces.

Because of the designs of the *gluing* stations (i.e. as opposed to *labelling* stations), these machines are generally limited to the application of a relatively narrow vertical strip of adhesive to the container surface. This strip is designed to coincide with the overall height and position of the wrap-around label. Containers are constantly rotated (i.e. about their own axes) at a controlled speed, so that the glue strip is coincident with the leading edge of the label in the magazine. Owing to the tack of the adhesive and the continuous rotation of the containers, the label is removed from the magazine, starting with the leading edge, and is progressively wrapped around the container surface. As the label is removed laterally from the magazine, the trailing edge is pulled across a small-diameter driven glue roller which – during normal continuous operation – is constantly in contact with the label stack, thus applying a second vertical strip of adhesive which subsequently forms the overlap bond.

Unlike the indirect-transfer application, most of the rear surface area of the label remains unglued and, for most applications, only 5–10% of the label area (i.e. the strips on the leading and trailing edges) is covered with adhesive. It should be noted that this labelling concept allows little flexibility with regard to glue patterns. Compared with 'patch' or 'spot' labels (i.e. as commonly applied by indirect-transfer machines), full wrap-around labels have a large

256 CARBONATED SOFT DRINKS

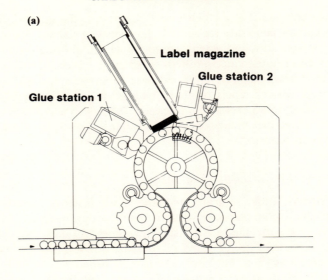

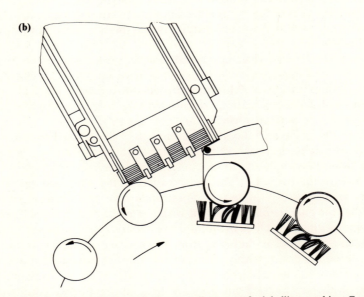

Figure 12.6 (a) Plan view of a typical rotary, direct-transfer labelling machine. Containers are glued directly (using hot-melt adhesive) and thus become their own label removal mechanism. (b) Detail of stationary label magazine. Individual labels are removed owing to the constant rotation of the containers to which a vertical strip of hot-melt adhesive is applied for initial label pick-up. As labels are removed laterally from the magazine, a strip of hot-melt adhesive is applied to the trailing edge which provides the overlap bond.

surface area and since the surface area of adhesive (in the form of the narrow vertical strip) is small in comparison, the tack of the adhesive is vitally important since it must be adequate to overcome the resistance of the label magazine retaining fingers. In practice, this means that (other than for very low output requirements), all types of cold glue are inadequate since the initial bond strength is insufficient to guarantee efficient label removal. Accordingly, with few exceptions, direct-transfer machines are designed for use with hot-melt adhesives. Hot-melt glues are non-aqueous (i.e. contain no water content) and while there are many different formulations, the most important property of hot-melts is their ability to achieve a very quick, permanent and strong bond resulting from the rapid chilling of the adhesive after application.

Historically, it has only been possible to use hot-melt glues for non-returnable containers, since hot-melt glues are difficult (if not impossible) to remove by means of conventional container washing techniques. However, this situation may change in the future since a new generation of water-soluble hot-melts is emerging although, at the time of preparation of this chapter, their effective use under production-line conditions has not been firmly established.

Compared with indirect-transfer machines, direct-transfer labellers are generally simpler in construction although this advantage must be balanced against the significantly lower versatility and flexibility. As with indirect-transfer machines, in-line and rotary machines are available, but the choice is most likely to be determined by operating speed requirements than by the nature of the labelling application. In-line machines are limited to operating speeds of approx. 400 cpm (primarily due to bottle stability reasons), whereas rotary machines are available for outputs up to 1000 cpm, dependent upon application.

12.4 Special container decoration applications and systems

Compared with some other industries, the container decoration requirements of the soft drinks industry have been simple and unsophisticated, for the most part being limited to the application of simple paper labels to plain, standardised containers.

Until recently, the entire labelling spectrum for the soft drinks industry could be covered by machines capable of applying patch body or full wrap-around body labels (in the vast majority of cases of a simple square or rectangular shape) plus shoulder/neck labels. However, owing to the constantly increasing sophistication of the marketplace, coupled with fierce competition, innovative and creative marketing policies have led to a significant diversification of container designs and the manner in which these containers are decorated in readiness for the point-of-sale. An additional important factor is the increasing diversity of product types which are marketed under the general 'umbrella' of soft drinks (i.e. products other than

258   CARBONATED SOFT DRINKS

carbonates and squashes, etc.). It follows that to meet this changing market scene, application machinery must be versatile and flexible and, in many cases, special-purpose equipment is needed to fulfil a particular requirement. Without doubt, the range of equipment available to the soft drinks industry today is far more comprehensive and diverse than ever before and, while by no means a totally comprehensive survey, the following review of special applications and systems includes all those likely to be of interest to the purchaser of capital plant for a modern soft drinks packaging hall, either currently or in the foreseeable future. In some cases, the equipment reviewed below has been specially developed for the soft drinks industry, while other systems have previously been used in other industries.

12.4.1 Combination labelling machines

From the foregoing, it can be seen that, generally, indirect-transfer machines are best suited for the application of 'patch' or 'spot' labels (e.g. front, back, shoulder, neck labels, etc.) while direct-transfer machines are designed primarily for full wrap-around body labels with overlap.

With the sudden and dramatic arrival of the PET bottle on the soft drinks scene in the late 1970s, the industry found itself needing a new type of labelling concept in order to apply much larger labels than those traditionally used, since, primarily for sales and marketing reasons, the 'norm' for this new generation of package was a full wrap-around body label. Since PET bottles were designed for larger volumes (e.g. 1.5 and 2 litres), the necessary wrap-around labels were correspondingly much larger.

Bottlers quickly realised the inadequacies of existing machinery for such label sizes and it was necessary to purchase new machinery to meet this need. However, in many cases this caused problems, especially prior to the introduction of lines dedicated entirely to the production of PET bottles. When combining PET (i.e. with wrap-around labels) with conventional production (e.g. glass bottles with body or body/shoulder labels) two different labelling machines would be necessary, and this resulted in layout difficulties due to space limitations, etc. To overcome this specific difficulty, a range of 'combination' machines evolved, incorporating (for the first time) both indirect- and direct-transfer labelling systems. Inevitably such machines were larger than conventional machines in order to accommodate both systems in tandem, and to incorporate a carousel large enough to incorporate all the relevant bottle turns for both types of application. The initial design concept of combination machines was that only one of the two labelling systems would be used at any given time; however, further development has resulted in the ability of both systems to be used *concurrently* for special label applications. The prime example of this is for the application of a full wrap-around body label (i.e. by the direct-transfer method) plus a shoulder or neck label (i.e. by the indirect-transfer technique) in *precise register* with the body label. This is

CONTAINER DECORATION 259

achieved by use of a specially designed carousel incorporating a drive system to produce bottle turns compatible with the two label pick-up points.

Other types of 'combination' machines available include those which combine wet glue and self-adhesive labelling stations, the latter normally being used for special purposes such as occasional sales promotions, price stickers, etc.; and also machines which combine two different *gluing* systems on one single labelling station. A good example of this type of application is when U-shaped or L-shaped strips are applied over the bottle closure (i.e. for security purposes) when a normal cold glue (e.g. casein) is applied to the area in contact with the glass bottle and hot-melt or synthetic glue applied to the area in contact with the metal or plastic closure.

12.4.2 Reel-fed labelling

Over recent years there have been steadily increasing demands from the bottling and packaging industries for the development of efficient, high-speed labelling from the reel. At the time of publication of this book, reel-fed labelling is an area of specific and growing interest to the soft drinks industry. There are several reasons for this, a prime one being, of course, economics since most types of square and rectangular labels can be produced more economically in reel format than in the traditional pre-cut method. Apart from initial manufacturing economies, there are also significant savings in packaging costs when supplying labels in large-diameter reels rather than in small, individually packaged lots of 500 or 1000 pre-cut labels. There are also other economies resulting from various operational advantages, but, without doubt, a primary

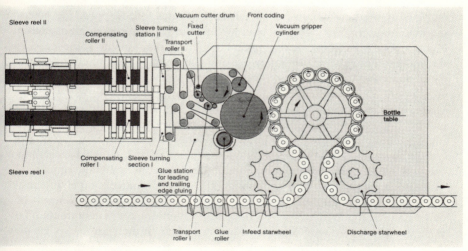

Figure 12.7 Plan view of a modern fully automatic, reel-fed labelling machine. Note the dual reel-feed arrangement, which is linked to an automatic web-splicing system to eliminate downtime during changeover.

reason for the requirement for reel-fed machinery in the soft drinks industry is its ability to handle efficiently the various types of plastic film labels that are now available, many of which cannot be efficiently handled in pre-cut form on more conventional application machinery.

Having mentioned briefly the economics of reel-fed versus pre-cut labels, a cautionary note should be added in this respect. Reel-fed labelling (at least as far as the drinks industry is concerned) is very much in its infancy and, accordingly, a realistic overall picture of the cost comparisons of reel-fed and pre-cut labels has not yet emerged. Currently, as the usage volume of reel-fed labels is very low and very few label printers are ideally equipped for their production, it is certain that the real benefits have not yet been fully realised. To underline this point, a recent detailed study highlighted the fact that, for some applications, reel-fed labels can cost up to 25% less than pre-cut labels of a similar size and material specification. It seems certain that the appeal of reel-fed labels to the soft drinks industry will increase in the future and, with this in mind, the following summary of advantages and disadvantages should be of interest to the reader:

Advantages of reel-fed labelling

Cost. Initial production costs are reduced (guillotining/punching of individual labels replaced by a simple reel splitting operation). Packaging costs – labour and materials – are eliminated (no paper/elastic bands, shrink film, polystyrene trays, cartons, etc., as required when supplying pre-cut labels in packs of 500/1000).

Reduced operator involvement. Reel-fed labels can be supplied in large-diameter reels (up to 1.5 metres), and with a twin-reel system (with automatic 'flying splice'), machines can operate for long periods (in some cases – dependent upon label width and gauge of the material – up to a full shift) thus eliminating the need for constant operator attendance at the label magazine.

Operational advantages. Generally speaking, higher production efficiencies can be obtained compared with pre-cut systems, owing to the elimination of many of the common stoppages that result from material problems associated with pre-cut labels. More consistent dimensional tolerances are achieved, and the absence of magazines eliminates the need for constant running adjustments to accommodate varying label sizes. Double or multiple label 'picking' (i.e. resulting from adjacent labels in a stack sticking together as a result of cutting burrs, inadequately cured inks/lacquers, static, etc.) is eliminated. In turn, this prevents 'loose' labels causing stoppages in the transfer system.

Ability to handle various substrates. Most reel-fed labelling machines are capable of handling a wide variety of materials including conventional paper stock, paper/plastic laminates and the diverse range of plastic films that is now available for manufacture of container labels. In this respect, it should be emphasised that there are significant difficulties and limitations in handling plastic labels in pre-cut form, and these have contributed greatly to the appeal of reel-fed systems.

Plastic labels. These behave completely differently to paper labels when pre-cut and stacked in a magazine, which makes them very difficult to control satisfactorily. In turn, this leads to inconsistencies of glue-film application and subsequent inefficiencies. In addition, there is the significant problem of static between adjacent labels, which can be extremely difficult to overcome.

Plastic labels have become very popular in the soft drinks industry in the USA, and it seems very likely that they will achieve similar acceptance and popularity in Europe and elsewhere in the future. Accordingly, it is worth analysing briefly the advantages of plastic labels and some of the reasons for their widespread acceptance:

(a) *Graphics.* Plastic labels appeal to the marketeers as a result of the very high quality graphics that can be achieved. Plastic film can be printed by flexographic or gravure methods, dependent upon individual requirements, although there is a cost implication in this respect. Generally speaking, the added brilliance of the colours and the high gloss finish achievable with plastic film exceed the limits of conventional paper labels. It is becoming increasingly common, when using plastic labels, to leave part of the label area completely unprinted so that the natural colours of the product are visible through the clear film, thus enhancing the graphics of the label and the overall effect of the product on the shelf.

(b) *Elasticity.* This is an extremely important advantage especially when applying full wrap-around labels to PET bottles containing carbonated drinks. Severe problems have been experienced (especially when products are filled at low temperatures) owing to subsequent expansion of the pressurised bottles. It is quite common for the diameters of large-capacity bottles (e.g. 2 and 3 litre) to increase by several millimetres. Accordingly, tremendous stresses are induced in the label and this can result in the splitting of paper labels or failure of the adhesive bond at the overlap position. By and large, this particular problem can be overcome by the use of plastic labels owing to the inherent elasticity of the material.

(c) *Strength.* Compared with paper, most types of plastic labels have greater tensile strength, and accordingly are less prone to damage (e.g. tearing, scuffing, etc.) after initial application.

(d) *Shrinkability.* Several of the materials used for manufacturing plastic films are heat-shrinkable and these can be used for labels that extend beyond the cylindrical body portions of the bottles to which they are applied. After initial application on the machine, the upper and lower edges of the labels can be carefully shrunk to conform to the curvature of the bottle. This has the

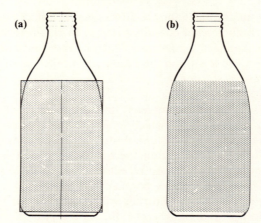

Figure 12.8 Shrinkable-film labels for non-returnable bottles. This is an economic alternative to the application of shrink-sleeves: (a) initial label application; (b) label after heat shrinking.

advantage of increasing the size of label that can be applied, which is often very valuable for sales and marketing reasons.

(e) *Impervious to moisture.* Unlike paper labels, plastic labels are totally impervious to moisture and this eliminates several problems when the product is stored under adverse conditions. After absorbing atmospheric moisture, paper labels can 'bubble' or 'ripple' and are easily susceptible to damage. None of these problems relates to plastic labels.

(f) *Overlap adhesion.* In the case of full wrap-around labels with overlap, plastic labels can either be bonded using conventional hot-melt glues (as used for paper labels) or they can be 'welded' by use of a solvent. In many cases, this 'solvent welding' technique will produce an overlap bond stronger than a glued bond and the costs can often be lower than when using hot-melt adhesive.

Disadvantages of reel-fed labelling

While significant advantages can be gained from a reel-fed application, there are also several disadvantages that should be carefully considered when choosing new capital plant for container decoration. Among these disadvantages are:

Limited flexibility. When compared to machines for applying pre-cut labels, reel-fed labellers are relatively inflexible. In the main, they are designed for the handling of body labels only and for application to vertical (i.e. cylindrical) surfaces. Furthermore, most reel-fed machines on the market operate with a knife 'cut-off' system, which means that only square or rectangular labels can be produced. No doubt continued development will result in reel-fed machines being capable of applying labels to non-vertical surfaces (i.e. neck and shoulder labels, etc.) and profiled labels but this will require sophisticated die-cutting techniques which have not yet proved viable at production line speeds.

Cost. Generally speaking, reel-fed labellers are more sophisticated than precut machines and initial capital cost tends to be higher for any given application. This extra sophistication has cost implications with regard to the level of maintenance, adjustment, etc., and the training and skill levels of the operating and maintenance personnel.

Speed limitations. In the main, the output of a reel-fed labelling machine is determined by the maximum web speed attainable and this, in turn, governs the maximum label speed in relationship to label size. Although reel-fed machines are available for most output requirements, operational speeds cannot generally match those attainable by pre-cut label application machines.

Label re-loading. Most designs of reel-fed machines can accommodate large-diameter label reels in order to minimise the frequency of change-overs. Such reels, however, can be as large as 1.5 metres in diameter and are extremely heavy and some form of mechanical lifting mechanism will normally be necessary to facilitate reel loading onto the machine.

12.4.3 Tamper-evident devices

Owing to what are, for the most part, negative reasons, the term 'tamper-evident' has become particularly well known and has assumed increasing importance during the 1980s. It is a sad fact of life that a very small number of cranks and extremist pressure groups have been able to cause such widespread concern (particularly in the food and drinks industries) by means of their unscrupulous actions in 'tampering' with products in supermarkets in order to exact revenge or to register, in the most dangerous of ways, some form of protest.

In addition, it appears that a significant sector of the consumer market is not averse to opening products on display in order to evaluate them by means of smell, or even taste (a practice commonly known in the USA as 'grazing'). Clearly, as a result of these unfortunate developments, manufacturers of consumable products have become anxious to ensure that their products are 'tamper-evident' when displayed at the point-of-sale. In some branches of the drinks industry (notably spirits, liqueurs, etc.), ROPP (Roll-On, Pilfer-Proof) closures have long been used for security reasons and, more recently, these and similar closures have found increasing acceptance in the soft drinks industry. Metal ROPP closures are now quite commonplace, as indeed are their pre-moulded plastic counterparts which are extensively used for PET bottles.

However, for various reasons, such closures do not always provide the best solution to this problem and alternative means of tamper-evidence have to be found.

In the context of this chapter, therefore, it is important to note that several interesting possibilities exist within the 'container decoration' spectrum. Various types of machinery are available for handling special labels, strips, foils, neck bands, etc., all of which can act as security devices. Where such devices are used they can – indeed should – be carefully designed to complement and enhance the basic container dressing (i.e. body, shoulder labels, etc.) so that they not only act as a security measure but also assist in the overall visual appeal of the container at the point-of-sale. Examples of such tamper-evident devices are:

Neck-strip labels. These can be applied in either a 'U' or an 'L' format over the bottle neck/closure area. U-shaped strips have long been used in the spirits industry to denote payment of Excise Duty although they do, at the same time, provide a highly effective tamper-evident device. The strips can be of a simple rectangular shape or elaborately profiled, within certain limits. The strips can be applied either in-line with, or at 90° to, the centre-line of the main body label. L-shaped seals are normally applied to the front of the neck area (i.e. aligned centrally with the body label) and the short 'leg' is pressed down on to the top of the closure. While both types of strip can be applied using only cold glue, for most applications it is strongly recommended to incorporate hot-melt adhesive locally, in the areas where the strips contact the plastic or metal closures, to ensure maximum security.

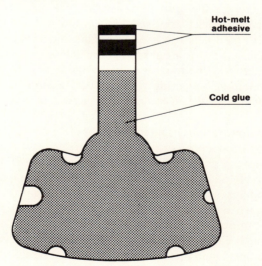

Figure 12.9 Combined shoulder/tamper-evident label. The glue pattern involves the use of cold glue for areas of the label contacting the bottle and hot-melt for the contact area with the closure.

Specially designed neck or shoulder labels. Such labels, which are normally applied in close proximity to the closure, can be re-designed to incorporate a strip or 'tab' projecting from the upper edge of the label which is then adhered to the side wall (or side wall and top surface) of the closure (again, often using hot-melt adhesive for security), which will readily tear if there is any attempt to unscrew the closure. Such labels are becoming increasingly common throughout the European Mineral Water Industry.

Deep-cone neck-around labels. Subject to the suitability of the geometry of the bottle neck/closure area, it is possible to apply a deep-cone neck-around label (i.e. full wrap-around with overlap) which encompasses both the upper neck area (which must be cylindrical or conical) as well as the side-wall of the closure. Often, such labels are perforated at a point just below the lower edge of the closure to facilitate removal. It should also be noted that aluminium neck foils (but *not* capsules), as described in the next section, can also serve as an effective method of tamper-evidence.

Shrinkable neck bands. An interesting alternative to the application of glued labels is the possibility of applying a shrinkable plastic (usually PVC) band applied to the closure area. By the use of carefully designed graphics, such neck bands can be a very attractive feature of the overall presentation. While various types of machines are available for applying such bands, a relatively recent innovation is a system for the application of these bands within the configuration of a labelling machine. A special dispensing unit (which cuts the bands to the required length from a reel-fed system) is located above the labelling machine infeed and the incoming containers pick up the cut bands in similar fashion to bottles entering a capping machine. The bands are subsequently positioned accurately (i.e. vertically) by means of specially centring bells on the labeller carousel and are then shrunk tightly to the neck area by means of a shrink unit located in the labelling machine or on the downstream conveyor, dependent upon application. Such neck bands can incorporate horizontal or vertical perforations to facilitate removal.

12.4.4 Metallic neck decoration

For many years, some products within the drinks industry (notably premium beers, champagne, sparkling wines, etc.) have been decorated with a form of metallic neck finish and this has proved to be an extremely important marketing ploy. Without doubt, the 'lustre' of a metallic finish conveys a deluxe, high-quality image and this can be an important factor in achieving effective sales of certain products.

More often than not, this metallic finish is achieved by the application of aluminium neck foils on a specially equipped labelling machine. While once, perhaps, the exclusive preserve of the brewing and champagne industries, neck foiling is becoming of significant interest to the soft drinks industry for certain premium products.

The traditional method of achieving a metallic neck finish on a bottle has been by application of pre-formed capsules made of lead or aluminium. Such capsules are fed to the bottle necks from a mechanical dispenser and are securely attached to the bottles by means of a special 'crimping' operation (i.e. no adhesive is used). The use of pre-formed capsules has declined in recent years, a major reason being the advent of labelling machines capable of applying glued aluminium neck foils concurrently with the bottle labels, which give a similar finished effect.

By comparison with foils, capsules are extremely expensive and the application machinery is limited to relatively low speeds. Capsuling requires a separate special-purpose machine, which takes up valuable floor space and necessitates an additional operator.

The applications of neck foils on labelling machines was first introduced in the early 1970s, since when, as a result of constant development and evolution, the application of foils can be considered almost as simple as the addition of an extra label. Metallic neck foils (manufactured from aluminium with a gauge of 12–15 μm) can be either pre-cut or reel-fed. Pre-cut foils are more common, the prime reason being that of flexibility. Pre-cut foils can be profiled to give a horizontal lower edge (i.e. to simulate a capsule) or to give a central point or radius that can be centralised with the body labels. Conversely, because of the rotary knife cut-off system used, reel-fed foils (although offering the cheapest solution) can only be used for straight, capsule-type finishes. In both cases, the foiling application can be integrated within a rotary-type labelling machine, subject to the additional foil application equipment being supplied.

Basically, foils are handled in the same way as conventional labels, at least as far as initial gluing, removal from the magazine and transfer to the containers are concerned. The one main difference, however, is in the final application. As a result of the manufacturing process, foils are extremely 'malleable' and in view of the total absence of a grain direction, they can readily be made to conform to the compound-curved areas of a bottle neck. For optimum visual appeal, the foils must, of course, be applied very neatly and tightly to the neck areas and this is achieved by means of a 'foil finishing unit' located in the machine discharge starwheel. A special design allows bottles to be rotated at speed during their transfer through the starwheel, during which period the foils are brushed tightly to the bottle-neck contours by means of a double-rotating brush system. In the vast majority of cases, the neck foils are designed to encapsulate the container closure. In their natural state, aluminium foils are bright silver in colour but often these are lacquered to give a gold or coloured finish. An important advantage is, of course, the ability of foils to receive graphics and this often takes the format of a central motif, crest, logo, etc.

12.4.5 *Self-adhesive labelling*

Without doubt, the self-adhesive label has an important role to play within the very broad labelling spectrum. Like all other systems, however, there are advantages and disadvantages and for the soft drinks industry the latter are such that they dramatically limit the appeal of self-adhesive labels. The main disadvantages are:

Cost. Although it is dangerous to generalise, self-adhesive labels of the sizes commonly used for soft drinks containers are significantly more expensive than conventional paper labels.

Operating speed. Self-adhesive labels must be applied from the reel, and the application speed (i.e. in terms of cpm) is a function of the maximum possible web speed and the label size. In many cases, the line production speeds preclude the use of self-adhesive labels.

Container surface condition. To ensure efficient label transfer and satisfactory adhesion, the container must be completely dry without the risk of formation of even the slightest condensation, and these conditions are very difficult – if not impossible – to achieve on a filling line.

While these three main disadvantages mean that self-adhesive systems are not used for 'mainstream' labelling of soft drinks containers, they can play an important role in certain specialised applications. For example, one advantage of self-adhesive labels over wet-glue labels is that better handling can sometimes be achieved for labels of very small dimensions. Accordingly, very small round or oval 'medallion' labels are often best applied by self-adhesive means. With the influence of the supermarket business, self-adhesive labels can play an important role in the application of price stickers, special promotion stickers, bar code labels, and labels that convey information to the consumer.

In recent years, 'combination' labelling machines (i.e. for applying both wet-glue and self-adhesive labels) have become quite common since this type of configuration increases the flexibility of the machine, particularly if there is a need to apply an extra or special label at very short notice. In this respect, self-adhesive labels have an additional advantage since, being handled in web format, using the 'direct transfer' concept, no handling parts are necessary. Conversely, wet-glue labels require individual handling parts that must be designed and manufactured, hence there is usually a lead-time of several weeks.

In summary, although self-adhesive labelling systems have a part to play in the soft drinks industry, in view of their limitations and lack of flexibility, it is most likely that they will only be used for very special needs and be complementary to – rather than be an alternative to – other labelling systems.

12.4.6 *Container pre-labelling*

Although it accounts for an extremely small percentage of the total production, the possibility of purchasing pre-labelled or pre-decorated bottles (i.e. from the bottle manufacturer) should be mentioned. In most cases, containers are pre-decorated by means of a plastic shrink sleeve or stretch sleeve.

It should, however, be noted that there are several major disadvantages with this concept, all of which have significant cost implications. Apart from the fact that the bottler is paying for an operation that he can normally undertake more economically himself, the shrink sleeves and stretch sleeves used for this decorating concept are more expensive than conventional labels or, indeed, plastic labels. For sleeving applications, the gauge of the plastic film has to be of minimum thickness to withstand the stresses induced during the application processes and this, obviously, is a direct cost factor.

In the case of shrink sleeving, if the label is to extend to the curved areas above the cylindrical body section (which is normally the case) high-duty shrink tunnels are required if a completely smooth, crease-free shrink is to be achieved and this results in high energy costs. Apart from the cost aspects, however, the use of pre-labelled containers can cause production problems. They also allow very little flexibility in the handling of important and unexpected new business since it takes much longer to obtain pre-decorated bottles for such a requirement than to have additional labels supplied. Any changes to the labels resulting from new formulations, special offers, price changes, etc., can prove prohibitively expensive if containers are pre-decorated.

12.4.7 Can labelling

While canning is becoming more important to the soft drinks industries in many countries (and has long accounted for a significant proportion of total production in the USA and the UK), there has, to date, been little or no requirement for can labelling, as is the case in the food industry. However, owing to changing circumstances and the evolution of new label materials and application systems, can labelling is likely to become attractive to soft drinks producers, at least for low-volume brand production.

It is unlikely that labelling will replace the printed can, especially in the case of very high volume products, but there are distinct cost and operational advantages on those canning lines which must produce a large range of products and flavours in relatively small batches. With the economics of can manufacture/decoration being very much geared to volume, many can makers impose minimum order quantities that are difficult to justify financially for a flexible canning line. In these circumstances, a soft drinks canner is faced with the major problem of stocking extremely large volumes of cans over prolonged periods with the obvious financial strictures that this imposes, and there is also the major inventory control problem and the lack of flexibility resulting from necessary changes to the decoration graphics. Under these conditions, the labelling of beverage cans is certainly well worth serious consideration since this operation can be achieved in a highly efficient manner either by the use of pre-cut or reel-fed labels in paper or plastic.

12.4.8 Container orientation

Although not a traditional requirement in the soft drinks industry, there has been an increasing need in recent years (as a result of the product and package diversification mentioned in the introduction to this section) for containers to be automatically orientated prior to labelling/decorating to ensure that all the labels are accurately positioned with regard to some design feature of the container (i.e. moulded crest, logo, etc., in the glass; special closure – e.g. swing-stopper; pre-applied capsule, etc.). Space limitations make it impossible to give a detailed description of all the possibilities but in view of the increasing importance of this special aspect of package decoration to the soft drinks industry, it is well worth recording that a wide range of automatic container orientation systems (both mechanical and electronic) is available for use on many types of modern labelling machine. It is recommended that the labelling machine manufacturer is fully consulted with regard to the capabilities of any given model prior to introducing any type of container or decoration that requires some form of orientation.

12.5 Factors influencing the efficiency and quality of container decoration

Before making a large capital investment in purchasing new machinery for package decoration, most companies will carefully assess all the available equipment prior to determining the one that is best suited to their needs. A number of factors will influence the final decision, including technical specification, flexibility, size/layout, operational efficiency/economy and capital cost. However, it must be emphasised that selecting the best available machine for any specific task – even if this also means, in turn, paying the highest price – is, in isolation, no guarantee of operational success when the machine is installed in the production line. It is important to appreciate that the 'application machine' is only one element (albeit a very important one) in the equation and it is necessary to consider the correlation between the machine and a number of key factors in the container decoration operation.

Many factors combine to influence the overall efficiency and quality of package decoration, the main ones being summarised as follows:

- Specification of packaging materials (i.e. containers and labels/foils, etc.)
- Adhesive specification
- Plant layout
- Machine condition.

Without doubt, the entire space allocated for this chapter could be devoted to one of several items in this section, notably 'Adhesives' and 'Specification of label papers', such is their importance to the subject. However, space constraints obviously determine that only a brief summary of the most important aspects can be included.

It should, however, be noted that while most machinery manufacturers have a good working knowledge of adhesives and papers, etc. (and will be most willing to give basic guidance and recommendations), they are not normally in a position to offer the same level of professional expertise as the manufacturers of these materials and the suppliers should certainly be consulted fully before embarking upon the design/introduction of a new package to the marketplace. Virtually all the leading suppliers of adhesives and labels/foils, etc., issue detailed specifications for their products, which can be referred to the machinery manufacturer for endorsement. Additionally, they also often make recommendations for correct usage of their products, and supply documentation incorporating 'trouble-shooting' guidance.

12.5.1 *Specification of packaging materials*

Containers. When considering the containers to be labelled in a soft drinks filling line, three main factors are important:

(a) *Material.* In the vast majority of cases the containers to be handled will be manufactured from glass or plastic although, as mentioned earlier, the labelling of metal cans may be of increasing importance to soft drinks producers in the future. In the case of glass bottles, it is important to differentiate between returnable and non-returnable containers, especially if hot-melt adhesives are to be considered in view of the difficulty in removing these in the bottle washer (see earlier comments). The exact nature of any type of surface treatment of glass bottles should also be established with the bottle suppliers since this can directly affect the choice of adhesive.

With regard to plastics, PET is relatively simple to label using synthetic resin adhesives (e.g. PVA glues) or various types of hot-melts, dependent upon application.

(b) *Condition at point of labelling.* A careful assessment should be made of the prevailing conditions at the label application point, i.e. will the container be cold or hot, wet or dry? These factors are mainly related to the choice of adhesive, but they can also be relevant to the specification of the label paper.

(c) *Container geometry.* The size and shape of the containers to be labelled must be carefully assessed. As a general rule, paper labels should only be applied to cylindrical or conical sections of the containers in order to eliminate the risk of creasing. Application of labels to compound curved areas should be totally avoided. Label size is important, and the labels should not be too big for the application area (a common error). Although there are exceptions, as a

CONTAINER DECORATION 271

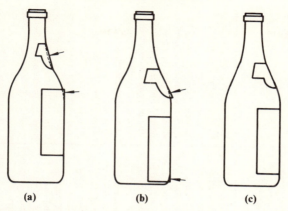

Figure 12.10 Importance of label position relative to bottle contour: (a) unsatisfactory; (b) unsatisfactory; (c) ideal.

general rule, body, back and shoulder labels should not exceed 180° of the container circumference when applied by an indirect-transfer type machine (i.e. patch labels). Aluminium neck foils are, of course, applied to the compound curved areas of the container necks, but this is only possibly because of the special 'worm-grain' finish of the foils which totally eliminates any graining and gives an extremely high degree of 'malleability'.

Plastic labels, as with paper labels, should generally be applied only to

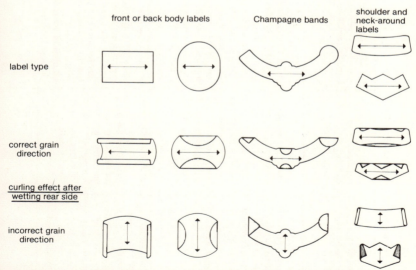

Figure 12.11 Correct grain directions for various types of label. The grain direction may be quickly and positively tested by wetting the reverse side of the label and observing the direction of curl; the axis of curl is parallel to the grain direction.

272   CARBONATED SOFT DRINKS

Table 12.1 Properties of a good label paper

Label material type and production method 1	Properties required by label manufacturers 2	Properties required for label handling system			Comments 6
		Labelling machine 3	Package conveying/storage 4	Label removal in bottle washer 5	
A — basic labels from unrefined (uncoated) papers (natural papers with and without wood pulp for normal requirements.)	good offset printability: minimum problems due to punching/cutting burrs, adequate tensile strength: lies in a flat condition: adequate wet strength when used for returnable bottles	Lies in flat condition when stored correctly: burr-free when punched: adequate tensile strength: conforms well to bottle contours due to low stiffness: adequate moisture absorption capacity: can be accurately cut/punched to size.	Label paper must have adequate wet strength and adequate wet opacity commensurate with the handling stresses and operating speed of the filling line. The label paper and its graphics should be adequately smear and scuff resistant. Colours and graphics should not fade nor be discoloured due to excessive penetration of glue through the paper.	Label papers used on returnable bottles should be suitable caustic-resistant to avoid 'label pulping' in the bottle washer (this inhibits the washing efficiency and drastically shortens the useful life of the detergent solution.) Conversely, the paper should have adequate caustic-permeability to facilitate efficient and prompt label removal from the bottles.	Varnishes/lacquers should not be too thickly applied, otherwise labels are rendered too stiff. Gold 'bronzing' by means of linseed oil/varnish should not penetrate too deeply into the paper structure otherwise moisture absorption capacity is reduced to an unacceptable level. This results in prolonged setting-times for the adhesive which can cause "skidding" and lifting of the label edges. Metallic finishes obtained using gold and silver gravure inks are preferred to other more problematical methods.
B — refined label papers (one side calendered or machine-coated: papers with or without wood pulp, also separately coated chromo-paper)	as A, but also suitable for gravure printing if nature of surface is appropriate: good–very good print reproduction: high wet strength.	Labels must not show excessive curling tendency (which impedes the labelling function) following the absorption of the required amount of water by the rear surface.			

C	high gloss, cast-coated label papers	as A and B: brilliant print/colour reproduction: usually wood pulp-free and with high wet strength.		
D	metallised paper labels	as A and B.		Bonding agent between paper and aluminium coating must have adequate caustic permeability.	Normally no problems, even with returnable packages.
E	aluminium foil/paper laminated labels	as A and B	as A and B	For returnable bottles, only paraffin-wax should be used as a bonding agent between the paper and aluminium foil.	Potential problems with returnable bottles. Washer must be of correct size/type and high detergent strength normally needed. Possible effluent problems.

cylindrical areas. However, some plastic films used for labels are heat-shrinkable and this gives some flexibility for extending the labels slightly above and below the cylindrical body section.

Decoration material. Three main materials are used for the dressing/decoration of containers: paper, metal (i.e. aluminium foil) and plastic (in various grades and specifications). In some cases, the decorating medium can be a combination of these materials.

Since the vast majority of soft drinks labelling applications involve labels made from paper, and since, by its very nature, paper is more variable in its specification and behaviour, the main comments here are related to conventional paper labels.

What properties should a label paper possess? To answer this question, it is necessary to consider all the important individual criteria and to summarise them in order to assess the various advantages and disadvantages of a given paper type, and to reach a decision with regard to its suitability for a specific requirement. The following are the most important properties to be considered:

- tensile strength
- wet-tear strength
- flexural stiffness
- textures of front and rear surfaces
- moisture absorption capacity
- permeability
- wet opacity
- grain direction
- density (weight/thickness ratio)
- 'printability' (suitability for accepting inks, lacquers, varnishes).

Table 12.1 underlines the importance of choosing a label paper with optimum physical characteristics to suit individual requirements.

(a) *Paper label testing.* Many test procedures for various standards have been established in order to determine the different physical properties of paper labels. While many of these tests require relatively sophisticated equipment (which label manufacturers and labelling machine manufacturers should both possess), a number of the tests can be carried out by the bottler using very simple pieces of apparatus. Full details of these tests should be obtainable via the label supplier.

(b) *Label storage.* Label storage is of paramount importance and can have a very dramatic effect upon operational performance. Money spent on a suitable, humidity-controlled label storage room will repay itself many times. Storage conditions for labels should be such that the labels remain flat at all times. Curled or very dry labels should be stored in conditions that allow them to regain their original moisture content and thus assume their original flat condition. It is essential that labels do not lose their moisture content during storage, since dried-out labels become very stiff, have reduced tensile strength and have a strong tendency to wrinkle when applied to the containers.

Paper labels should be stored at a temperature of 18–22 °C and in a relative humidity of 60–70%. Storage conditions for foils can vary, but the essential requirement is that the foils should remain flat during storage in a minimum relative humidity of 55%. Labels and foils that are subsequently transferred from such storage conditions to the humid atmosphere of a bottling hall will not create any undue problems. The only exceptions to this rule apply in countries with a very dry average annual climate.

Label stocks should be rotated, on a 'first-in, first-out' basis.

12.5.2 *Adhesives*

In many respects, it can be argued that the adhesive is *the* most important element in the vast majority of container dressing applications that are relevant to the soft drinks industry.

Since it has been previously highlighted that most modern labelling machines (whether of the direct- or indirect-transfer types) utilise a 'glue-picking' concept (i.e. removal of individual pre-cut labels from a magazine), it follows that the function of the adhesive is not simply to stick labels to containers, but is a vital element in the sequence of mechanical operations that transfer labels from the magazines onto the containers. *Accordingly, the importance of correct adhesive selection cannot be over-emphasised.* As much care should be taken in choosing the optimum grade of adhesive for any specific application as is spent in assessing the type of application equipment.

All bottlers are strongly urged to consult fully with their adhesive supplier and the manufacturer of the application machinery in order to determine the type and grade that will give the best performance, which should include efficiency (i.e. 'machinability' on the equipment to be used), adhesive properties for the specific application, consistency of quality, 'yield' (number of labels applied per unit weight of adhesive) and cost. *It is important that, in no circumstances, should adhesives be chosen on the basis of cost alone. Unfortunately, this is often the case in practice.*

The factors that determine the choice of adhesive grade to be used are summarised as follows:

- type of application machine
- production line speed
- container material
- nature of container surface (e.g. treated/untreated glass, etc.)
- label material (paper/plastic/metal foil)
- label material specification
- container condition at labelling point (e.g. wet, dry, cold, hot, etc.)

Table 12.2 Comparative properties of labelling adhesives

	Dextrins	Vinyl dextrins	Starch	Non-casein	Caseins	Dispersion/emulsions/latexes	Hot-melts
Outstanding features	Good initial tack; fast setting; limited adhesion range	Very fast setting; fluid consistency.	High initial tack; medium to slow setting speed; cohesive.	High initial tack; medium setting speed	High initial tack; medium setting speed with fast setting on cold bottles.	Fast setting; low initial tack	100% solids; needs heated applicators; very fast setting; excellent adhesion
General machine performance	Good, but low cohesive strength can result in splashing and throwing on high speed equipment	Good, but needs special attention to run low viscosity.	Excellent	Good–excellent	Excellent	Fair	Excellent
Adhesion to coated glass and flame-treated polyethylene	Poor	Better than dextrin but should be tested	Good–excellent	Good–excellent	Good–excellent	Good–excellent	Excellent, but should be tested under end use conditions
Adhesion to plastics	None	None	None	None	None	Excellent, some products available for un-flamed polyethylene and polypropylene	Excellent
Water resistance (for ice-proof labelling)	None; easy clean-up and label removal	None; easy clean-up	Fair (semi-ice-proof)	Good (generally fully ice-proof)	Fair–good (can be fully or semi-ice-proof).	Fair–good; can be more difficult to clean up	Non-water soluble
Major usage	Older, slower speed machines; not suitable for coated glass	Low viscosity a limitation; better on coated glass than dextrin	All types of glass bottle labelling; most suited to high-speed labelling equipment; also widely used for adhesion to flame-treated polyethylene; non-casein types are not generally suitable for returnable bottles	As starch	As starch	Generally restricted to use on plastics but can help solve problems with 'springy' labels	Used for all types of can, bottle and jar labelling; extensively used for wrap-around work on all containers

* All comments refer to general performance. Specific tests should be conducted to determine actual results on the adhesive to be used. (Reproduced by kind permission of National

CONTAINER DECORATION

- returnable or non-returnable containers
- ambient conditions in the packaging hall
- production line layout
- subsequent warehousing/product storage conditions
- anticipated shelf-life of product

The main types of adhesive in commercial use in the bottling industry today are: casein, starch jellygums, dextrin, synthetic resin emulsions, latex, hot-melts and vinyl dextrin. Table 12.2 lists the comparative properties of these adhesives with respect to container labelling, while Figure 12.12 gives an indication of the suitability of various adhesives for handling different package materials/surface conditions.

12.5.3 Plant layout

After the best machine has been selected for any specific duty, it is vital that it should be installed in a plant layout that is totally compatible with the machine operating concept in order to achieve the best production efficiency from the unit. Despite the tremendous improvement in the quality of packaging materials over the years (i.e. labels, adhesives, etc.) the 'golden rule' that labelling machines should run as continuously as possible (with a

Package material/surface condition	Adhesive types						
	Dextrine	Modified Dextrine	Casein	Starch	Modified Casein	Synthetic	Hot-Melt
Glass – without "cold-end" treatment	◐	●	●	●	●	●	●
Glass – with normal "cold-end" treatment	○	◐	●	●	●	●	◐
Glass – with extra "cold-end" treatment	○	○	○	○	◐	◐	◐
Cans	○	○	○	○	○	◐	●
Polyethylene (flame-treated)	○	◐	●	●	●	●	●
Polyethylene (non flame-treated)	○	○	○	○	○	◐	◐
PVC	○	○	○	○	○	◐	◐
PET	○	○	○	○	○	◐	◐
Polypropylene	○	○	○	○	○	◐	◐
Polystyrene	○	○	○	○	○	◐	◐

KEY

● = very good   ◐ = good   ◐ = relatively good   ○ = unsuitable

Figure 12.12   Adhesive suitability chart for various container materials.

minimum of stoppages) still applies. Careful design of the production line layout can contribute significantly towards this end.

Most modern labelling machines are equipped with some form of 'speed modulation' which automatically adjusts the machine's speed dependent upon the prevailing production line conditions (i.e. the volume of containers on the line before and after the labelling machine). If the machine is to be 'free-standing' (i.e. not forming part of a monobloc unit), adequate accumulation conveyors must be incorporated both upstream and downstream to ensure a smooth, efficient and continuous flow of containers under all conditions, and these conveyor areas should also provide location points for the line switches/sensors that control the machine's speed modulation.

If container flow at the labeller infeed ceases for any reason (e.g. a filler stoppage), the layout and machine control should permit 'controlled idling' conditions. In other words, the machine should be allowed to continue to operate at a low speed (without containers passing through and without label application) in order to keep the adhesive under optimum control and to ensure that containers can immediately be fed to the machine infeed (i.e. on resumption of flow from the filler) without any problems. At all points in the line, containers should be handled smoothly, gently and with minimum pressure but this is particularly important in the area between the labeller discharge and the infeed of the packaging machinery where everything possible must be done to avoid damage to the newly applied labels.

Direct contact with the newly applied labels should be eliminated, and this means that all conveyor guide rails should be carefully positioned above and below the labelled areas of the containers. Similarly, the infeed systems of the packaging machinery used should be designed to avoid label contact.

The layout should allow adequate 'drying time' in the area between labeller discharge and packer infeed for all 'wet glue' labelling applications. The drying time will, of course, be determined by the length of accumulation conveyors although, conversely, it follows that the longer these conveyors, the more risk there is of label damage owing to container-to-container contact after labelling. The amount of drying time can vary enormously dependent upon many factors including output, label specification, container material, adhesive used and ambient conditions.

12.5.4 *Machine condition*

A most appropriate alternative title for this paragraph would be 'The human factor' since, without doubt, much depends upon the enthusiasm and diligence of the operating and maintenance personnel responsible for handling the application machinery. No matter how much time, effort and money is spent on selecting the machinery and defining the specifications of the materials to be used, the end result can very easily fall well short of expectations if the machinery is not adequately and correctly operated and maintained. Great

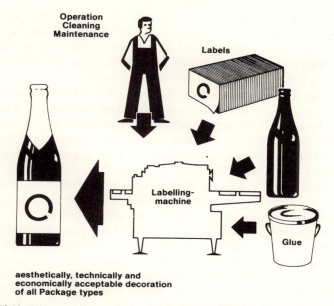

Figure 12.13 The co-ordination of key elements in the container decoration operation.

care must be devoted to cleanliness, correct settings, and regular and efficient maintenance. All reputable manufacturers of capital plant provide appropriate documentation to assist these activities, and many offer comprehensive training 'packages'. Training is often a neglected area but, without doubt, money spent in this direction is very quickly recouped via higher levels of operational efficiency.

Throughout this chapter an attempt has been made to emphasise the interdependence of machine and materials and to underline the importance of the correct correlation between the various elements in the container decoration process. These points are graphically illustrated in Figure 12.13.

12.6 Future developments

Throughout the 1970s and early 1980s the emphasis when planning new bottling lines for the beverage industry was often placed upon quantity – i.e. there was a prolonged trend towards faster production lines, but this was often at the expense of operational efficiency and package quality. In recent years the emphasis, without doubt, has shifted from quantity to quality and nominated line speeds have been more moderate with greater stress placed upon higher line efficiencies, quality of product and product presentation.

As the markets for soft drinks become fiercer and more sophisticated, it appears certain that the 'container decoration' requirements will become more important and more demanding. The trend towards the use of plastic film labels (the advantages of which are outlined in this chapter) is gaining pace very rapidly and before long the use of such labels will almost certainly be commonplace throughout the beverage industry.

As bottling lines generally will need to be more flexible to cater for a wider range of products, container types and end-of-line packaging, it is more than likely that the diversity of container dressing requirements will be such that the installation of more than one labelling machine or 'decorating' machine will be necessary. This factor will, of course, have a significant impact upon bottling line design.

A key issue that inevitably will have to be addressed in the future is the likelihood of increasing pressures being brought to bear in some market sectors to revert from non-returnable to returnable containers as a result of the environmental lobby. Should such a trend come to pass, it is unlikely to be accompanied by any reversal of the move towards more diverse and sophisticated container dressings and this will mean certain challenges for the industry. For example, some container dressings can only be satisfactorily achieved using hot-melt adhesives which, hitherto, have basically been incompatible with returnable bottles. Much research is currently being carried out with a view to producing commercially viable water-soluble hot-melts which can be used for returnable containers. Similarly, ways and means of dealing with plastic film labels applied to returnable containers will have to be found.

13 Container-inspection equipment
P.W. BINNS

13.1 Introduction

In today's market place, the consumer demands product liability, and the governments of the world are producing more and more legislation to protect the consumer. On-line container-inspection equipment is becoming the norm on packaging lines and in the future will be an essential part of every production line. A summary of inspection equipment past, present and future is provided in this chapter. The two main areas of inspection equipment – i.e. empty-container and fill-level detection with filler management – will be covered in greater detail.

Before 1960, there was very little inspection machinery available for the packaging industry, and so the only inspection that took place was based on visual examination. In those days this was sufficient for the relatively slow production speeds in use. The inspecting operator would be looking for 'foreign' bottles, damaged bottles (in particular, chipped sealing surfaces) and bottles containing any solids or liquids that should have been removed by the bottle-washing process. In the early 1960s, with the advent of automation, packaging line speeds increased; this was achieved through bottle standardisation and the introduction of the automatic packer and unpacker, palletiser and depalletiser, pre-mix filling units and rotary labelling machines. The introduction of the can in the beverage industry produced further justification at this time for automatic inspection equipment. The first two items of equipment to be introduced were the gamma-level detection system and the empty-bottle inspection system. Both systems enabled internal examinations to be carried out, the first on closed cans where the fill level was invisible and the second on washed, empty bottles to check for dirt or foreign object residue.

These items of equipment were first introduced in the USA and later brought to Europe. The introduction of these inspection machines brought controversy to the packaging halls and therefore studies were undertaken in the USA in the middle 1960s to compare the use of these first simple inspection units against the human visual inspection. The results were basically similar for a period of approximately 20 minutes, after which time human inspection deteriorated quite considerably. It was found at speeds of 300 bottles per minute (bpm) that after 20 minutes the human inspector started to miss obvious rejects, and to carry out visual inspection at these speeds it was therefore necessary to rotate the human inspector every 20 minutes; it was also found that these human inspectors were not totally capable of doing the job again for a period of three hours.

As line speeds increased above 300 bpm the conveyors had to be split so that more and more visual inspection could be added. The cost of the extra conveyors and sighting staff boosted the introduction and development of automatic inspection machines, although they were unsophisticated and unreliable by present-day standards.

Through the 1970s line speeds increased up to 800 bpm and 1250 cpm (cans per minute); these speeds, together with the introduction of more stringent legislation requiring minimum contents and customer protection, brought forward the necessity of faster and more efficient empty-bottle inspectors and fill-level detectors to ensure the product was being filled into a clean container and the correct amount of product declared on the label was present.

At the end of the 1970s and in the early 1980s micro-technology was introduced into packaging lines and this created potential for better line control thus allowing higher speeds and operating efficiencies to be achieved. Owing to these higher speeds, and to the introduction of new regulations in the USA and the EEC, product security became a most important part of the production line. No longer must the customer, in the case of an accident, prove that something was wrong with the product – the supplier has to prove that he had taken all reasonable care and attention to ensure the product was safe. Faced with the possibility of costly legal action, the security features of quality-control equipment became increasingly important. Empty-bottle inspection, residual-liquid detection, fill-height inspection (especially the control of overfill), monitoring burst bottles and checking of seams on cans were all identified as critical areas warranting increased quality control.

The foregoing covers the history of container inspection to the present time and the reasons for its growth. From this point onwards a modern production line will be examined with respect to inspection equipment which could and should be incorporated.

Figure 13.1 shows a typical bottle-filling line with eight positions where containers are inspected; each item of inspection machinery will be explained individually and, where applicable, the appropriate equipment relating to a canning operation will be included.

13.2 Empty-container inspection

13.2.1 *Pallet inspection*

After the depalletiser (position 1 in Figure 13.1) and before the palletiser it is important to inspect the empty pallet to ensure it has no broken slats, is

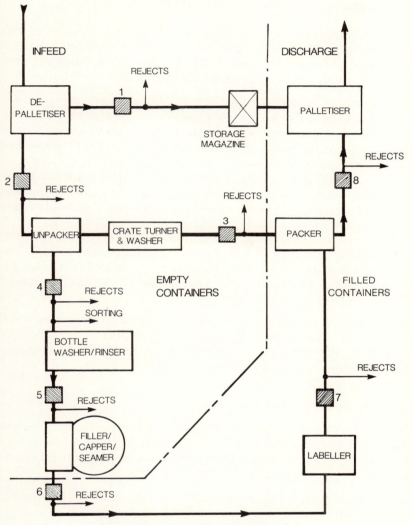

Figure 13.1 Container inspection positions.

dimensionally correct and within specification. This is normally carried out by the use of photo cells to check individual slats and overall dimensions. If the pallet meets the required standard it is counted and allowed to carry onto the palletiser ready for the finished product to be palletised. If not, the pallet is removed for refurbishment or scrap, thus ensuring the palletiser is always ready to run at maximum efficiency.

13.2.2 *Bottles in crates*

This is the inspection of the crate to ensure that the crate being presented to the unpacker has the full complement of bottles of the correct type and specification (position 2 in Figure 13.1). The types of inspection equipment that can be incorporated at this point are as follows:

(a) Sorting for different coloured bottles using a colour camera: e.g. a 24-bottle crate of flint bottles could also contain some green bottles of the same size; if the number of green bottles is above a predetermined percentage, this crate could be rejected before reaching the unpacker by using a special majority logic programme.

(b) Detection of different height, upside down or lying bottles in the crate using ultrasonic measuring devices; these crates could be removed before reaching the unpacker as it is possible that they would create a stoppage at that unit.

(c) The presence of plastic closures or crown corks can be detected by using inductive proximity switches or special reflective light barriers. These crates can be removed before reaching the unpacker to stop damage to the unpacking heads. This requirement is normally specifically requested.

If any of the above faults arise, it is possible to remove the crate prior to the unpacker, or to marshal the faulty articles into lanes prior to the unpacker, for hand sorting and replacement on the line or for palletising and subsequent reprocessing through the depalletiser.

13.2.3 *Empty-crate inspection*

Normally after the unpacker crates are inverted to remove foreign objects and undersize or foreign bottles that have been left in the crates before the crates are washed (position 3 in Figure 13.1). After the washer and before the packer, the crates can be inspected for the following:

(a) Check for foreign objects, broken bottles still in the crates and remove, if required.

(b) Check the crates for colour and physical size to ensure the correct crate is repacked; these can be rejected if not to specification.

(c) Inspect for logo printed on the crate or private labels to ensure the correct crates are repacked; if the wrong logo or label is present, these crates can be removed.

13.2.4 *Bottle inspection – sorting*

Bottle sorters (position 4 in Figure 13.1) have become a more important part of the production line due to the introduction of a wider range of containers to

Figure 13.2 Inspection of empty bottles in crates. (Courtesy of Heuft Ltd, Atherstone, England.)

suit marketing needs. Dependent on the line layout and the number of containers to be handled, bottle sorters can range from a simple two-way sorter to a multi-way sorting system. The monitoring unit consists of a detection bridge which is normally a CCD (Charge Coupled Device) camera system which checks for height and shape of the bottle. The detector grades the bottle height information into different classes and the shapes of the bottles are then transformed into special shape vectors which identify one bottle type from another; the unit compares the actual shape vector with programmed vectors of the desired 'good' bottle and triggers the appropriate rejection device accordingly. The system can also detect different colours of bottle and by incorporating the camera system it can separate, for example, the difference between crown-cork and screw-thread finish bottles.

In other applications special cameras can also detect the presence of ACL labels so that ACL and non-ACL can be separated and with extra camera systems the differences between ACL labels can also be identified.

286 CARBONATED SOFT DRINKS

In automatic bottle sorting, bottle handling is especially critical; it is recommended that prior to a bottle sorter a pressureless combiner should be incorporated into the production line to assist in the removal of fallen bottles and other foreign objects, although the rejection equipment of a good bottle sorter can reject fallen bottles. The pressureless combiner can deliver the bottles in a 'touching' but pressure-free sequence, and this is an advantage for bottle sorting because the empty bottles are generally unstable and transporting them in a touching mode ensures a more stable stream of bottles; additionally, the speeds of the conveyors can be minimised, thus reducing bottle noise. Owing to the instability of empty containers it is imperative that the correct type of rejection system is incorporated to suit the containers to be handled; no matter how sophisticated the inspection equipment becomes, the empty-bottle sorter is only as efficient as the rejection system (see Section 13.5).

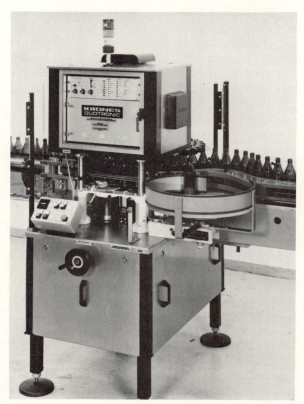

Figure 13.3 Krones 'Duotronic' empty bottle inspector. (Courtesy of Krones UK Ltd, Manchester, England.)

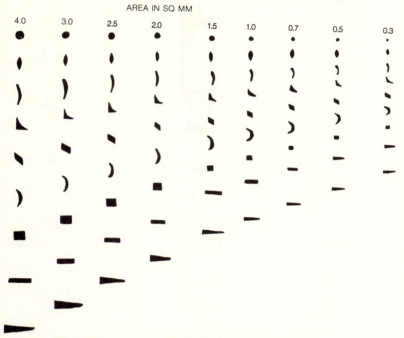

Figure 13.4 Comparison of differing shapes having same area.

13.2.5 Empty-bottle inspection

As this is one of the main areas of container inspection (position 5 in Figure 13.1), more detail will be given on the subject. There are three main inspection areas on a bottle:

(A) Bottom (base)
(B) Finish (chip neck)
(C) Side wall

Before describing each section, the most important item to take into consideration is the definition of a defect (Figure 13.4). This diagram shows defect size in square millimeters and the interpretation of shapes that can be put to a certain size in area.

The inspection principle of bottom and finish are basically the same; therefore, the principles described below cover both (A) and (B) (see Figure 13.5).

Type (i) – Single diode and scanner. One of the first systems to be used for bottle inspection; the basic principle of this system is shown in Figure 13.6. A strip mirror is rotated to scan the base of the illuminated bottle, giving different

Figure 13.5 Bottom inspection.

views of the base of the bottle per revolution; the signals are directed to a diode. If an object or defect is on the base of the bottle, the normal signal received by the diode is changed to a higher signal. This is used to activate a rejector to remove the container. The problem with this system is the use of only one diode, which means the same sensitivity is used for the whole base area thus making it impossible to pick out small objects because the normal base markings must be allowed for, i.e. mould marks, knurling, air bubbles, etc.

Type (ii) – Four diodes and scanner. This is another system that was used at about the same time as Type (i) for bottle inspection. The basic principle of the system is shown in Figure 13.7. In place of the rotating strip mirror a dove prism is rotated to scan the base of the illuminated bottle; working with the prism are four diodes. This principle gives approximately the same results as Type (i) with the exception of the ability to control the sensitivity of four areas of the base.

CONTAINER-INSPECTION EQUIPMENT 289

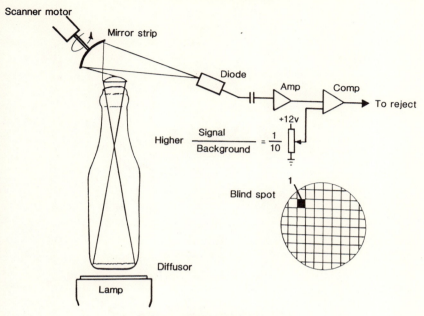

Figure 13.6 Single diode and scanner.

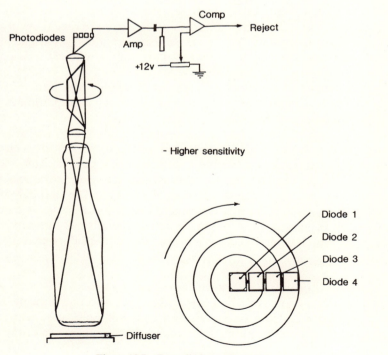

Figure 13.7 Four diodes and scanner.

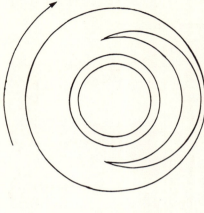

Blind spot in the middle
- beam splitter
- pre triggering

Figure 13.8 Disadvantages: rotary scan.

The above, while still used today, were the only systems available up to the end of the 1970s, and their disadvantages are illustrated in Figure 13.8. There is a blind spot in the middle; doughnut-shaped opaque rings and moon-shaped areas are very difficult to detect. To overcome these problems various refinements have been tried, i.e. beam splitters, pre-triggering and faster scanning; however, these problems are still an inherent fault on these types of inspection systems.

In the 1980s the laser scanner and the CCD camera came into commercial use, and these systems offered much greater accuracy in inspection than had previously been possible.

Type (iii) – Laser scanner. The basic principle of this system is shown in Figure 13.9. In this case the bottle is rotated, unlike Types (i) and (ii). While the bottle is rotating, the laser beam is directed onto a rotating hexagonal mirror which transmits the laser across the base of the bottle. The laser light beam is collected in a prism for comparison; and if a fault is seen in the base the laser light beam is interrupted. From the length and number of interruptions the size of the fault is determined; if it is of a size to be rejected a signal is given to the rejection system.

Type (iv) – Matrix camera system: Inspection principle. This offers a high-speed inspection system of a modular design. The inspection principle is based on special grey scale image-processing techniques. Solid-state matrix-array cameras are used as receptors. Each matrix-array camera used provides an

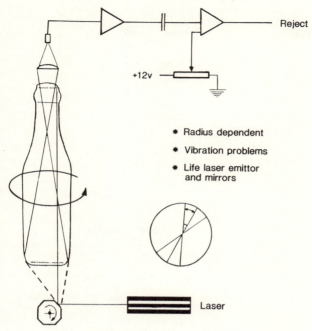

Figure 13.9 Laser scan.

effective grid of approximately 180 000 light sensors, called picture elements or pixels.

Vision principle. The intensity of light received by each pixel is first converted into a voltage level after which it is transformed by the image processor into a digital value, corresponding to that intensity of light, or shades of grey, at the point of image. This principle, in combination with a special algorithm for rapid information processing, enables the unit to detect at high speeds very small contrast differences (differences in shades of grey) between pixels. While the concept is simple, the technology is highly advanced; the principle of image processing can, to a certain extent, be compared to human vision, as illustrated in Figures 13.10 and 13.11.

Inspection features. The described principle of vision enables:

- high inspection speeds up to 100 000 bottles per hour
- high precision systems allowing inspection of very small defects down to 0.2 mm^2
- adjustable sensitivity and defect indication by modulated flashpoints on a monitor
- high reliability (due to micro-electronics) achieved by specially designed matrix-array cameras and the use of stroboscopic light to avoid any motion blur

292   CARBONATED SOFT DRINKS

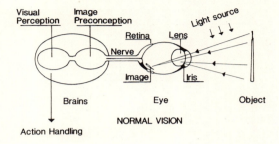

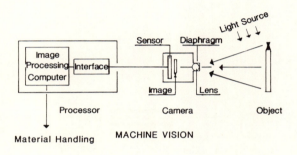

Figure 13.10 Comparison of matrix camera operation with normal vision.

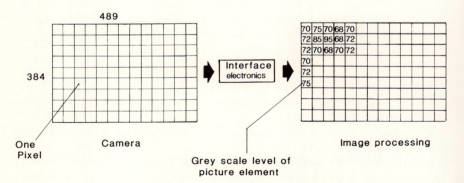

Figure 13.11 CCD Camera-image processor.

- no problems from colour variations owing to the grey shade analysis principle
- flashpoint quantity control by counting the number of flashpoints.

Most systems are broken down into a modular inspection as follows:

Module 1 – Base inspection
Module 2 – Residual liquid inspection
Module 3 – Closure finish inspection
Module 4 – Side wall inspection
Module 5 – Rejection.

Individual modules or any combination can be obtained, depending upon the requirements.

13.2.6 *Empty-can inspection*

The inspection of empty cans (position 5 in Figure 13.1) on the infeed to the filler is an area that is now being considered as very important for efficient filling. For inspecting cans in this area only the CCD camera system can be used. The same principle as for bottle inspection with CCD cameras is used, except the illumination is from the top. The main areas for inspection are the base of the can for foreign objects, the flange for splits and knock-down flanges, the side wall for dents and missing lacquer or lacquer blisters.

13.3 Filled-container inspection

13.3.1 *Bottle inspection*

Fill level control can consist of a unit just checking for under-filled bottles; these days it is important for this unit to be more than just a level-control unit but a unit monitoring the complete operation of the filling and closing equipment (position 6 in Figure 13.1). These units are normally complete systems comprising under-fill, over-fill, missing closure, filler and closer management system. For the measuring of bottles, the fill level can be controlled with an HF (high frequency) measuring bridge. This bridge can monitor high and low fill levels using one bridge. More recent versions of these bridges can give analogue read-outs of the actual fill height so that an average fill-height read-out of the total day's production can be obtained. Furthermore, it is also possible to use the actual fill-height analogue read-out of every bottle together with a valve identification to give statistics for each single valve performance, i.e. statistics incorporating average (QN) and standard deviation for every valve. The same principles apply to gamma inspection; in addition to the normal high- and low-fill detection bridges, quantitative analogue gamma bridges have been developed to give read-outs for fill-height control in exactly the same way as the high frequency. These gamma units, however, are mainly used on canning lines. In this application the use of a wider gamma beam with a larger detection area is used precisely to register the fill height of each single container. In general, it can be said that analogue read-outs are not as precise as the absolute high and low read-outs of normal gamma bridges, but using statistics on a total production day gives a very precise read-out for QN and also average fill height on each single valve. This same remark applies also to the high-frequency bridge. It is normally accepted that the standard accuracy of a gamma bridge is ± 1 mm. The value of QN on an analogue system would definitely be better than ± 0.1 mm.

Both high-frequency and gamma systems relate to the conveyor top as the reference point. This means they do not measure the air space (headspace), which is usually measured with a special scale since bottle tolerances are defined and measured from the top of the bottle.

The introduction of a new generation of systems for fill-height control using CCD cameras has now been developed. While not in common use at the moment, it can measure the actual headspace, is independent of the bottle and conveyor tolerances and can be incorporated into an analogue system as with the high-frequency gamma. This CCD system can only be used on clear bottles but gives a more accurate QN and standard deviation.

Filling machines and closing machines are always blocked together. The filling machine will have either a crowner, a capper or, in the case of cans, a seamer blocked with it. Therefore, with the level-detection equipment a closure-inspection device is always incorporated making the unit a level-and-closure detection system. For the detection of metal closures – i.e. crown corks, caps and can lids – a special inductive type sensor is used; for plastics and other materials a special infrared light cell is used. These sensors detect the presence of the closure; if the closure is missing, a signal is relayed to the rejection system which will remove the container from the production line.

13.3.2 Filled-can inspection

Apart from the use of gamma to inspect cans as mentioned in Section 13.3.1, a new type of inspection device for cans (position 6 in Figure 13.1) is now available. This device uses a sonic inspection system which is installed between the filler and the seamer. The principle of the sonic system is that a signal is directed into the can from a fixed point and when this signal contacts the liquid the signal is returned to the original point. The time taken for the signal to return is measured, thus giving the level of liquid. Using this principle the headspace in a can is measured. At the moment there are certain drawbacks with this system. The accuracy of the system is dependent on the surface of the liquid; the smoother the surface the more accurate the system; bubbles and foam will affect the accuracy. The system can only be installed between the filler and the seamer as it can only operate on cans without lids. However, where gamma-level inspection is not acceptable, this system is a very good substitute.

13.3.3 Labelled-bottle inspection

Correct bottle dress (position 7 in Figure 13.1) is a requirement that is becoming more important owing to legislation regarding the contents of the container. This is now obligatory in many countries and as the information relating to product constituents and filled quantity is usually printed on the label, it is therefore important that each bottle is labelled correctly. Two types of label inspection are used:

1. Simple 'presence-of-label' inspection using infrared cells stating whether the label is present on the container or not; this inspection can take place either within the labelling machine or on the conveyor after the labeller.
2. The use of CCD cameras has also been brought into this element of inspection as these cameras can check for label position. This means that the bottle dress can be checked for twisted labels or labels that are too low, too high or out of position; it is also possible to check for 'flagging' labels or labels that have dog-ears or have been badly scuffed. In certain circumstances, it is possible to check that the correct label has been applied to the bottle and has not been placed upside-down. These latter checks are very useful to monitor the performance of the labelling machine and are greatly appreciated from the marketing viewpoint.

Normally, with all types of level detector a closure detection system is incorporated to ensure that the correctly filled container has a closure. For the inspection of metal closures, inductive sensors are used and for plastic closures a special infrared light cell is used. The signal from this system is sent to the same rejector as the label-inspection system.

13.3.4 Crate/pack inspection

This is a unit (position 8 in Figure 13.1) that can be used on both bottle and can lines. Normally after the packer, the crate or pack is inspected for a full complement of containers.

1. Full bottle crate inspection ensures that the correct number of bottles and closures are present in the crate as it leaves the packer; if not, this crate can be rejected.
2. Full pack inspection of cans or non-returnable bottles ensures that the correct number of containers are in a shrink-wrap pack and that the shrink film is formed correctly; if not, the pack can be rejected.

With either of the above inspections it is also possible to check for under-fill (using gamma absorption) to ensure the container has not been damaged in

Figure 13.12 Fill level and cap monitor. (Courtesy of Heuft Ltd, Atherstone, England.)

the final stages of packaging and has not lost part or all of its contents.

The inspection for bottles or cans is carried out by the use of either photocells or proximity switches, dependent on the type of closure used. If a closure is not in the pre-programmed position the pack will be rejected. The gamma inspection to check for gross under-fills in the pack is carried out by sending a gamma signal of certain strength, and over the correct number of bottles per row a predetermined amount of gamma will be absorbed. If the amount of gamma received is greater than expected, a container is either broken, low-filled or missing.

13.4 Filler and closure management

Filler management is a tool that can be added to the level- and closure-detection system. Filler management performs two functions: the first is the collection of all relevant data, i.e. the number of good containers produced, the number of faulty containers, the reason for these faulty containers, and the efficiency of the filler, closure machine, etc; and the second is preventive maintenance, safety and quality control.

The maintenance function is filler-valve and closure-head identification, which enables the maintenance department to know which filling valve or closing head is causing problems. This operation is carried out by the registering of each faulty container (whether under- or over-filled) to the valve from which it came. With the analogue system each bottle that is produced is identified with the valve that produced it; this same principle is applied to the closing machine.

The safety aspect has a burst-bottle feature, which is a method of associating a bottle that burst inside the filling machine with the valve where the burst occurred; on that revolution the bottles from the valve before and after (number free-programmable) are removed. On the following revolution of the filler the bottle which is filled on the valve where the burst occurred on the previous revolution is removed along with bottles each side, if required. This process can be repeated to suit quality-control requirements.

The quality-control feature is a sampling facility. This is a process where a complete revolution of the filler or closing machine can be sampled, allowing quality control or maintenance to inspect the operation of the filling and closing machine under production conditions.

There is one other quality control feature – the inspection of the seams being applied to the can, which is monitored in operation. This can be done by either a simple method of checking the average seam thickness which looks for a trend of the seam going out of its normal tolerances or a more in-depth checking of the seam on both first and second operations using a double-seam tightness monitor which actually checks the complete seam on each operation and gives wear factors of seaming rolls or broken chucks. The first method gives a warning that something has *started* to go wrong on the seamer and needs to be rectified; the second method is a preventive maintenance tool that gives warnings in advance.

13.5 Rejection systems

Container-inspection (quality-control) systems are only as good as the rejection unit. Therefore this section, while not about container inspection, must be incorporated as each of the foregoing items of inspection equipment will have a rejection system of some type. There are many rejection systems and a complete chapter could be spent just dealing with this subject; therefore this section will only deal with the principles that are necessary for a good rejection system. There are two types of rejection systems: the single segment and the multi-segment.

Single-segment systems are used for two purposes:

1. The removal of containers for scrap; for this purpose a high-speed single-segment ram would be used since a container which is rejected with this system is not required afterwards.

Figure 13.13 Multi-segment rejector for bottles. (Courtesy of Heuft Ltd, Atherstone, England.)

2. If a container is required, then the single-segment rejection would consist of a fixed-curve guide changing the static friction of the container into a glide friction. At the end of the curve is a single-segment rejector driven by a high-speed cylinder/solenoid valve allowing very smooth diversion. For the removal of crates or cartons a pusher would be required, and for this duty it must be of robust construction guaranteeing trouble-free operation under harsh environmental conditions.

Multi-segment rejectors are of varying designs as they must smoothly divert containers of all types, empty or full, glass and plastic bottles (0.1 litre up to 3 litre) and two- or three-piece cans. The construction of the system must be robust with high precision, guaranteeing maximum performance with minimum maintenance. For this the latest micro-electronics control the rejector precisely on fixed and variable-speed conveyors. The designs must be compact, allowing them to fit into tight areas. The operation must be smooth, with minimum wear and tear and low energy consumption. The ability to reject touching containers has an advantage in noise reduction. Taking all the above features into consideration, the choice of the multi-segment rejector is very important and depends entirely on the stability of the container to be handled.

13.6 Future developments

Following the introduction of microprocessor-controlled quality-control and container-handling equipment, the sophistication available to the machinery manufacturer has increased dramatically, providing many differing functions and possibilities; in many cases, these welcome advances are severely limited, not by the electronics themselves, but by the inability of the user to handle the sophistication to the best advantage. Unfortunately, the quality of the technical and operating staffs on modern soft drink production lines may not always be of the standard required for the latest high-technology equipment now available.

The electronic machinery to which reference has been made in this chapter covers the following: bottle sorters, container laning machines, inspection of empty bottles and cans, closure or seam control, sealing-surface inspection, fill-level monitoring for bottles and cans with filler management tools (valve location, bottle-burst detection, sampling, etc.), label-position control with vision technology and carton and crate inspection/control. Generally, all these units are single, independent developments, probably with an assortment of microcomputers from different suppliers and carrying varying types of control panels and fault-finding aids.

This lack of standardisation is costly, inefficient and bewildering to the maintenance and operating personnel; a revised approach on a system basis is required which will produce a uniformity of concept, leading to the following advantages:

- Good, reliable mechanisms, minimising change parts and change-over times and providing automatic adjustments (where practical) and economy of spare parts.
- Standardisation of multi-processor hardware and software with modular techniques allowing a basic computer to be adapted to suit a particular application.
- The ability to upgrade a range of machines to meet new developments and legislation.

The 'system concept' will also allow a positive integration of the individual units so that, for example, the fill-level monitor may instruct the ink-jet coder to mark only those containers that have passed inspection; the fusion of all the units mentioned previously is possible at the present time and adaptable systems are available for the benefit of forward-thinking companies.

14 Secondary and tertiary packaging

R.E. LEIGHTON and A.J. MITCHELL

14.1 Introduction

This chapter deals with the 'packaging' of finished carbonated products and covers the machinery and materials necessary to receive the filled bottles or cans (the 'primary' packages) from the production line and prepare them in such a way that they may be safely and economically transported to the warehouse and, from there, via the distribution network, to the customer. The principles involved in handling both returnable and non-returnable products will be examined and, although some of the operations are similar, it is always prudent to select machinery to suit the precise purpose for which it is intended.

The packaging of carbonated soft drinks is a vast and interesting subject involving techniques that are constantly under development; within the scope of the chapter and the space available, it will only be possible to highlight a little of the background of packaging and to explain some of the systems currently in use.

14.2 Returnable bottles and crates

For many years the operations of the soft drinks industry were based on robust glass bottles which were returned to the factories for refilling many times in their working lives. The number of trips that a bottle made before it was lost, broken or damaged to an extent that it could not be used, was extremely variable and somewhat indeterminable, but an average was believed to be around twenty. To safeguard the bottles on their journeys to and from the retail points required some form of strong container and it was inevitable that the chosen material would be timber. At the time, this was a first-class choice since timber was readily available, relatively cheap, easy to cut and, when nailed and wired, formed a suitable enclosure which was also easy to repair if damaged; the boxes, known as 'crates', could be printed with the bottling company's name or logo and then varnished to produce a durable and weather-resistant finish. However, there were several disadvantages with wooden crates; even with the use of simple jigs to cut the sections, the finished dimensions were very difficult to control. Some crates were designed with wooden divisions to separate the bottles but the internal structure was weak and susceptible to damage during the loading operation of large, heavy bottles. Most important of all, there was an extremely serious fire hazard when large quantities of the crates were stored on bottlers' premises.

When stacked on pallets, these wooden crates were not very stable but securing the top layer with string or tape was sufficient to ensure that the load was not displaced during the relatively slow and short distribution movements that were current in the years immediately following the Second World War.

There was little standardisation on crate sizes for the range of bottles in use, but all large bottles above about 10 fl oz capacity (285 ml) were invariably packed in a 4 × 3 configuration, thus perpetuating the 'dozens/hour' term when specifying the outputs of bottle-handling machines – now having been mostly superseded by the more acceptable 'bottles/hour' style.

For small bottles, wooden crates were designed to allow the bottles to be placed on their sides with the rows alternating to achieve a nesting pattern; although this arrangement produced a high utilisation of the crate volume, there were also problems with damaged labels and if one bottle leaked, the contents would be distributed over several bottles in the crate. Attempts were made to mechanise the loading operation but without real success; it was not until the 'lay-down' crate had been replaced by the 'stand-up' type that automatic equipment could be utilised to assist the packing operatives.

The advent of plastics and the development of moulding technology allowed the plastic crate to be introduced to the industry in the early 1960s; a far superior crate was produced, being virtually indestructible and available in various attractive colours. By careful design the plastic crate could be made immensely strong even with wall and base sections much thinner than the previous wooden structure; hence during the change-over period when plastic and wooden crates were used in mixed proportions there were great problems in stacking stability and, where applicable, in mechanical handling owing to the variations in size of the two types. However, once wooden crates had been eliminated, the preponderance of plastic crates encouraged the development of automatic bottle-crating machinery catering for the full range of outputs.

The design of a plastic crate is a complex matter and is a balance between the requirements of the unloading and loading operations and the practicalities of injection moulding. Some of the design criteria would be:

1. Maximum number of bottles per square unit of base area.
2. Provision of internal divisions to allow each bottle its own pocket.
3. The divisions should provide a 'lead-in' for easy bottle entry.
4. The top surface of the crate should be above the level of the bottle closures.
5. The base of the crate should be perforated as much as possible to allow debris to fall out and to promote full drainage during a washing operation.

302　CARBONATED SOFT DRINKS

6. Convenient handholes should be positioned so that free and easy access is possible even with bottles in the crate.
7. The underside of the base should include projections to allow one crate to stack on top of another both in a uniform stack (column stacking) and when one crate is turned at 90° to the other (interlocking bond); these projections should locate within the top rim of the lower crate to provide good lateral support.
8. The outside dimensions of the crate (adjusted by the sizes of the upper and lower rim sections which also give added strength to the plastic structure) should be chosen so that the desired interlocking bonded layer pattern may be obtained without gaps in the final formation. Additionally, the layer pattern should almost fill the area of the selected pallet to afford maximum stability when loaded pallets are stacked side by side.
9. The overall strength of the crate should allow loads of filled bottles to be stacked four pallets high with no possibility of distortion or instability.

Although there is a strong technical case for producing plastic crates with internal divisions, there are many crates in use without this facility; ribs are usually added to compensate for the loss of the strengthening effect of divisions (very considerable even when they are only half-depth) and other sections are normally increased to prevent distortion. The advantages of a divisionless crate are basically cost, ease of cleaning and simplified loading of bottles; however, at the decrating point, if the full complement of empty bottles is not present in the crate then the bottle pattern is lost and the decrating heads cannot locate on, and extract, the bottles.

14.3 Automatic recraters

Hand loading of bottles into crates is strenuous and labour intensive and to alleviate this situation various attempts were made to mechanise the operation. The first units were semi-automatic machines that used compressed-air cylinders to raise and lower a suitable assembly of bottles; the bottles were collated on a conveyor table and a plate holding a series of air-operated chucks (same number and pattern as the pockets in the crate) lowered by hand control to engage on the bottles. The assembly of bottles was lifted, transferred a short distance to the crate and lowered into position. This was a slow process and only suitable for the smallest plants.

The first fully automatic recrater employed the principle of 'drop packing', a system where the bottles were marshalled on a slat conveyor in a configuration to suit the crate pocket arrangement (Figure 14.1). At the end of the conveying section, modules of bottles (equal to the crate pattern) were fed onto a grid system; the empty crate was positioned and raised under the grid which then opened to allow the group of bottles to drop into the crate. These machines could cycle at between 3.5 and 4.0 seconds and were therefore suitable for

SECONDARY AND TERTIARY PACKAGING

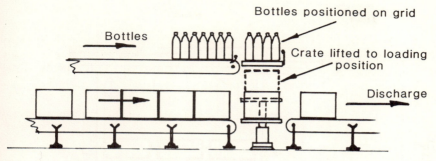

Figure 14.1 Diagrammatic operation of 'drop' type of packer.

operating speeds of approximately 1000 crates per hour; the principle was used successfully for many years but the harsh treatment of the bottles, label damage, noise and product agitation generated the need for a 'pick and place' system and, as line production speeds increased, multi-head machines were developed.

14.3.1 The modern recrater

Nearly all manufacturers of recraters now follow similar principles in the method in which the bottles are collated, lifted and placed into the crates, although the geometry of the mechanisms varies considerably. Assembling the bottles into the necessary collations and then lifting and transferring them into the waiting crates are operations that cannot be hurried – whether the bottles are large or small; the generally accepted minimum time for the complete cycle of movement is 7.5 seconds and this promotes both smoothness and reliability of the operation. Hence, knowing the number of bottles in a crate, it is a simple calculation to determine the size of machine required for a particularly output.

Example Cycle time: 7.5 seconds, i.e. 8 movements per minute
Operating speed: 45 000 bottles per hour
Crate size: $9 \times 4 = 36$ bottles

$$\text{Number of heads (lifting modules)} = \frac{45\,000}{8 \times 60 \times 36} = 2.6$$

Therefore, a *three-head* machine should be specified.

The arrangement of a high-speed recrater is shown in simplified form in Figure 14.2 and a Crown-Baele recrater can be seen in operation in Figure 14.3. The bottles are marshalled on a multi-slat conveyor system running at right angles to the path of the crates – which can travel in either direction according to the requirements of the layout. At all points where the mass of bottles is being handled, it is important to ensure that lane guides, etc.,

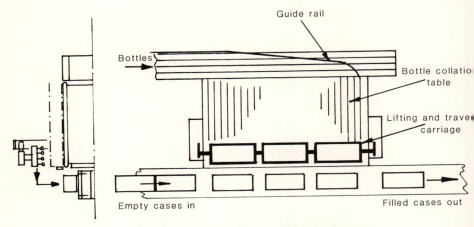

Figure 14.2 Working principles of three-head recrater.

are fitted with label protection ribs to make certain that only glass is in contact with guides. The marshalling lanes are arranged to align with the rows of bottle pockets in the crates; at the end of each lane a stop bar positions the rows of bottles ready for the lifting-head assembly to descend and grip the bottles. A series of mechanical flags or proximity switches positioned prior to the pick-up point senses that the full complement of bottles is available for transfer.

The bottle lifting-head assembly comprises a number of modules on which gripping devices are mounted to suit the bottle configuration in the crate; the grippers may be of the mechanical type or compressed-air operated. They are flexibly mounted on the module to allow for discrepancies in bottle position owing to variations in diameters. As the lifting-head assembly prepares to move upwards from the table, the conveyor is stopped for a short time to ensure a smooth lift without interference from the pressure of the remaining bottles; the carriage moves over the crate conveyor and either deposits the bottles in the crates or waits for the necessary crates to be positioned. In some instances, the bottles gripped in the lifting heads may have to be moved apart or closed-up according to the pattern in the crate and, in such cases, air-operated links will adjust the positions of the head grippers during the transfer movement. The empty crates are metered into the machine, normally by a flight bar conveyor which also positions the crates accurately to receive the bottles; simple infeed jigs are usually provided to guide the bottles into the crates at the points of entry. The most important aspect of the modern recrater is that bottles are *gently placed* into the crates, thus minimising bottle/label damage and noise.

The example quoted previously specified a three-head machine which is depicted in Figure 14.2; if larger bottles of only 12 per crate had been

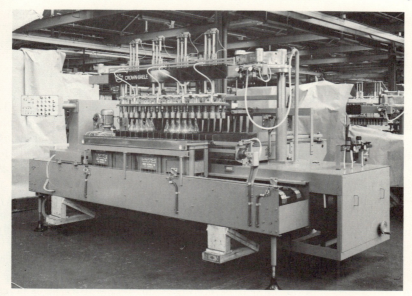

Figure 14.3 Three-head recrater. (Courtesy of Crown Cork Company Ltd, London.)

considered, the required machine size would be *eight-head*. This number of heads would make the bottle-lifting carriage too long for practical purposes and the machine must be constructed to handle the crates in a double row; consequently, the recrater would constitute two rows of four lifting-head modules per row. It is not unknown for such large machines to lift almost 200 kg of glass and product at each cycle and hydraulic mechanisms are invariably used to handle these unusual loads.

The choice of orientation of the crates during the loading operation (short side leading or long side leading) depends upon both the output required and the configuration of bottles in the crate; the 'dead times' in the recrater operating cycle are:

(i) the forward movement of bottles in the lanes to position them for the next pick-up; and
(ii) the discharge of filled cases and the introduction of the next batch of empty cases to the drop-in point.

The correct balance must be obtained between these two factors, and the machinery manufacturer, when acquainted with all the facts, can give the best advice.

As with many complex machines, recraters perform most efficiently when arranged to operate under one set of conditions; it is possible to interchange lifting-head modules, guide-rail assemblies, etc., to suit other sizes of bottles and crates but unless change parts are correctly fitted and necessary adjustments meticulously carried out, the subsequent performance of the machine will be seriously affected.

14.4 Non-returnable container packaging

The rapid expansion of supermarket activities from the early 1950s radically changed the packaging strategy of soft drink businesses; retailing returnable bottles involved supermarkets in charging and crediting deposits, utilising floor space for returned empties and sorting and checking empties for return to the bottling factories – all of which cost money. A primary package was required that was economically disposable and possessed all the advantages of the returnable glass bottle; although cans were enjoying increasing popularity, they could not supplant this well-established, re-sealable, 'see-through' container. The lightweight, non-returnable (or no-deposit) bottle, carefully designed and treated, was conceived and quickly developed and marketed; it was an instant success and, in addition to the inaugural small sizes of bottles, a full range of containers eventually became available.

A 'throw-away' bottle had to be contained in 'throw-away' packaging; during the introductory stages of the new packages, the bottles were supplied to the soft drinks factories in open cardboard boxes from which the bottles were decanted at the feed end of the production line and into which the filled and labelled bottles were repacked by hand, sometimes with the addition of cardboard divisions to prevent bottle-to-bottle contact. This system was somewhat rudimentary but at least non-returnable bottles could be processed using the existing production plant originally designed and installed for returnable bottles and the duality of operation was a situation that had to be accepted and mastered.

As the success of the new package became evident, production requirements justified high-speed, dedicated bottling lines and the non-returnable bottles were supplied to the factories in bulk, assembled in layers on pallets. This development transferred the responsibility of providing cartons from the bottle supplier to the bottler; the cartons were supplied in flat form to reduce transport costs and a labour-intensive operation of erecting the cartons, loading the bottles by hand and sealing the cartons was instituted, a similar procedure to that adopted on canning lines at that time. The actual loading of bottles into the cartons was sometimes mechanised by adapting the returnable bottle recraters described in Section 14.3 but the inherent flexibility of the cartons created some problems.

The extremely competitive nature of the supermarket business demanded ever tighter control of packaging costs and promoted the introduction of a plastic secondary packaging material; this was a transparent polymer film which possessed a 'memory' to reduce in size when subjected to heat. A pack of bottles could be loosely wrapped in the shrink-film and, after passing through

Figure 14.4 Examples of PET bottles shrink-wrapped in trays. (Courtesy of Medway Packaging Systems, Maidstone.)

a heating tunnel for between 5 and 7 seconds, the film would tighten to produce a reasonably firm and stable pack. In order to retain the bottles in the desired grouping during the wrapping and shrinking operations, they were usually loaded into a shallow cardboard tray which also served to protect the underside of the film from being punctured when the packs were placed on top of each other during the palletising operation (Figure 14.4).

The shrink-film developed for this application possessed a 'two-directional shrink characteristic'; it was designed to contract at a much greater rate in one direction (i.e. around the pack) than in the other (i.e. at 90° to the main shrink). Combined with the choice of film width (the length of the 'sleeve'), this ensured that the overlap on the short sides of the pack, when shrunk, securely supported the tops and bases of the end bottles and also formed a 'window' to allow air circulation (preventing condensation which causes label mould) and which, with care, could be used as a handhold for lifting purposes.

The development of shrink-wrapping containers in trays was also applied to soft drinks can production and progress in this field closely followed the styles adopted for non-returnable bottles; however, owing to the superior dimensional control on can sizes and their greater stability, operating speeds of the can-packaging machines were usually in excess of those handling bottles.

The success of the shrink-wrapped pack in the soft drinks industry is irrefutable; it is economical, protective to the product, a deterrent to pilfering and the disposal of the used packaging material is no real problem. In the last twenty-five years the technology has been refined, effectiveness improved, costs contained and the application machinery developed to a high standard of sophistication as described in later sections.

14.5 Tray erectors, tray loaders and shrink-wrappers

This section deals with the various types of the above machines which are usually manufactured by different companies but may be co-ordinated and linked together to form a packaging group; the distinction must be made between such an assembly of machines and the purpose-built, integrated units that have been developed for high-speed operations.

14.5.1 *Tray erectors*

There are two methods by which trays may be formed. One is by a separate tray-erecting unit which feeds trays to a second machine for loading with cans or bottles. The other method forms the tray around a collation of containers and is usually associated with the integrated machines described in Section 14.7. The main disadvantage of a separate tray erector is that the internal dimensions must be such that the containers may be introduced without difficulty; therefore, the tightness of collation leaves much to be desired and cannot compare with the 'grip' of a tray formed around, and pressed onto, the containers.

A tray erector comprises three main sections: a magazine to hold the flat tray blanks, a forming box or die and a ram. An oscillating arm equipped with vacuum suckers removes one tray blank at a time from the magazine and places it over the vertical hole in the forming box; adhesive is applied to the required points on the flap sections of the blank and the ram descends to push the tray blank down through the forming box. During this latter operation the flaps are bent to create the side walls of the tray; when the vertical ram reaches the bottom of its stroke the tray is fully formed and the ram returns to the top position to await the arrival of the next tray blank. The completed tray is held in the forming box until the next tray ejects it onto a conveyor; during this time the adhesive sets and forms the corner bond.

The intermittent motion of this type of machine limits the speed of production to approximately 25 trays per minute although small trays with a shallow wall height can be erected at a faster rate. Successful operation will depend upon the design of the tray blank, quality of materials and the storage/handling conditions of the cardboard; dimensional accuracy of the blank is crucial and die-cut articles are well worth the extra cost. To give the maximum stability to the group of containers in the tray, it is important that the corners are formed to either a chamfered shape or, preferably, a radius

SECONDARY AND TERTIARY PACKAGING 309

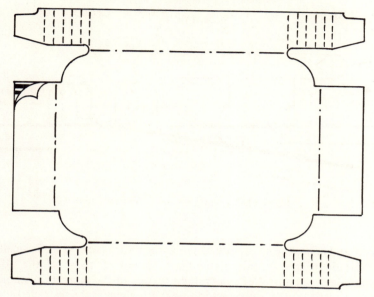

Figure 14.5 Typical tray blank with radiused corners.

equal to that of the container. A well-designed tray blank suitable for holding a twelve 0.5-litre bottle pack is depicted in Figure 14.5.

The specification of the cardboard is one for discussion with the supplier but usually 'B' flute corrugated board (or a variant) is satisfactory; the transportation and storage requirements must be strictly followed since no machine can efficiently operate with damp, distorted, bowed or damaged blanks.

14.5.2 Tray loaders

Similar to bottle recraters, the basic choice with these machines lies between 'pick and place' and 'drop' packing. The 'pick and place' operation, as with packing into cartons, is an adaptation of the multi-head bottle recrating machine. The problems are aggravated by the low height of the tray wall and the requirement to make the bottles as tight a fit as possible in the tray; the light weight of the tray also renders movement and careful positioning difficult to control. A useful modification was the addition of lifting platforms to raise the trays to a centring frame through which the bottles were fed; lowering the loaded tray back to conveyor level allowed the packs to discharge from the machine by travelling underneath the frame. Despite careful attention to all stages of the operation, this method of packing into trays is never as efficient as a continuous operation tray loader.

The use of the 'drop' pack type of machine also creates problems although the elevation of the tray at the point of bottle introduction reduces the degree

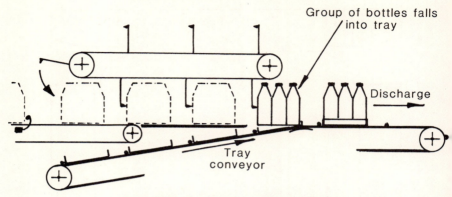

Figure 14.6 Continuous operation tray loader.

of 'drop' and hence the possibility of damaging the fragile bottles. Basically, the other disadvantages of this method are similar to those of the adapted recrater – lack of control of empty, light trays and the inherent instability of the larger bottles in the trays; an additional drawback is the restricted speed of operation – about 750 packs per hour.

The 'start–stop–start' mode of packing by the 'drop' machine was overcome by retaining the feed arrangement of bottles travelling above the trays but incorporating two major modifications: the bottle and tray conveyors were made to run continuously (and substantially at the same speed) and the trays were raised to a point as near as possible to the bases of the bottles (Figure 14.6). As the bottles approached the feed point, they were divided into groups by means of timed gates which momentarily interrupted the flow; a flight-bar conveyor (on an overhead chain system) was timed to enter the gaps thus created and propelled the collation of bottles across a deadplate. The trays, also transported by a flight-bar conveyor, were raised at an angle and timed to meet the bottles at the edge of the deadplate. The bottles dropped a distance equal to the height of the tray wall and this type of machine was therefore suitable for loading small, stable bottles into a tray with a low wall height, for example 10–25 mm. Owing to its continuous method of operation, the output of this type of tray loader is superior to that of the drop packer and a speed of 1800 packs per hour is attainable, dependent upon the actual bottles and trays.

14.5.3 *Shrink-wrappers*

Shrink-wrapping consists of two separate elements:

(a) wrapping and sealing the film;
(b) shrinking.

SECONDARY AND TERTIARY PACKAGING

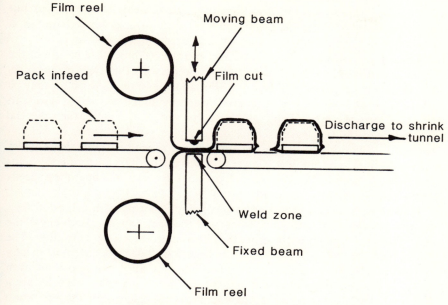

Figure 14.7 Principles of sleeve-wrapping.

Most shrink-wrapping machines work on the principle of being supplied by two reels of film, one mounted at high level (above the pack conveyor) and the other below the pack conveyor. The ends of the two reels are 'welded' together to form a vertical curtain through which the pack is pushed. 'Welding' is accomplished by placing two layers of film together and applying pressure and heat; in practice, the two ends of film are trapped between a stationary, lower beam (carrying an electrical element) and a reciprocating, upper beam which applies the pressure (Figure 14.7). The packs are separated and spaced as they enter the machine and when each pack has pushed into the film screen and has passed the 'weld zone', the moving beam descends, pulling more film from the top reel, and locates on the lower, fixed beam. The welding operation takes place and the pack is now wrapped in a sleeve but with the two webs of film from the reels attached. The double thickness of film must now be cut to allow the sleeved pack to be discharged and leave a curtain of film for the next pack; this is done in a variety of ways but a common method is to use a hot wire cutter placed in the centre of the welding beam which causes film separation by a controlled burning operation. The temperature of the welding bar must be synchronised with this function and this is achieved by arranging a constant base temperature which is slightly below the film fusion point; at the appropriate time for welding and film cutting the temperature of the welding bar is raised for the minimum amount of time to effect a weld.

Machines vary considerably in detail and some designs operate with only one reel of film; this has the advantage of only one welded join per pack but requires complex mechanical cam and lever systems in order to maintain a positive hold on the film after the cutting operation.

The discontinuous mode of operation and the reciprocating nature of the welding/cutting bar do not permit high speeds with this type of machine. Dependent upon pack stability, outputs of up to 1800 packs per hour may be obtained. Higher speeds are attainable by arranging the conveying system to feed two rows of packs side by side through a wide film; a cutter then slits the film between the two packs. Other designs incorporate a series of welding beams on a continuously moving flight-bar arrangement and this enables higher outputs to be obtained; however, it was not until the integrated systems (described in Section 14.7) were developed that the very high operating speeds of the mid-1980s were achieved.

Having contained a group of bottles safely in a tray with a loose-fitting, continuous sleeve of film wrapped around the pack (open at both ends), the next stage is to apply heat to shrink the film and produce a tight envelope. The packs pass through a tunnel which is heated by banks of thermostatically controlled electrical elements; fans circulate hot air to promote an even temperature and avoid localised 'hot spots'. The shrink characteristics are a function of time and temperature; therefore, the success of the operation depends upon the temperature of the tunnel and the time of travel through the tunnel. Variation of the conveyor speed will allow the best condition to be obtained but it is vital to provide sufficient space between consecutive packs to afford the necessary degree of air circulation. The film thickness and the width must also be taken into account to produce a tight pack, evenly shrunk and with no 'bunching' of surplus material at the open ends.

On discharge from the tunnel, the shrinking process will be complete but the film will be extremely soft; although a cooling fan is usually mounted above the outfeed conveyor to assist in hardening the film, great care must be exercised in conveying the packs until such time that the film has fully cooled. If hot packs touch each other they will stick together and damage the film; conveying packs in soft film along roller conveyors with the rollers spaced too far apart will cause the bottles to rock and the top section of the pack will deform, harden to a loose condition and produce a slack pack.

14.6 Wrap-around carton machines

Although the shrink-wrapping operation has enjoyed tremendous popularity in the last decade due to its low basic cost, acceptability by the supermarket trade and visual attractiveness of the transparent pack, there are still some applications appropriate to the use of cardboard secondary packaging. Outside the carbonated soft drinks industry this would apply to high-value products (whisky, gin, etc.) in glass bottles; some soft drinks are also packed in cartons, particularly if the goods are being transported over long distances and the subsequent handling and distribution methods may justify a type of packaging more robust than the shrink-wrapped pack.

For these purposes the wrap-around carton has been developed. As its name suggests, the carton is formed around a collation of bottles (with or without cardboard divisions) and a number of variations on the basic pack are available:

- Full and total coverage of the bottle group.
- Showcase carton in which the bottles may be seen through 'windows' in the sides and/or ends of the pack.
- Incorporation of easy-open panels.

The design of the wrap-around machine varies considerably but the principle of operation is to place the collation of bottles on, or next to, a cardboard blank and then fold the blank around the bottles, hot-melt glue being applied to the appropriate points. Subject to a well-designed and good quality blank, a very tightly finished and protective pack may be achieved. The compact, intermittent-operation machines can cycle at about 1500 packs per hour whereas the continuously moving, larger machines may attain a speed of 2400 packs per hour, dependent upon actual pack size.

14.7 Integrated shrink-wrap packaging machines

The development of bottling and canning lines capable of very high operating speeds (50 000 bottles per hour and 120 000 cans per hour) forced machinery manufacturers to reassess the existing capabilities of the secondary packaging part of the process; it became unacceptable to consider a multiplicity of tray loaders and shrink-wrapping units with the attendant costs of bottle division, pack combination, extra operators, greater floor space, etc. The action of the then current range of machines created a limitation of about 1800 packs per hour, and to break this speed barrier required units which operated in a continuous style, dispensing with the intermittent motion mechanisms and finite dwell times associated with these separate tray-loader/shrink-wrapper combinations hitherto utilised. Specialised packing machine companies adopted a radical approach and the composite, integrated units were conceived and developed.

These units handle both bottles and cans and in the latter instance the inherent stability of the filled can allows speeds of 6000 packs per hour to be obtained. The shapes and stabilities of bottles greatly affect the optimum output and the recent introduction of the one-piece PET bottle with the multi-point base has created some problems – particularly the notorious $1\frac{1}{2}$-litre size with its large height/diameter ratio; however, the industry has risen to the challenge and by suitable modifications to the handling mechanisms these bottles may now be satisfactorily processed at high speed and high efficiencies.

As with all complicated machinery, individual designs vary considerably but a popular model is shown in diagrammatic form in Figure 14.8. The complete operation may be divided into five distinct stages:

- Formation of the container group.
- Placement of the group on the tray blank.
- Erection of tray walls around the containers.
- Sleeving of the pack with film.
- Heat shrinking of the film and subsequent cooling.

All these stages proceed in a continuous motion with no pause in the movement of the containers and packs.

The mass of containers is directed onto the infeed conveyor system which is fitted with a series of lanes equal to the required number of containers in the length of the pack (which is across the *width* of the machine). The continuous rows of containers in the lanes are separated into collations by means of vertical pins at conveyor level which rise between groups of bottles, creating gaps in the pattern; these pins are then lowered from contact with the containers and the individual groups are propelled by means of an overhead flight-bar conveyor. Each flight bar pushes a collation of containers over a thin deadplate and onto the tray blank which has been fed on an inclined conveyor from the magazine at low level. At the point where the container group meets the tray blank, first the leading wall of the tray is lifted and then the trailing edge is raised to form, in effect, a clamp around the containers; the flight bar moves upwards to leave the pack which is then transported by carrier chains fitted with spring-loaded supporting plates locating on the front and rear walls of the tray. Hot-melt adhesive is then applied to the appropriate parts of the side walls (still horizontal) and these are quickly raised by deflection bars and retained firmly in position until the adhesive sets. The result of these positive operations is that the containers are now held firmly in a tight-fitting tray ready for shrink-wrapping; at this stage, a production code may be added to the tray by inked rollers or any other desired method. The sleeve-wrapping section of the integrated machine departs from the weld-bar system and is able to effect the desired pack wrap by dispensing controlled lengths of film cut from one reel. The pre-cut piece of film (equal to the vertical periphery of the pack plus about 50 mm overlap) is held on a perforated elevating conveyor by vacuum and timed to project its leading edge through a gap in the pack conveyor as a pack straddles the gap; the film is trapped between the pack and the conveyor and a rotating arm quickly lifts the trailing edge of the film over the pack and down through a second gap to form a complete sleeve. This ingenious mechanism can operate smoothly at very high speeds and forms the nucleus of the new generation of wrappers which now permit the present-day prodigious outputs to be obtained from integrated units.

The sleeve-wrapped assembly of containers is then conveyed through the

SECONDARY AND TERTIARY PACKAGING 315

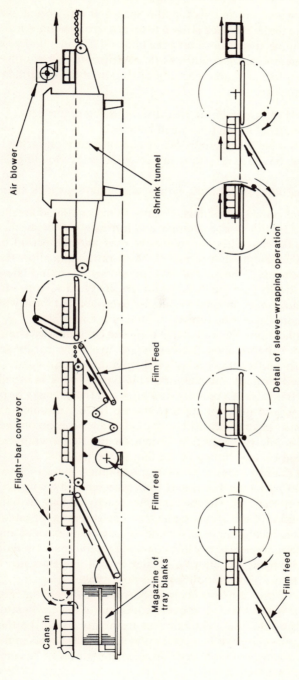

Figure 14.8 Arrangement of integrated shrink-wrap packaging machine.

shrink tunnel where not only is the film shrunk in the normal way but the overlapped join under the pack is fused to form a strong bond. An air fan at the exit from the tunnel assists in cooling the film and it is only under abnormally high ambient temperature conditions that refrigerated air is utilised.

In addition to the enhanced speed of operation of the integrated units, the other advantages are:

- Controlled tightness of the tray around the containers.
- Maximised use of film (one small overlap).
- Unit is suitable for microprocessor control and monitoring.
- Very large board magazine at low level for ease of loading (30 minutes' supply).
- Low level film reel location with dual spindle for quick change-over.

Inevitably, packaging groups must be capable of handling different bottles in different packs and the more complex and intricate are the mechanisms, the more difficult the machines are to change over and the period of 'settling-down' can be unacceptably protracted. Microprocessors allow the involved timing requirements to be altered by program selection and latest developments automatically position (by air cylinders, etc.) guides, stops, etc., not only to simplify change-over procedures but also to ensure that the new settings are exactly as required and not subject to human opinion.

In some applications, the walls of the trays can be eliminated completely and the containers supported on flat cardboard. Even the cardboard may be dispensed with and containers in small quantities (particularly cans) are shrink-wrapped without any support. This type of pack is usually used to promote multiple sales at a reduced unit price and the film may be pre-printed with the logos and appropriate advertising; in this instance, the printed information may have to be displayed at a certain position on the pack and this requires a colour-recognition system on the sleeve-wrapper to ensure correct registration and timing of the film to the pack.

Despite the ability of the modern integrated unit (Figure 14.9) to operate at speeds of 6000 packs per hour, it is sometimes advantageous to divide the throughput between two slower machines, each working at 65/75% of its maximum output; in the event of one machine stopping for any reason, the other can automatically increase in speed to the maximum value and maintain operation of the complete line before container reservoir accumulations fill up. Such an arrangement will afford a greater overall line efficiency but can be expensive in terms of capital outlay and floor space.

No doubt further refinements will be made to these integrated units, possibly in the fields of automatic replenishment of the cardboard magazine and film reels; in addition, the inherent instability of tall bottles could be improved by providing overhead clamps at vulnerable points in the handling cycle, thus allowing running speeds on these containers to be comparable to those obtained with cans.

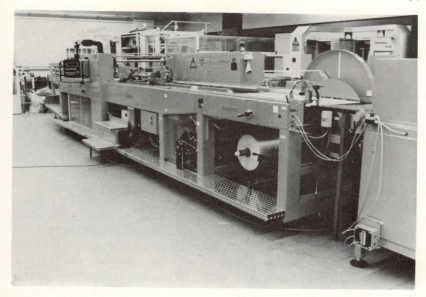

Figure 14.9 Integrated shrink-wrap packaging machine. (Courtesy of Medway Packaging Systems, Maidstone.)

14.8 Multi-packs

Although a comparatively small sector of the packaging scene, the multi-packing of bottles and cans has been popular for many years; it involves combining containers in small quantities of between two and twelve units – but more commonly in groups of four or six. The purpose is a commercial strategy to encourage the sale of more than one container – usually at a slightly lower unit price; multi-packs are also an aid to merchandising when products may be displayed to advantage and the packaging material itself used for promotional purposes. For transportation and normal distribution, the 'mini-groups' are usually combined in collations which are then shrink-wrapped in trays or packed into cartons.

The methods of multi-packing vary considerably, dependent upon the container and the nature of the retail outlet. One of the simplest systems is applied to cans and comprises plastic loops joined together which hold either six or four units; the application machine collates the cans as necessary, the loops being fed from an overhead reel and placed around the group of cans.

Bottles may be multi-packed by means of cardboard clips located over the necks or, more widely used, cardboard envelopes around a group of four or six bottles with the necks protruding through holes in the top surface and a convenient carrying device designed into the pack.

Shrink-wrapping has its place in multi-packing, as has been indicated previously in Section 14.7, and this process may be satisfactorily applied to both bottles and cans.

The special-purpose machinery for multi-packing bottles or cans may, in most cases, be supplied and maintained under a rental agreement from the companies providing the cardboard or plastic raw material.

14.9 Palletisers

Pallets have been in use in the soft drinks industry for so long that it is difficult to imagine factory and warehousing operations without them; previously, moving quantities of wooden crates and cartons involved mobile conveyor systems, trolleys, hand trucks – and a great deal of hard work. Before automatic machines were introduced, the operation of palletising had to be carried out by hand; although strenuous and labour-intensive, the procedure conveniently allowed the operatives to take into account the inevitable variations in size of the supposedly 'standard' wooden crates and, by careful selection and arrangement, reasonably stable stacks were produced. The gradual introduction of plastic crates necessitating, for a period, a mixed operation of wood and plastic, exacerbated the situation and created many problems with stability of the finished pallet as well as the difficulties in decrating and recrating as mentioned elsewhere.

The crates were arranged on the pallet either in a 'column stack' system or in an 'interlocked' or 'bonded' pattern (see Figure 14.10). Column stacking is straightforward, can accept variations in crate sizes (particularly heights) and, since it is very difficult to omit one or more crates from the centre of the stack, ensures that the full quota of crates is on the pallet; this latter factor is very useful when cases of returned empty bottles are checked by the incoming goods department. The disadvantage of column stacking is the lack of stability and the practice of tying the top layer with cord or a rubber strap certainly assists but is not the full answer. The bonded arrangement, even with wooden crates, produces greater stability although it is advantageous to secure the top layer for safety; the plastic crates, correctly designed with base projections locating into the crates below, form the perfect stack from the stability viewpoint.

In an effort to reduce the manual workload during palletisation, the crate conveyors were often elevated to a level approximately 2 metres above the floor. A mechanical or hydraulic hoist lifted the empty pallet to conveyor level and the palletising operatives, working on a staging at that level, commenced loading the pallet; as each layer was added, the pallet was lowered to allow the operatives to load subsequent layers at the same height. This greatly reduced the height through which the heavy crates had to be lifted, particularly the upper layers which would normally be at head height. From this principle were

SECONDARY AND TERTIARY PACKAGING 319

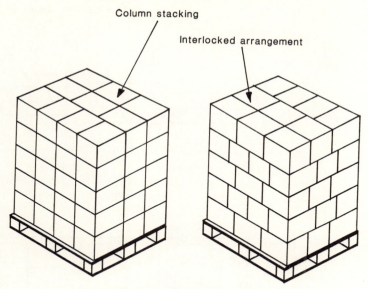

Figure 14.10 Types of stacking methods.

developed the automatic palletising machines of today, which comprise three basic designs:

- low-level sweep
- high-level sweep
- low-level hook and lift.

The decision on which type of machine should be specified depends upon plant layout and speed of operation: a high-level machine can be operated at a faster rate but a low-level machine may be controlled by an operator at floor level who is therefore able to perform other duties.

14.9.1 Low-level sweep palletiser

As the name suggests, the infeed crate conveyor will be approximately 0.5 metre above floor level; since the top of the loaded pallet will be about 2 metres above floor level (dependent upon the crate height dimension and number of layers), there must be a means of lifting the crates to the required height; this action is carried out by the 'elevating table' on which the crates are assembled in the required pattern and then transferred to the waiting pallet load.

Figure 14.11 shows the arrangement of a palletiser in its simplest form; crates are fed into the machine by a metering conveyor, turned through 90° as required by the pattern and then pushed in rows onto the pattern-forming table. As each row is formed, the incoming crates are halted to allow the sweep

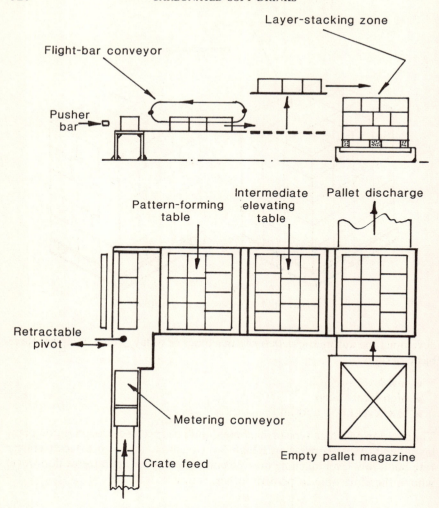

Figure 14.11 Arrangement of low-level sweep palletiser.

bars to push the row of crates clear of the transverse conveyor; the sweep bars then return either over or under the next row of incoming crates in order to speed up the operation and avoid 'dead' or waiting times. Each row, when being pushed onto the pattern-forming table, pushes the previous row until the full pattern is formed and can be transferred onto the intermediate elevating table; as this movement takes place, the sweep bars push the first row of the next pattern onto the pattern-forming table. Holding a complete layer of cases, the intermediate elevating table now moves vertically (either up or down) until the underside of the table is slightly above the level of the top face of the pallet

SECONDARY AND TERTIARY PACKAGING

(if the layer is the first to be placed on the pallet) or, alternatively, slightly above the top edges of the crates of the previous layer (for a part-loaded pallet).

The table (and crates) now move horizontally to a point vertically over the pallet, a restraining bar is lowered behind the layer and the table retracts to allow the crates to drop onto the pallet or previous layer. This degree of drop is minimised by designing the elevating table as *thin* as possible, either polished stainless-steel plate or free-rotating, small-diameter rollers. Having returned horizontally to the loading position, the table moves vertically to the position where it can accept the next layer which, on a well-designed unit, will now be available.

This procedure continues until the pallet is loaded with the required number of layers and transfers to the discharge point for removal by fork truck; a new pallet is dispensed from the magazine and moves to the loading position.

14.9.2 High-level sweep palletiser

On this design of machine the crates enter the unit at about 2 metres above floor level, at which height the layer patterns are formed in a similar manner to that adopted in the low-level palletiser. The pallet is fed into the machine at low level and, in the loading position, is raised to the underside of the intermediate table as shown in Figure 14.12; the completed layer is dropped onto the pallet as the intermediate table retracts, after which the pallet and cases lower to a position ready to accept the next layer. Since the pallet hoist indexes each time by an amount equal only to the height of the crate, very little time is taken for this manoeuvre and this advantage, coupled with the elimination of one horizontal movement of the completed layer, allows a faster cycling time on the high-level unit compared with the low-level type. The section of the cycle where time may be lost is when the empty pallet must be raised from pallet conveyor level to a position underneath the intermediate table; as the weight is minimal a high-speed motor can effect this movement in 4–5 seconds thereby ensuring that the empty pallet is in position to receive the first layer without halting the layer-assembly operation.

14.9.3 Low-level hook palletiser

This machine utilises the crate handholes to lift the assembly of crates and dispenses with the sweep operation used on the other types of palletiser. The crate feed and turning systems are the same as in the sweep-type machine, as is the pattern-forming arrangement; when the formed layer is ready for transfer to the pallet, an overhead lifting frame cantilevered from a vertical column lowers a series of hooks which are arranged to enter the handholes of the crates. The whole layer is elevated to a position where the lifting frame can rotate through 90° and place the layer either onto the pallet or the preceding layer of crates (on a part-loaded pallet). This method of palletising provides a

322   CARBONATED SOFT DRINKS

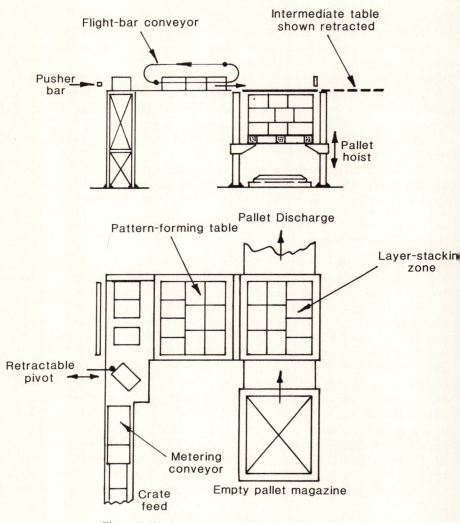

Figure 14.12 Arrangement of high-level sweep palletiser.

relatively high output while retaining the benefits of low-level operation from the machine attendance point of view; it does, however, rely upon absolutely standardised crates with easily accessible handholes to allow free entry of the hooks.

14.9.4 *Palletiser applications*

Although the previous sections dealt mainly with the palletisation of wooden and plastic crates, the machines are also used to handle cartons and shrink-wrapped packs of both bottles and cans; for these non-returnable packages obviously the hook palletiser cannot be used and the choice of machine lies between the high- or low-level sweep units. Cartons are usually constructed from good-quality board and, when stacked, the load of the upper layers is spread sufficiently to prevent damage to the cartons at the base of the stack; however, shrink-wrapped bottles (even in trays) suffer from point loading on the closures of the bottom layers which may cause, at best, indentation of the trays or, worse, penetration of the trays which will lead to the collapse of the pallet. To overcome this problem, intermediate cardboard layer pads are inserted between layers of packs and most palletisers can be adapted to fit these automatically, selecting the pads from a magazine store; this operation is carried out immediately after a layer of packs has been deposited on the pallet stack and, in a well-designed machine, does not affect the output.

The actual loading pattern of the packages on the pallet must be carefully considered; the most important requirement is stability, closely followed by utilisation of the pallet area. In the case of plastic crates and cartons, particularly the larger sizes holding large bottles, the choices of pattern are not very great; however, with shrink-wrapped packs of small bottles collated in low numbers, there may be several patterns available and these must be thoroughly analysed and discussed with the machinery manufacturer before a decision is made.

With some combinations of pack and pallet size, it is possible that the only layer pattern available will have spaces between adjacent packages; provided these are not too great, a satisfactory bonding effect may still be obtained but modifications will be required to the pattern-forming mechanisms on the assembly table. Either overhead retractable wedges or rising stops between the conveyor rollers are incorporated to create and maintain the spaces while the layer is formed. Sometimes a section of the pattern forming table is fitted with driven rollers to 'open out' packages and create a desired gap.

Turning crates or packs through 90° in order to produce the required pattern is carried out at the infeed to the machine immediately after the metering conveyor which not only counts the packages and activates the subsequent handling operations but also separates the packages to allow turning, where necessary. When turning is required, an air-operated stop locates off-centre on the leading edge of the package and the latter then rotates around this pivot; the operation is one that takes time – far longer than if a package is allowed to enter the machine unrestricted. To avoid undue delays (which would reduce the overall output of the machine), an assessment must be made of the proportion of packages which have to be turned to achieve the desired pattern; if this is well in excess of 50%, it would be advantageous to turn *all* the packages on the conveyor system before the palletiser and then *re-turn* under 50% of the packages as they are indexed into the machine. The turning operation may be speeded-up by adding a swinging arm to the pivot mechanism which positively twists the package as required.

As the pallet loads are built-up, some crates or packs may not be positioned

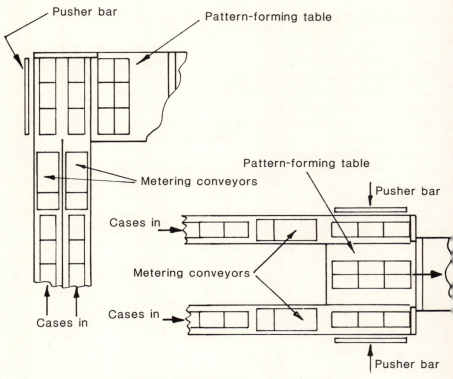

Figure 14.13 Examples of double case feeds to palletiser.

as accurately as desired; this could be due partly to the necessary running clearances of packages between guides and partly to the occasional difficulty of locating one plastic crate over another. On some machines an additional facility is provided to compress each layer in both directions horizontally after it has been deposited on the pallet load; air-operated clamps momentarily grip the layer pattern to straighten any out-of-position crates or packs and make the loaded pallet tidy and stable.

To cater for the recent increases in can- and bottle-filling speeds, higher outputs on palletisers may be obtained by adopting a double feed of packages; dependent upon the layer pattern, this feed arrangement would be either longitudinal or lateral in direction, as shown in Figure 14.13. The modification allows a high speed of pallet load formation yet minimises the linear speeds of packages as they are assembled in the layer pattern; it is particularly beneficial when handling unstable packs such as 3×2 collations of tall bottles.

The modern palletiser is custom-built to handle particular packages in particular patterns at a specified speed; adaptations to other requirements are rarely satisfactory. The intermittent type of movement which is inevitable with

SECONDARY AND TERTIARY PACKAGING

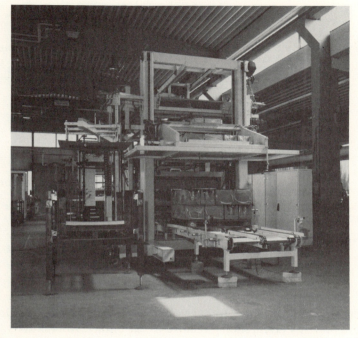

Figure 14.14 High-level sweep palletiser. (Courtesy of Schaefer GmbH, Munich.)

these palletising machines, coupled with relatively heavy packs (e.g. 6 × 2-litre or 3-litre bottles) demands robust mechanisms and great emphasis must be placed on the structural elements of the machine; only a ruggedly built unit will produce the accuracy of placement of packages that is necessary for overall pallet stability.

Microprocessors have greatly improved the reliability and operation of palletisers, permitting the variable-speed movements and 'soft starts' that are so conducive to good package handling; programming allows a speedy change-over of package and pattern combinations with minimal mechanical alterations and settings which can self-adjust according to the program selected.

14.10 Over-wrappers

The problems of the stack stability of loaded pallets have been mentioned several times in this chapter. Plastic crates in an interlocking pattern are extremely stable and suitable for transportation by fork truck and road trailers. High-quality, closed cartons are reasonably stable, particularly if spots of special adhesive are added to the top faces of the packs; this adhesive

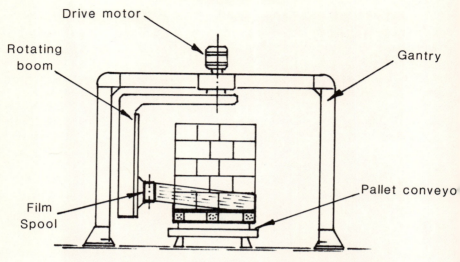

Figure 14.15 Diagrammatic arrangement of pallet stretch-wrapper.

has considerable resistance to shear but easily separates in tension and therefore prevents layers sliding relative to each other. Alternatively, plastic straps may be fitted in a vertical plane to secure the load to the pallet.

Shrink-wrapped bottles and cans present a different problem; strapping, to be effective, distorts the packs (particularly bottles) and cuts the shrink film unless additional cardboard pads or trays are added. The solution to this problem is to overwrap the complete pallet load immediately after the palletiser; one method is to fit a pre-formed shrink-film bag over the pallet of goods and contract the film by the application of hot air, either in a tunnel or by a hand unit.

Alternatively, the pallet load may be covered in stretch film (approximately 500 mm wide) spirally wound around the assembly; for low outputs this operation can be carried out by hand or by means of a motorised robot unit which travels in circles around the pallet and applies the film to a pre-set program. Higher speeds require automatic units and the machines available fall into two types: those which rotate the pallet load adjacent to a fixed spool of film and those which rotate the film around the stationary pallet. The latter type is the more popular and is shown diagrammatically in Figure 14.15.

The pallet load is positioned by proximity switches or light beams and the free end of the transparent cling film pressed against the pallet at low level; the boom then rotates at approximately 25 rpm, winding the film tightly round the load, the tension being adjusted to suit the packages. The spool rises and descends to produce the required number of 'wraps' after which the film is cut and the free end pressed against the load. Microprocessor control initiates and monitors all movements, allowing a program to be selected appropriate to the pallet load being handled.

This process of tertiary protection produces a tightly enclosed pallet load which prevents movement and distortion of the packages during warehouse and distribution handling; it also deters pilfering of individual containers and packs and allows removal of one or more layers while retaining the remainder as a stable load.

15 Effective application of quality control

D. McDONALD

15.1 Introduction

Although the organoleptic qualities of soft drinks and some of their major ingredients present specific problems in the attainment of consistent high quality, this challenge has been successfully met by the introduction of comprehensive quality control (QC) and quality assurance (QA) systems incorporating advanced laboratory and in-line instrumentation aided by the application of statistical and microbiological techniques relevant to high-speed packaging operations.

The effects of increasing legislation, added to the growing consumer insistence on safer products and higher quality standards in a competitive market, have augmented the prime objective of the modern quality technologist – the achievement of *consistent* product quality within company standards. Supportive expertise must now be provided to reduce manufacturing costs by tighter controls on raw material quality and utilisation, and to improve production line efficiencies.

This chapter reviews the evolution and growing importance of both QC and QA in the soft drinks industry. Particular emphasis is accorded to the practical problems of establishing and operating effective quality systems in three types of business, viz. small-to-medium, national and international; not surprisingly, the wider scale of quality supervision demanded by both national and international operations receives principal attention, although the particular issues relevant to smaller companies are also highlighted.

As production speeds increase and equipment becomes more complex, it is particularly important that the QA systems are geared to *prevent* defectives as distinct from the QC systems applied in-plant in order to *detect* defectives.

Finally, the chapter examines some of the important technical factors that are likely to influence the industry's growth and development, emphasising the contribution required from quality technologists both now and in the future and stressing the prospect of challenging and exciting careers for newcomers to the soft drinks industry.

15.2 Evolution of QC in the soft drinks industry

15.2.1 Concept of quality

Before beginning to describe the development and increasingly important role of QC in soft drinks manufacture, it is important to first lay down the ground rules for achieving consistent first-class quality, as this requires a multidisciplined approach within each company in order to be successful.[1] Put simply, manufacturers produce their various product flavours to formulations which incorporate predetermined quality standards governed by consumers' expectations of a consistent, good-flavoured and refreshing drink, at a reasonable price. We therefore have three critical areas of quality to consider.

- quality of design – what we believe our consumers want
- quality of manufacture – our best efforts at making it
- quality of marketing – what the consumer actually gets.

These shall now be examined in more detail.

Quality of design. This is an R&D and marketing responsibility, to quantify what the consumer wants (no mean task) and develop formulations that match this expectation. Raw material sourcing, costs and availability, compositional legislation and processing requirements all feature in this important development stage. As there continue to be many notable failures with 'new' soft drink products, it must be assumed that this area of quality continues to prove most difficult to quantify accurately. QC input at this stage is normally limited to establishing the necessary tolerances for the quality parameters to be used in production control and verifying that the processing requirements can be met.

Quality of manufacture. The bulk of this chapter will be devoted to the application of QC during the manufacturing process for carbonated soft drinks, requiring the combined disciplines of chemistry, physics, engineering, statistics, microbiology – and common sense! With high-volume production and using statistical sampling methods, the achievement of 100% product fully within specification is not normally possible. A more practical target is to ensure the maximum expected quality according to the process capability of the production line, which should have been selected to meet the company's quality standards.[2]

Quality of marketing. In addition to their role in the development of new products (or re-formulation of existing products), the marketing function has an important responsibility for ensuring that their company's product range reaches the market in the same condition as it was when produced. Through co-ordination and liaison with sales, production and distribution, marketing can help ensure that products reach the consumer well within shelf-life and are competitively priced and packaged. The effects of age, heat, sunlight and dampness during storage before sale, the interaction of some ingredients plus possible microbiological activity in the product, all combine to reduce the 'factory-fresh' product quality. Although soft drinks suffer far less from these factors than many other products and have, in most cases, a shelf-life of up to one year, products containing light- or heat-sensitive ingredients such as ascorbic acid, quinine, aspartame and certain food colours, can deteriorate appreciably and lose their palatability and attractive appearance. This can be a significant problem in certain overseas markets.

These three key areas of quality must be borne in mind when setting up a comprehensive quality system extending from raw material supplies right through to final point of sale.

15.2.2 Evolution of soft drinks QC

A number of key factors in the development of the industry[3] after the Second World War helped to accelerate the evolution of soft drinks technology and the need for in-plant QC. These included

- increased demand for international branded soft drinks – particularly colas
- introduction of soft drinks legislation covering product composition, contents (volume), labelling, ingredients and prescribed container sizes
- significant new product and package developments, including the introduction of comminuted fruit drinks, low-calorie drinks and one-trip containers – particularly PET bottles.

These factors brought in train the introduction of the pre-mix filler design (where the finished product and not carbonated water, is handled in the filler bowl) and the coagulation chemical treatment of the raw water, which required more technical supervision in-plant. Although, initially, laboratories were staffed by either qualified chemists (who tended to be laboratory, rather than plant, oriented) or trained production personnel (as in the USA), evaluation of the procedures and QC approach used by the major international franchisors enabled these to be selectively applied in the industry, demanding the employment of multi-disciplined technologists with direct plant experience. As their contribution to the business increased, so did their status in the industry. The following sections review the application, in practice, of quality systems in three different levels of business:

- The small-to-medium business
- The national manufacturer with multiple plants
- The international (franchise) business.

15.3 The small-to-medium-sized business

Many of these were family owned businesses supplying local sales and with strong brand loyalty. After many years of heavy dependence on experienced, reliable production personnel for quality supervision, the need for closer technical control of production became increasingly apparent as production speeds increased, formulations became more complex and one-trip packaging was introduced. These companies were also competing with the high-quality branded products supplied by national companies and had to reassess their quality of design, manufacture and marketing to stay fully competitive. With increasing dependence on the new QC function and its vital contribution in the control of raw material utilisation and costs, QC became more firmly established in the management team and progressively assimilated the necessary techniques to apply the new skills of microbiology and statistics in their quality plan. Where two or more plants were operated, a central QC co-ordinating role became necessary to ensure common standards, procedures and quality performance, and these responsibilities were coupled with product and packaging development. QA systems began to be developed to prevent production of defectives as well as further improvement in QC procedures and equipment for the detection of defectives.

15.3.1 *Contract packing*

As larger companies turned to contracting out production as an alternative to building new, expensive plant, this development became an important catalyst for smaller businesses with latent expansion plans. The high plant and product standards demanded by the contractors frequently required up-grading of plants, resources and, in particular, the QC function. Marks and Spencer have provided a good example of the growth potential for their suppliers through contract packing in the UK – provided that the packers recognise the critical role of quality in this type of operation.

15.3.2 *Setting up a cost-effective system for QC*

As close control of overheads became increasingly necessary in the highly competitive drinks market, there were major constraints in the smaller companies on the introduction and expansion of QC and progress was somewhat slower than in larger companies. Basic tests for Brix, carbonation and contents were initially introduced, but as product development became increasingly important, more experienced chemists were engaged to handle both QC and product/packaging development. This also enabled a more professional approach to be taken and prime attention was given to the principal sources of substandard quality by setting tighter Brix and carbonation standards and by checking these key quality parameters at regular intervals throughout production. Similarly, procedures for ingredient processing and accuracy were improved through the introduction of Brix and acidity standards and closer laboratory supervision of flavoured syrup preparation.

Finally, the introduction of benzoate-preserved comminuted bases for fruit drinks, replacing juice-based drinks preserved with the more effective sulphur dioxide, highlighted the need for more stringent hygiene procedures and routine microbiological control.[4]

It is interesting to note that until QC had proved itself in many organisations, initial reporting relationships were to production management and this is still the norm in many US plants where trained production personnel handle the QC responsibilities. Independent reporting to general management developed as the value of effective QC became more apparent.

In summary, therefore, with limited resources available, the operating costs of QC were more than off-set by the savings in ingredient utilisation during processing, i.e. through less waste and by tighter control of the syrup/water proportioning during the pre-filling blending and carbonating process. Reductions in line rejects and in consumer complaints were other areas of cost benefit through the introduction of QC. Finally, elimination of trade spoilage (which could affect considerable quantities of stock in the trade) was achieved through more frequent intensive sanitation procedures, including the introduction of hot sanitation techniques before and after production of the more sensitive beverages containing comminuted fruit bases.

15.3.3 *Product and packaging innovation*

Companies in this category had a significant advantage over larger companies in being able to launch new products or packaging more quickly and with less investment risk. This greater flexibility enabled the more alert companies to capitalise on current consumer trends with minimum advertising spend, i.e. through local launches. Where necessary, the company's own technical resources could be augmented by specialist consultancy support and a number of examples exist of important innovations to the industry introduced in this way and adopted later by larger companies. A major contribution to the success of such innovations was made by the chief chemists, quality managers or technical managers of these smaller operations.

Some companies also had the good fortune (or wisdom) to have acquired natural spring or well-water sources which could be more fully exploited with the dramatic growth in bottled waters.

Although the soft drinks industry's concentration in recent years has sadly led to the closure of many smaller businesses, those that have survived through shrewd marketing, contract packing, or by becoming low-cost producers, have also contributed to the industry's technological development and owe this, in some measure, to the efforts of their QC staff.

15.4 National operations with multiple plants

15.4.1 Impact of industry concentration

The increasing concentration of the international beverage business, through acquisitions, joint venture and equity share, has significantly reduced the number of soft drinks manufacturers in Europe and the USA. In the UK, for example, there are less than a dozen major carbonated drink manufacturers operating high-speed, low-cost production units – generally on a multi-shift basis. This has required some adjustment in approach to the control of quality through up-dating of test procedures, sampling plans and use of in-line inspection equipment. Through improved plant design, both processing and filling plant have increased reliabilities but the sheer scale of the operation can mean major cost penalties when things go wrong. With 1000 bottles per minute (bpm), bottling lines capable of producing over eight million cases of product per annum and canning lines now operating much faster than this, QC response to line problems has to be immediate, the problem rapidly assessed and remedial action implemented without delay.

Reduction in operating plants has, however, eliminated some of the problems of product quality variability between plants – a significant issue in the past in some organisations.

15.4.2 Organisation of QC at plant level

A key requirement for plants operating shift systems is the achievement of equal quality performance on all shifts. This is normally a people-related problem requiring considerable company effort in the selection, training, motivation and supervision of key line personnel – particularly as automation reduces the number of line operators, all of whom make some contribution towards product quality.

Basic requirements for an effective QC system include

- comprehensive raw material, processing and finished product specifications;
- a quality plan (endorsed in a company technical manual) identifying the sampling frequency and source, tests required and test equipment to be used (see Figure 15.1);
- well-equipped laboratory facilities, ideally located close to both production floor and key processing operations and including separate microbiological laboratory;
- a clear understanding of the QC function's responsibilities and level of authority by both plant and central management;
- a positive, planned programme for the training, re-training and motivation of the QC team.

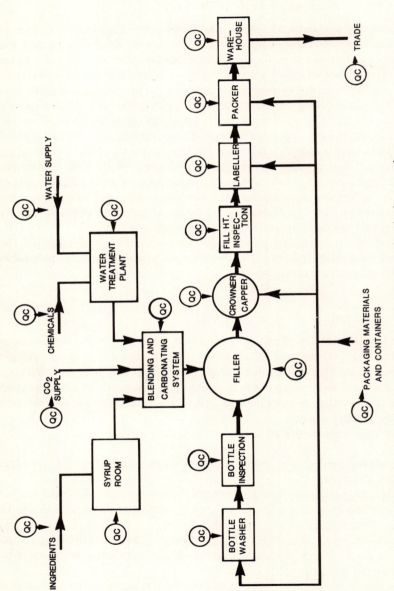

Figure 15.1 QC sampling plan incorporating process and packaging operations.

15.4.3 Centralised organisation for quality

In a multi-plant operation, co-ordination and standardisation of quality systems are fundamental to the achievement of consistent quality standards. This responsibility may be vested in a quality controller or manager, with a small support team for statistical data processing, development of new test procedures and a hygiene specialist (or microbiologist) for plant sanitation audits. Regular auditing of all company plants will be a key responsibility of the central quality function to assure senior management that quality is, indeed, under control. This is therefore a *quality assurance* responsibility, rather than one of quality control and helps to differentiate the respective roles within the quality organisation.

In a larger organisation, the central quality function may be included with engineering and R&D, and be placed under a technical director with full board authority. Clear functional and reporting responsibilities of the plant QC manager with both plant and central management must be defined. One favoured structure is for the QC manager to report to the factory manager with functional responsibility to the central quality controller.

The central quality function would also have responsibility for the preparation and updating of all technical manuals, for liaison with both R&D and marketing in the development and introduction of new products, packaging and ingredients and with R&D and engineering in the development of new process systems.

15.4.4 Bottling versus canning QC requirements

Although a number of routine tests are common to both canned and bottled products (e.g. Brix, carbonation, pH or acidity, fill contents and microbiological status) there are certain important differences in either test procedure or emphasis that need highlighting. Owing largely to the design and material of the modern can, higher production efficiencies are generally obtained on canning lines than on glass-bottling lines and, with fewer stoppages, quality tends to be more consistent. The quality integrity of the can and its contents depend heavily, however, on close control of the end seaming operation and effective headspace air removal through flushing with CO_2 or nitrogen immediately prior to can end application. Nitrate levels in process water must be low as can corrosion may develop with high nitrate content. Sugar for canning must be free from sulphur dioxide preservative (still used in some parts of the world) and it is essential that any proposed change in the design, composition or specification of the can or end component is advised to the canner by the supplier and, where considered necessary, thoroughly checked out before adoption. This is particularly relevant to the internal lacquering system used and the PVC compound on the end component. Can contents are normally checked by an in-line fill-height inspector which employs a radioactive isotope source.

336   CARBONATED SOFT DRINKS

Figure 15.2 Combined carbonation and headspace air content test on canned beverages. (Courtesy of Pepsi-Cola International.)

In bottling operations, high-quality glass or PET containers are vital for good line performance. While crown-closure application is quite simple and presents few quality problems, close control of cap application (both plastic and aluminium caps) is necessary for good carbonation retention yet easy consumer removal. Contents variability between bottles can occur with low-calorie and some foaming products at the narrow neck fill-point of the bottles, making it difficult to balance equipment settings to provide both good fill and standard carbonation.

Efficient utilisation of ingredients through tight control of fill contents and Brix (or acidity for low-calorie drinks) is a significant QC responsibility in both canning and bottling and will avoid major cumulative losses on high-speed units particularly. Consistent high quality normally means maximum raw material conversion with minimum wastage.

15.4.5 *Equipment selection for quality*

In view of the major investment involved for modern high-speed bottling and canning lines (e.g. £4 million for a 1000 bpm returnable bottle line) the capability of the line to meet all of the company's quality standards must be carefully assessed before purchase. Each equipment component contributes to the final product and package quality, and it is important for QC to participate in equipment selection from the outset through to commissioning and final assessment of the plant's process capability against the quality tolerances demanded by the company.

15.4.6 Development of in-line quality-monitoring equipment

Following the major increase in production speeds in recent years, periodic on-line sampling followed by laboratory testing proved to be inadequate as action response to substandard quality was too slow. Development of in-line test equipment has accelerated in both Europe and the USA and alternative equipment is now available to monitor key parameters such as Brix, carbonation, pH, colour, clarity, etc. Although earlier equipment was designed to monitor the appropriate quality parameters, the latest designs transmit the test results to the process control mechanism, ensuring that any out-of-standard situation is rapidly corrected. Prime interest in Brix and carbonation has provided equipment available from a range of manufacturers including Terris, Anacon and GAC in the USA and also Maselli[5] and Embra in Europe. Density monitoring in-line, as an alternative to Brix, is preferred by some soft drinks manufacturers and equipment is available from Paar Scientific Ltd, London (Figure 15.3). Increased popularity of diet drinks (containing no added sugar) has required alternative control parameters such as acidity, and in-line instrumentation using infrared spectroscopy is now available.

In addition to the basic tests for carbonated products, water-treatment plant control can be readily exercised by the use of various types of in-line instrumentation for monitoring pH, total dissolved solids, alkalinity, residual chlorine, etc. Other well-established systems include in-line empty-bottle inspection (after washing) utilising new camera techniques, bottle washer detergent strength monitoring by conductivity and fill-height inspection for both bottles and cans.

Recent developments include multiple-label inspection (an increasingly important requirement on high-speed bottling lines as manning levels are reduced) and in-line microbiological sampling linked to rapid test methods for evaluation of finished product stability before release to the trade. Clearly, this has become a major feature in the quality control of high-speed soft drink production lines demanding familiarisation of QC personnel with the operating principles and techniques of this equipment.

15.4.7 Potential quality problem areas

The prime sources of variable product quality in a typical large production plant will now be examined in detail. These problems tend to stem from five

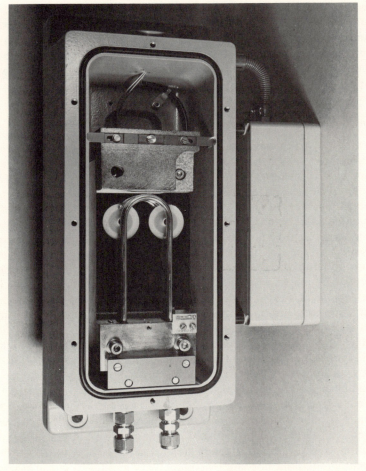

Figure 15.3 In-line densimeter for Brix monitoring of finished product. (Courtesy of Paar Scientific Ltd.)

principal sources:

- Poor material quality
- Process malfunction through either equipment failure or human error
- Ingredient omission through either equipment failure or human error
- Inadequate sanitation
- Malfunction or inadequate control of filling/carbonating/proportioning equipment.

Raw material quality. From the range of ingredients used, chemical additives produced to BP or equivalent standards are unlikely to provide quality problems and the quality plan should therefore focus on sugar, CO_2 and fruit materials. Although these particular ingredients should be tightly specified and quality-controlled during manufacture, they are more likely to provide problems in some parts of the world compared with other ingredients. The quality plan should, therefore, include batch sampling and at least physical examination of the fruit materials,[6] taste and odour checks on CO_2 and Brix, colour, taste and micro checks on each batch of sugar. Water, as a prime ingredient and liable to seasonal quality variation, requires particular QC attention in respect of both incoming water quality and control of treatment used. See Section 15.4.9 which covers this vital area in more detail. Packaging materials can present short, sharp, serious problems on the production line and inspection policies differ from one company to another. Some companies develop a comprehensive vendor rating scheme which includes regular inspection visits to suppliers' plants by QC staff from the bottling or canning plant, with free access to the suppliers' quality records. Alternatively, incoming goods inspection schemes can be established for statistical sampling and examination of incoming packaging for approval (or rejection) before use.[7] Selection of the best system will depend on the reliability of the supplier and availability of the packaging – a real problem in certain countries. In view of the critical importance of container dimensions for good filling and handling on the production line, spot sampling (at least) for dimensional checks is a worthwhile insurance before the containers are filled. Unfortunately, in many countries, serious packaging defects continue to be discovered either on the production line or, worse still, in the trade, through lack of appreciation of the value of pre-use inspection schemes.

Process malfunction. Although a malfunction or operator error can occur at any time in either water treatment or flavoured syrup preparation, morning start-up (particularly Mondays or after holiday shut-down periods) tends to be the most vulnerable period when QC needs to be on guard against unusual problems developing. Coagulation water treatment plants can be notoriously unreliable after shut-down periods and when they are subjected to peak early morning demands. In flavoured syrup preparation, some products require special processing such as filtration, pasteurisation or homogenisation which, under peak pressure conditions, can be inadequately processed with disastrous effects on the finished product in the trade. Each important stage of the ingredient processing operations has to be highlighted in the Quality Plan, with appropriate supervisory checks by QC staff and the status of the process logged for each day's production.

Ingredient omission. Although effectively part of the previous section on process malfunction, omission of vital ingredients continues to be a costly vulnerable area of processing operation in many companies. This is recognised by the major international franchisors who normally supply their bottlers with

340   CARBONATED SOFT DRINKS

a 'unit pack' which includes all the key ingredients for flavoured syrup preparation except sugar, making the syrup process much simpler and more reliable. Where unit packs are not used, formulae need to be clear and unambiguous, with the ingredients added in an optimum sequence – as certain ingredients can interact. Various systems are used to verify the accurate addition of *all* the ingredients, including operators double-checking each other's measurement and addition of every ingredient and logging these in a batch report. While the successful application of modern laboratory techniques such as HPLC (Figure 15.4) has enabled analytical verification of ingredients such as saccharin, benzoate preservative and caffeine, other ingredients such as flavour (essence) content, fruit juice or comminuted fruit content, quinine, aspartame, sodium citrate, etc., cannot be readily checked before the syrup is required for production. Fortunately, more advanced syrup or finished product blending systems are being developed and are already operating with apparent high reliability, including in-line monitoring instrumentation. In most world-wide plants, heavy dependence continues on reliable operator processing, backed up by independent QC checking. In addition, the important organoleptic check on the flavoured syrup prior to bottling creates a related problem, particularly in large operations producing flavoured syrups on a batch basis as the considerable number of batches produced daily places a heavy load on the sensory capacity of the QC staff. The industry continues therefore to seek a suitable instrumental 'sniffer/taster' to deal with this problem, although development of highly reliable processing

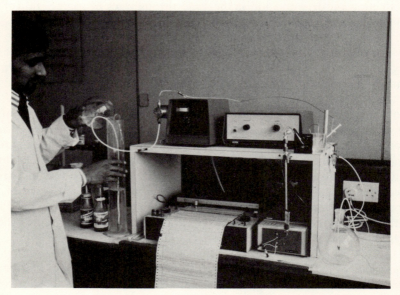

Figure 15.4   Determination of benzoic acid and saccharin levels in finished beverage. (Courtesy of Cadbury Beverages.)

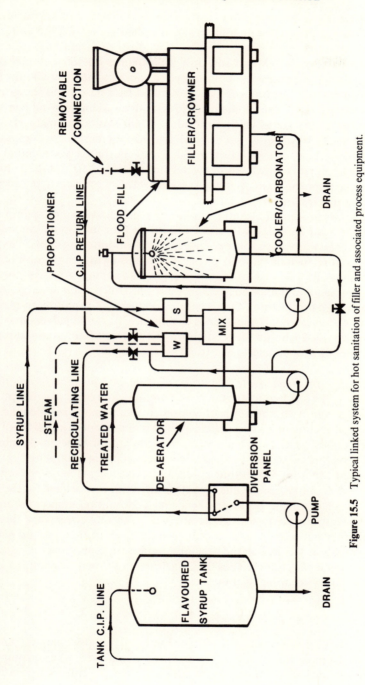

Figure 15.5 Typical linked system for hot sanitation of filler and associated process equipment.

plant is clearly a better option. Until then, it remains important for QC laboratories to include adequately trained, skilled tasters for the vital test.

Inadequate sanitation. Major soft drinks companies operate comprehensive plant sanitation programmes to ensure good product shelf-life without trade spoilage. The use of fruit juices or comminuted fruit bases, and the growing trend away from the use of chemical preservatives, demands the *consistent* application of stringent hygiene procedures. Hot sanitation methods using programmed CIP systems (Figure 15.5) have proved most effective employing, for example, 1% caustic solution at 80 °C or, alternatively, an initial cold detergent treatment followed by hot water sterilisation at 80 °C, for 20–30 min. In certain equipment systems, temporary removal of refrigerant is necessary before the hot treatment.

Suspect areas of plant cleaning tend to be pipeline 'dead-legs', pump rotor faces, filler-head springs and valves, i.e. where fruit-juice pulp can rapidly accumulate and provide a source of infection. QC monitoring of plant hygiene must, therefore, include periodic physical dismantling and inspection of plant after sanitation, followed by microbiological swabbing. As it is normally impracticable to hold finished product stock until micro clearance by QC (i.e. 3–5 days after production), the recent development of rapid micro methods using impedance and luminescence techniques has enabled results to be available in 12–24 hours (depending on the degree of contamination) and *before* stock is despatched to the trade.

Malfunction or inadequate control of filling/carbonating/proportioning equipment. This is frequently the most common source of variable product quality through a combination of equipment maintenance inadequacies or poor operator control. As described earlier, it is important to know the process capability of the plant (which may vary between products) before defining the quality tolerances to be applied. The frequency of sampling must be determined according to output speed and expected machine performance and, in some companies, the line operators are made responsible for the start-up of the plant to the necessary Brix and carbonation standards and for checking output quality at regular intervals. The QC checks are therefore confirmatory but include more comprehensive product analyses and packaging examination. Critical periods demanding close operator and QC attention are flavour or package changes and start-up/shut-down of plant, as errors are more likely to occur at these times.

15.4.8 Product recall

Product recall is not a decision to be taken lightly in any company and can seriously damage both company and brand name even when prompt and responsible actions are taken to resolve the problem. Recent UK legislation on product liability has underlined the importance of reputable companies having a product recall plan available and regularly up-dating this. It is essential that both management and staff are fully conversant with this plan and that occasional trial recalls are carried out to check their effectiveness. Product recall is normally generated by one of the following four categories of defective product occurring:

- Illegal product composition or volume of contents
- Product contamination creating a consumer health hazard
- Defective packaging creating a consumer safety hazard
- Product or package defect that would generate sufficient adverse consumer complaints without presenting any health hazard.

When a fault in any of these categories occurs, it is essential to identify rapidly the nature and source of the defect, the production date, the volume and number of varieties affected and a senior management decision should then be taken on the need for recall implementation. In parallel with this urgent action, the source of the fault must be eliminated before further production can continue and stocks that have been produced since the date of production of the defective batch should be quarantined, sampled and checked out by QC (and, where necessary, by either local authority or independent laboratories), after which a decision is taken on the stock's suitability for release for sale.

The frequency of recall announcements for a wide range of consumer products reported by the media highlights the seriousness of this problem – despite modern sophisticated technology. As high-speed manufacturers of a low-cost product, the soft drinks industry must give close attention to this area to maintain its good reputation.

15.4.9 Water quality and treatment

As carbonated soft drinks contain approximately 90% water, the quality of the treated water used and the processes required to achieve the necessary standards demand comment. Although, in the UK and many parts of Europe, water is generally of good potable quality and in plentiful supply, key quality parameters such as alkalinity, colour and mineral content can exceed those limits required for good-quality soft drinks and may affect product flavour and appearance during storage in the trade. Typical water-quality requirements are shown in Table 15.1. Various treatment systems are used to obtain water of the necessary quality but coagulation treatment employing cold lime softening, ferrous sulphate and chlorine has been the standard system used by many bottlers for some time. These plants do, however, require considerable space and the chemical dosage system needs close control and maintenance. As plants become larger, bottlers have to give increased attention to water-treatment equipment in terms of capital cost, reliability and space require-

Table 15.1 Recommended quality parameters for treated water for carbonated soft drinks

Characteristic	Maximum level	Effect on soft drink quality
Physical and Organoleptic		
Taste	Should be tasteless	Product off-taste
Odour	Should be odourless	Product off-odour
Colour	5 Hazen units	Poor colour and possible off-taste
Turbidity	1 mg/litre against Fuller's Earth standard	Poor colour and possible sediment
Sediment	Nil	Product sediment, off-flavour
Chemical		
Total dissolved solids	500 mg/litre	Saline effect on product taste
Alkalinity	50 mg/litre	Neutralises product acidity
Chloride	250 mg/litre	Salty effect on product flavour
Sulphate	250 mg/litre	May affect product flavour
Fluoride	2 mg/litre as F	Mottles teeth
Nitrate	10 mg/litre as N	May cause disease in young children
Organic Matter	1 mg/litre in 4 h at 27 °C (dissolved oxygen test)	Indicates organic pollution
Microbiological		
Coliforms	Must be absent	May cause illness to the consumer
Yeast	Maximum of 5/100 ml	May cause product fermentation
Total count	Less than 100/ml	Inadequate water sterilisation or infected equipment or source

General
In addition to the above, the water must be of potable quality and must not adversely affect the product in any way. Additional recommended – or legal standards exist for many other impurities in water, e.g. EEC Directives.

---

ments, and reverse osmosis has become a viable alternative for waters with low colour. De-mineralisation or electrodialysis are other well-established treatment systems for saline waters and equipment suppliers will tailor the alternative systems to handle any type of potable water.

Increasing concern over high nitrate levels from excessive agricultural use of fertilisers requires selective ion-exchange treatment. Ozone use as an alternative to chlorine has received fresh attention recently[8] and is certainly a highly effective colour remover and germicide, providing an alternative means for reducing trihalomethane levels in treated water – a matter of growing concern in the EEC countries at present.

In larger production plants, consideration may be given to the use of different quality grades of water to reduce costs. This requires clear identification of pipeline runs and close QC supervision to provide the necessary safeguards. In addition to full treatment of water for product use, additional treatments in a typical plant would include boiler-water treatment, ion-exchange softening for washer final rinse and pH correction/filtration of factory effluent before discharge to the main sewer or to a river.

The QC laboratory has therefore an important responsibility for monitoring the effectiveness of the various treatment processes in use in the plant and for ensuring that the quality of the treated water meets the standards required.

15.4.10 *Statistical QC*

The introduction of statistical techniques to the various aspects of high-volume production has provided another valuable tool for the assessment of quality performance at minimum cost and effort.[9] These techniques have proved invaluable in decision-making, when large quantities of product are quarantined for suspect quality; also in the evaluation of incoming packaging where large numbers are involved and defect levels may be low. On-line sampling, process control of finished product and evaluation of plant process capability have all benefited from the availability of applied statistical techniques. QC has been able to determine the extent of sampling schemes required according to the degree of risk acceptable, at minimum cost. It seems certain that statistical QC will have an increasingly important role to play in better quantifying complex in-plant problems for management decision (Figure 15.6).

15.4.11 *Microbiology*

The importance of microbiological QC in modern soft drinks manufacture stems from four important developments:

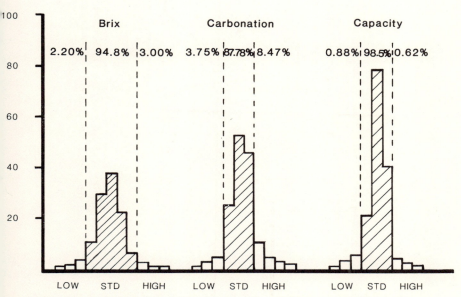

Figure 15.6 Statistical presentation of key beverage quality parameters by histogram.

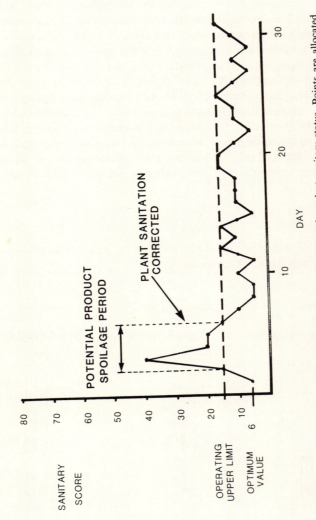

Figure 15.7 QC monitoring chart for assessment of plant and product sanitary status. Points are allocated according to microbiological results obtained against maximum limits. High counts therefore indicate poor hygiene.

1. Introduction of the pre-mix filling system, where the filler and carbonating equipment is in contact with the finished product and not carbonated water, as in previous systems.
2. Increased popularity of fruit juice and comminuted fruit-based carbonated drinks.
3. Major increase in bottling and canning output, involving much larger product quantities in any trade spoilage outbreak.
4. Increased complexity of modern processing and filling equipment demanding properly designed CIP sanitation.

A plant microbiological quality plan will include appropriate sampling and testing of

- ingredients – sugar, water, fruit juices or bases
- flavoured syrups and finished products
- containers – particularly washed returnable bottles
- plant inspection and swab programme after sanitation.

Pour-plate methods are still necessary for pulpy products and fruit ingredients, whereas membrane filtration of much larger quantities can be used for greater accuracy on water, sugar and clear-flavoured syrups and finished products. As aseptic filling techniques have not yet proved necessary for soft drinks bottling, complete 'operating theatre' sterility is not expected and it is therefore necessary to establish acceptable target levels for the items listed above, with the finished product target levels of prime importance. Regular sampling and tracking of results (as shown at Figure 15.7) will highlight adverse trends requiring investigation and remedial action.

Unfortunately, microbiology is a somewhat inexact science with plant results at times appearing illogical and difficult to interpret. It must be borne in mind, too, that yeasts and bacteria develop according to a classic pattern and may take two or three weeks to reach fermentation levels or bacterial spoilage. This can mean significant stock quantities reaching the trade and subsequently developing spoilage. Certain types of organism may be present initially in small numbers, within target levels, but develop rapidly, giving spoilage. The formulation of soft drinks has a direct bearing on their microbiological sensitivity, with pH, preservative, carbonation levels and the use of fruit materials being important factors in product stability as illustrated by Table 15.2. Certain flavour materials are known to contain trace amounts of natural preservatives and help maintain good shelf-life performance. Microbiology is clearly established as a key area of QC responsibility requiring special laboratory facilities, trained personnel and a systematic, comprehensive microbiological programme to be effective.

15.4.12 *Dispensed soft drinks*

This has become an important growth area for soft drinks, presenting specific quality problems requiring a quite different approach to QC. Post-mix has

Table 15.2 Key factors in microbiological stability of soft drinks

Formulation	Ingredients	Packaging	Processing	Sanitation	Shelf-life
Carbonation Brix pH or acidity Preservatives (natural) Preservatives (added) Nutrients (from ingredients) Redox potential Water activity	Sugar(s). Fruit juices or bases. Special ingredients: Milk products Caramel Beer (for Shandy) Wine (for Coolers) Chocolate base	Returnable bottle sterility (after washing); closure integrity and sterility.	Pasteurisation of either sugar syrup, flavoured syrup or finished product. Sterile filtration of clear, sensitive products, e.g. those containing beer or wine.	Hot or cold sanitation used. Frequency and effectiveness of sanitation.	Market temperatures. Length of shelf life required.
Stable product requirements Optimum stability with high carbonation, acidity, Brix and preservative levels. pH less than 3.5.	Optimum stability when fruit materials or special ingredients not used – unless either flavoured syrup or finished product pasteurised.	More care necessary with returnable bottle sterility. One-trip containers unlikely to provide micro problems.	Pasteurisation effective in eliminating microbial infection – particularly if sensitive special ingredients used.	Hot sanitation essential when sensitive products filled.	Reduced shelf-life may be required for sensitive products. 28–30 °C optimum temperature for rapid yeast growth.

become the more favoured dispensing system owing to its higher profit margin for the retailer. In this system, flavoured syrup is supplied to the outlet in either stainless steel tanks or in one-trip containers such as laminated bags for use in the 'Bag-in-Box' system. The outlet blends this syrup with carbonated water in equipment leased from the syrup manufacturer. In effect, each outlet is a mini soft drinks plant operated by the outlet's own staff but serviced by the equipment supplier. Correctly operated, a satisfactory quality product can be obtained, although simple filtration is the only water treatment possible and highly saline waters cannot be used for this system.

After careful assessment of a potential outlet's cleanliness, personnel and water quality, intensive training is provided for the outlet's staff after installation and commissioning of the dispensing equipment. Subsequent service and sanitation visits are made by the equipment field engineer, who will also check the quality of the dispensed drinks, i.e. Brix, carbonation, taste and dispensed product temperature. Equipment adjustments are made, if necessary, to meet finished product target quality standards. Although special field audits may be carried out occasionally by QC personnel from the syrup supplier, responsibility for maintaining product quality is firmly vested in the outlet, with the field engineer providing the independent check role assumed by QC in the plant operation.

As it is estimated that there are some 88 000 carbonated product dispensing outlets in the UK alone, this gives some idea of the logistical problems the branded syrup suppliers face in maintaining consistent product quality through field dispensing.

15.5 International QC and QA of soft drink operations

Maintaining consistently high product quality standards on an international scale is a complex but exciting challenge for any company.[10] This was pioneered by the US major franchise companies, who built up strong bottler networks around the world. Although some companies operate through subsidiaries with company-owned plants, franchising is by far the most significant type of operation and provides the main focus of this section. A typical organisational chart for supervision and liaison with a bottler network by an international franchisor is shown at Figure 15.8

15.5.1 The franchise system

Major international franchisors operate essentially the same franchise system in that they manufacture in-house their own flavours or concentrates, selling these to the bottlers and canners who process them into finished products and pack them, normally for local sales. The contract (or franchise agreement) requires the franchisor to provide both technical and marketing support to the brand through its bottlers. Assuming principal company focus on one major

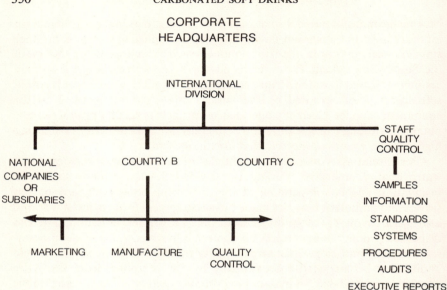

Figure 15.8 Organisation chart for international franchisor-bottler operation.

brand, e.g. a cola (and a diet version), a common product specification is applied world wide, subject to local legislative controls on product composition and packaging.

15.5.2 *Technical services*

To achieve the required quality standards, each franchisor employs a decentralised field technical services team backed up by a regional or central quality assurance laboratory and specialised engineering, packaging and data processing support. Each field technical manager is allocated a specific number of plants and countries and is accountable for product quality in these operations. Although new plant start-ups and new product or package launches form an important part of the technical manager's responsibilities, prime focus is directed at product quality and plant efficiency. Major factors such as quality and availability of raw materials, variable water quality and the influence of various ethnic cultures on the bottlers' way of doing business, demand a uniform but not inflexible approach to the application of QC. This is generally applied as follows.

New franchises. These must be carefully assessed *before* any commercial arrangements are finalised, to determine whether

- the plant is suitable for bottling (or canning)
- the plant is quite unsuitable for a franchise agreement

- subject to suitable modification and some capital investment, the plant will be able to meet the necessary franchise standards.

Initial trial production runs are carried out under the technical manager's supervision to assess plant process capability and to train the bottlers' staff in the techniques required by the franchisor. Subsequent full production will be closely monitored, as described below.

Technical supervision of existing franchises. Franchised plants are subjected to regular in-depth audits by field technical management to ensure that good manufacturing practices and quality standards are being maintained. Bottlers are advised of any deviations requiring correction, which may include an agreed period to complete the action required with some financial support from the franchisor. In addition to the plant audit, samples of treated water and finished product are required monthly for analyses by the regional or central laboratory. Samples of flavoured syrup, sugar and packaging materials may also be requested from time to time. Bottlers receive copies of the laboratory reports sent to the field technical manager. During each plant audit, the field technical manager will review with plant management the laboratory results to date on his production, his audit impressions and recommendations for improvement.

Finally, and of considerable importance, an independent trade sampling scheme is operated whereby samples of product are collected from trade outlets by either the franchisor's field management or by an agency, and these samples are sent to the franchisor's laboratory for analysis. By comparative evaluation of plant samples versus field samples analyses in relation to the plant audit reports, a good profile of the bottler's quality performance can be obtained, despite the relatively low sampling rate possible. In addition to his prime responsibility for quality, the field technical manager must provide a full technical service to his bottler. This includes support on new plant selection, the specification, supplier liaison and production trials of new packaging and the development through to production of new products such as the diet version of the core company product. The bottler, in this way, is receiving full technical service support to ensure that the business is being run profitably and kept up-to-date, and his product quality is being independently assessed to maintain the necessary high quality standards required in an intensely competitive market. This is, in effect, the cornerstone of a good franchisor/bottler relationship and is reflected in the excellent quality standards achieved by franchise bottlers in many parts of the world.

15.5.3 The International Quality Assurance Laboratory

Be it a regional or a central facility, the International Quality Assurance Laboratory provides an essential back-up to the field technical services team

352   CARBONATED SOFT DRINKS

and their bottlers.[11] In addition to routine testing of field product samples for Brix, pH or acidity, carbonation, volume of contents and general package quality, microbiological checks are carried out where necessary, although the value of these depends on the age of the samples. Water samples are checked for alkalinity, residual free chlorine, total dissolved solids (by conductivity), taste and clarity. The laboratory also assists in trouble-shooting investigations of unusual field problems, faulty raw materials or trade spoilage outbreaks. The laboratory is equipped with state-of-the-art sophisticated instrumentation, which is under constant review. Routine analyses are performed in a pre-programmed sequence for maximum efficiency and high sample throughput, with test results computerised, the data stored and reports promptly issued to the bottler and the field technical manager. The cost of a modern quality assurance laboratory will vary depending on location, but will be at least £50 000 for laboratory and computer equipment alone (Figure 15.9).

15.5.4   *Ingredient quality*

Water, not surprisingly, can provide some interesting problems in an international operation and careful assessment of the optimum treatment is required for both current and future demands. Seasonal variations in quality

Figure 15.9   International franchisor regional quality assurance laboratory. (Courtesy of Pepsi-Cola International.)

and supply from surface waters, well supply and boreholes all provide their specific challenge to bottler and technical manager and many excellent books exist[12] incorporating extensive practical experience in this critical area of ingredient processing.

Sugar is also an important (and variable) raw material in the world market and treatment may be necessary before use, particularly in clear, uncoloured beverages. The sugar may also require heat processing to destroy any organisms. The availability of High Fructose Corn Syrup (HFCS) in the '42' grade, or the higher quality '55' grade, has provided bottlers in a number of countries with an economic alternative in liquid form to sucrose. In the USA, blends of carbohydrate sweeteners (including invert syrups) provide further alternatives.

Carbon dioxide is not available with the necessary high-quality specification in a number of countries and therefore in-plant generating equipment is required. Whether generated locally or purchased from suppliers, carbon dioxide for soft drinks must conform to high purity standards and be regularly checked for taste and smell.

Although the major Cola franchisors supply all the key ingredients as unit packs, other types of beverages may require chemical additives such as citric acid, sodium benzoate, sodium citrate, saccharin and aspartame. These chemicals must be of BP or USP grade quality. Particular care must also be exercised in the selection and QC of fruit juices, comminuted fruit bases and approved food colourings as these ingredients have an important bearing on the organoleptic quality of the finished beverage.

15.5.5 *Packaging quality*

Although, in many countries, the returnable glass bottle with crown-cork finish is still the principal pack, the introduction of glass, metal and plastic one-trip containers considerably widened the technological scope of soft drink packaging. This has entailed developing specifications with suppliers, establishing sampling procedures and test facilities in bottling and canning plants, and training QC personnel. As previously described, some companies prefer to put emphasis on vendor rating schemes (with full quality accountability on the supplier) or enter into joint venture arrangements with suppliers – as some major companies have done in the UK with their can suppliers.

The impact of defective packaging on production efficiency and quality can be dramatic and therefore adequate QC time and resources must be allocated to this critical area in beverage manufacture. In some countries, restricted availability of packaging raises difficult operational decisions when packaging quality is marginal, stocks of finished product are low and sales are high.

Although principal emphasis has been placed on the importance of one-trip packaging quality, new returnable glass supplies are designed to have a trade life of up to ten years and therefore warrant close examination before use as the effects of any faults will recur as bottles are returned, washed, refilled and distributed. The logistical problems of incoming packaging material evaluation and re-assessment after use in the final product pack, present QC with quite different problems from that of ingredient QC. Double-seam control in canning, application control of roll-on or plastic closures, effective shrink-wrapping of finished packs – these all present a testing challenge to the QC team to ensure the final integrity of the pack in the trade.

15.5.6 *Trouble-shooting, the theory in practice*

The franchise bottler and canner will occasionally face unusual serious quality problems requiring good liaison with the franchisor's field technical manager and prompt, effective joint action to avoid major trade problems. Unfortunately, but understandably, these incidents are normally kept confidential by the companies concerned and others may not benefit from their experience. Although this topic could provide a fascinating chapter in its own right, the following few examples illustrate the wider-ranging demands of international quality control.

- Unusual silica levels found in a bottler's finished product were eventually traced to the bottler's unauthorised use of an anti-foam in flavoured syrup preparation in an attempt to improve poor filling performance owing to inadequate maintenance of the filling equipment.
- During a period of unexpected sugar shortage, a canner used sugar from a non-approved source. An unpleasant off-smell, accompanied by pinhole leakage, subsequently developed in the trade, generating many complaints and loss of goodwill. Investigation confirmed the presence of sulphur dioxide preservative in the sugar, producing a sulphur complex in the product and subsequent corrosion of the can.
- A batch of PVC can end liners was inadequately cured during processing by the supplier and residual monomer migrated into the product in the trade, providing a serious off-taste with adverse trade and consumer reaction.
- During a plant break-down over a holiday weekend, a carbon dioxide manufacturer produced a batch containing carbonyl sulphide impurity – a problem not unknown in the industry.[13] This was undetectable when received by the bottler, was used in production and the products were distributed to the trade. Some days later, the stock developed a strong off-smell, leading to trade returns and consumer loss of confidence in the brand name for some time.

It should be noted that none of these problems was detectable at the time of filling, highlighting QC's responsibility for quality extending through to ultimate consumption of the product. These practical problems also demonstrate the value of retaining laboratory reference samples should unusual trade problems develop.

In international franchising, therefore, the distinction between quality control and quality assurance can be readily seen with the bottler applying QC in the plant and the franchisor assuming a QA role.

15.6 The future

Quality control's continued importance in the further growth of the soft drinks industry can be confidently predicted but will take, increasingly, a *Quality Assurance* role, i.e. by establishing comprehensive programmes for monitoring each stage of the manufacturing process, including the use of modern sophisticated in-line instrumentation,[14] to *prevent* defective product being produced. Four key factors will demand this:

1. The continued development of more advanced process and high-speed filling equipment (see Figure 15.10)
2. Constant market pressure for new drinks incorporating unique ingredients or packaging, demanding technological innovation, e.g. aseptic filling of carbonated soft drinks.
3. Legislative changes (particularly in the EEC), requiring quantitative declaration of ingredients and reduction in permitted additives.
4. Computerisation of warehouse stocks and production output geared to

Figure 15.10 Modern continuous blending system for ingredient conversion to finished uncarbonated product. (Courtesy of Coca-Cola and Schweppes Beverages Ltd.)

both actual and projected sales demand will see a minimum 'dwell period' for warehouse stocks before distribution to the trade, i.e. production quality will need to be 'right first time'. This concept was seen by the author some years ago at Coors Brewery in Colorado, where production was loaded straight off the line into rail cars for immediate trade distribution.

15.6.1 *Influence of packaging*

As packaging design and material composition will have a significant bearing on new bottling and canning plant layout, recent introductions of the returnable PET bottle in Germany and the PET can in the USA underline the constant search for new materials to improve container performance and reduce costs. Future success of these or other containers, using perhaps laminated materials or composites, will require new engineering technology to fill, seal and handle on the production line and will also undoubtedly pose significant new problems for QC. Recent 'hole-in-the-wall' dedicated can supply to high-speed can fillers – introduced in the UK by two of the largest companies to guarantee can supply and reduce container costs – could become a more favoured concept, although this has already been used in the USA for many years in their highly competitive canned drinks market. In-line instrumentation for QC is already a feature of such lines.

Re-cyclability of one-trip containers and further EEC constraints on this type of packaging are other important factors that will help to shape future packaging and production-line design, particularly in Europe during the period leading up to full harmonisation in 1992.

15.6.2 *New ingredients, formulation and sanitation requirements*

The continuing search for novel (successful) products suggests the possibility of a new concept in carbonated drinks of more delicate flavours, sensitive to spoilage and demanding dairy or even pharmaceutical hygiene standards, including aseptic filling. Increasing popularity of tropical fruit flavours, natural mineral waters (including flavoured varieties), wine coolers (in the USA) and the potential for dairy product ingredients for soft drink formulations point the way to future trends. Automated CIP at elevated temperatures on a daily basis, with automatic in-line micro sampling and rapid stock approval are essential features of such systems.

Continuous blending of ingredients to provide a highly accurate flavoured syrup or finished product is already proving successful in a number of countries to replace the traditional syrup batching and proportioning systems and will probably become the standard system for major high-speed operations and, as capital costs reduce, will come within the range of smaller companies. Despite the exciting developments ahead for the large international companies, we must recognise and respect the continued success of many smaller companies around the world, where significant change in manufacturing practices or products will be much slower but where QC continues to provide essential support.

15.6.3 *Role of the soft drinks associations*

The important role of national soft drinks associations, such as the British Soft Drinks Association and the US Society of Soft Drink Technologists, must be stressed in supporting member companies (large and small) and the industry as a whole in the development of new technology to provide high-quality, low-cost soft drinks to consumers. By pooling knowledge and resources, major issues can be addressed, such as the establishment of optimum bottle-washer process conditions (a recent US study) – of considerable importance in future bottling-plant design, cost and efficiency.

15.6.4 *The final word*

The vital importance of Quality Control and Quality Assurance in the efficient production of soft drinks has been made quite clear in this chapter. Future developments in new types of ingredients, processing and packaging will impose increasing demands on quality personnel and the latter will require high standards of training in the latest techniques and equipment as they are introduced: ever-increasing production speeds and the probability of greater sensitivity of new formulations (owing to the use of natural ingredients) will leave no room for error. It will therefore be important for the industry's technologists in the major manufacturing countries to liaise with each other on relevant new developments; unfortunately, little formal world-wide co-ordination currently exists, although organisations such as the British Soft Drinks Association and the Society of Soft Drink Technologists in the USA offer excellent opportunities to their members to keep up to date on the latest advances in soft drinks technology. The single European market in 1992 will clearly demand closer co-operation between major manufacturers. In the UK, major interest has developed in the certification of manufacturers able to comply with the stringent quality requirements of British Standard 5750, which details the quality systems to be implemented to meet the Standard. Certification is carried out by qualified 'Lead Assessors' from the British Standards Institute or other qualified bodies. Although the food and drinks industry is just beginning to appreciate the advantages of certification, this important development has been instrumental in significantly raising quality standards in many manufacturing industries in the UK and BS 5750 has also been given status as a national standard by the Euro Committee for Standardisation (CEN), which should lead to implementation of the Standard in the Community's industries.

Quality is the common thread running through almost all of the preceding chapters and this factor demonstrates the contribution and impact made by the many facets of production on the excellence of the product. Consumers may be assured their demand for safe, healthy foods will be met by the soft drinks industry and that skilled, committed quality personnel play a vital role in that responsibility. This chapter is dedicated to past, present and future members of that professional élite.

References

1. P.J.M. Asselbergs, *European Brewery Convention Symp. on Quality Assurance, Zoeterwoude, the Netherlands*, Monograph X (1984), p. 76.
2. F. Price, *Right First Time*, Gower Publishing, Aldershot (1984) Chapter 2.
3. J.A. French, *Developments in Soft Drinks Technology*, Vol. 2, Elsevier Applied Science Publishers, Barking (1981), Chapter 1.
4. N. Tilley, *Developments in Soft Drinks Technology*, Vol. 1, Elsevier Applied Science Publishers, Barking (1978), Chapter 8.
5. J.O. Dierking, 'Determination of beverage carbonation by in-line pressure–temperature measurement', *Proc. 33rd Annual Meeting of Society of Soft Drink Technologists, Orlando, Florida* (1986).
6. W. Price-Davies and D. McDonald, 'Soft drinks' in *Quality Control in the Food Industry*, Vol. 3, Elsevier Applied Science Publishers, Barking (1972).
7. B. Moody, *Packaging in Glass* (revised edn), Hutchinson Benham, London (1977) Chapter 6, p. 122.
8. R. Rice, 'Application of ozone in soft drink bottling plants', *Proc. 34th Annual Meeting of Society of Soft Drink Technologists, Las Vegas, Nevada* (1987).
9. E.H. Steiner, 'Statistical methods in quality control' in *Quality Control in the Food Industry*, Vol. 1, Elsevier Applied Science Publishers, Barking (1967).
10. N. Tilley, 'How Coca-Cola assures quality with QUACS', *Soft Drinks Management Intl*, March (1988).
11. J.R. Engelland, ' A laboratory automation system', *Beverages*, 52, No. 4, Series No. 165 (1987).
12. C. Morelli, *Water Manual*, 'Beverage World', Great Neck, New York, September (1983).
13. R. Karelitz, T. Radford, D.E. Dalsis, 'Analysis of carbonyl sulfide and hydrogen sulfide in carbon dioxide', *Proc. 28th Annual Meeting of Society of Soft Drink Technologists*, Colorado Springs, Colorado (1981).
14. P. Fitzell, 'Approaching quality control automation', *Beverage World Intl*, April (1986).

Index

acesulfame K 57–59
acetic acid 96
acids 92–97
  storage 123
active chlorine (CIP) 131
adhesives for labelling 275–277
ADI (acceptable daily intake) 107
  colours 90, 100
  sweeteners 74
air
  contents 336
  effect on carbonation 210
  exclusion 219
air-driven conveyors 199–200
alitame 71–72
aluminium cans 157–162
aluminium closures 168–170
ambient filling 233–234
anti-oxidants 106
APV Gaulin homogeniser 85–86
artificial sweeteners see high-intensity
  sweeteners
ascorbic acid 96–97
aspartame 59–62

benzoic acid 101, 104
beverage consumption 7
bottle rinsers 194–195
bottles
  glass 142–148, 300
  plastic 148–156
bottle sorting 284–286
bottle unscramblers 181–182
bottle washers 189–194
Bran and Luebbe proportioner 220

can
  conveyors 202
  labelling 268–269
  rinsers 196
cans for carbonates 156–166
carbonated soft drinks

colours used in 98
components 91
future prospects 13–15
global scale 1
markets and growth 5–11
origins 5, 204
packages 9–11, 140
carbonation 204–212
  determination 208–212
  levels for various drinks 206
  measurement 208
carbonators 212–220
carbon dioxide (CO_2) 205–207
citric acid 93–94
cleaning-in-place (CIP) 125–132
  detergents for 131–132
  rate of flow in pipelines 130
  spray balls for 128
closures for carbonates 166–172
  basic types 141
coagulants 19
coagulation treatment 18–26
  monitoring plant performance
  24–26
colours 97–100
  ADI 90
  storage 125
container conveying 196–202
  decoration 243–280
  inspection equipment 281–299
containers and closures 140–173
contract packing 331
conveyors for bottles, cans 196–202
crate inspection 284
crates for bottles 300–302
crate washers 185–187
Crown-Century proportioner 222
crown corks 168
cyclamate 62–63

de-aerators 219–220
decapping machines 187–189

INDEX

decraters 182–185
  gripper heads for 183
density monitoring (in-line) 337
depalletisers
  boxes and crates 175–176
  bulk bottles 176–179
  bulk cans 179–180
detergents for CIP 131–132
dihydrochalcones 70–71
dispensed soft drinks 347–349
double seaming (cans) 160

ecology 3–4
empty bottle inspection 286–293
empty can inspection 293
empty container handling 174–202
emulsifiers 105–106
emulsions 84–87
  application 87–88
  evaluations 88–89
  manufacture 85–87
ends for cans (lids) 163–166
energy conservation 3
equilibrium pressure 207–208

filled container inspection 293–295
fillers 224–228
  largest and fastest 203
  requirements for 226
filling valves 228–242
fill level correction 240–241
filters
  carbon 23–24
  sand 21–22
filtration of syrup materials 117
flavourings 81–84
  application of 87–88
  creation and production 82–84
  evaluation 88–89
  legislation 81
  storage 124
fobbing and foaming 207
food additives (safety) 107
fruit juices
  storage 125
fumaric acid 96

gassing, undercover 240
gas volume chart 209
glass bottles 142–148
glucose syrup 40
granulated sugar 38
  batch dissolving 52
  bulk delivery 44–46
  continuous dissolving 52–53
  high-capacity dissolving 53–54
  storage 48–50, 123

heat exchanger, plate-type 120

high-fructose syrup 40
  storage 123
high-intensity sweeteners 56–80
  ADI 74
  approval and regulation 76–77
  future use 77
  storage 124
  use in carbonated drinks 56–57
high-level sweep palletiser 321
high-pressure liquid chromatography (HPLC) 340
homogenisation 85, 121–122
hot sanitation 341–342
hydrogen peroxide (CIP) 131
hydrological cycle 16

international soft drink operations 349–355
ion-exchange treatment 27–32

jetting arrangement in bottle washers 192

labelling machines
  direct-transfer 255–257
  indirect-transfer 247–255
  reel-fed 259–263
  self-adhesive 267
lacquering for cans 162–163
lactic acid 95–96
liquid sugar 39
  delivery 47
  storage 50–52, 123
low-level sweep palletiser 319–321

malic acid 96
metallic neck decoration (bottles) 265
microbiology 345–348
mixers for syrup tanks
  high-shear 112
  liquid jet 112
  systems 110–111
Mixomat M syrup plant 136–138
Mojonnier proportioners 221
monobloc filling units 224–225
multi-packs 317–318

nitrate removal 31–32

Ortmann and Herbst proportioner 223
osmosis, reverse 32–34
over-wrappers (pallets) 325–327

Paar densimeter 337–338
package types 140
packaging
  developments 2
  function 141
  non-returnable 306–308

INDEX

pallet inspection 282–283
palletisers 318–325
parabens 101, 104
pasteurisation 119–121
paracetic acid (CIP) 132
phosphoric acid 95
plastic closures 170–171
polyethylene terephthalate (PET) 149
  bottle production 150–156
preforms for PET bottles 150–153
preservatives 100–105
  storage 124
product recall 342–343
properties of label material 272–273
proportioners 220–224
pumps 115
  centrifugal 115
  positive-displacement 116

quality audit, definition 328
quality control
  application of 328–358
  definition of 328
  evolution of 330
  plant level 333
  sampling plan 334
quaternary ammonium compounds 132

raw materials checks 338–339
recraters 302–306
rejection systems 297–298
reverse osmosis 32–34
Reynolds Number 129–130
roller conveyors for bottles 198–199
roll-on closures 168–170

saccharin 64–66
saponins 106
secondary packaging 300–325
shrink-wrappers 310–312
slat-band chain conveyors 198
sorbates 105
sorbic acid 101, 105
spray balls for CIP 128
stabilisers 106
statistical QC 345

stevioside/stevia 66–69
stretch-blow moulding 153–155
sucralose 73–76
sugar
  granulated 38
  liquid 39
sulphur dioxide 101, 103
surface coatings for glass bottles 147–148
sweeteners, high-intensity 56–80
syrup room
  automation 132–135
  design 108–109
  equipment 109–122
  liquid measurement 116–117
  multiple component mixing 136–138
  operation 108–139
syrup tanks
  agitators 110–113
  storage and mixing 109–110

tamper-evident devices 170, 263–265
tartaric acid 94–95
tertiary packaging 325–327
thaumatin 69–70
tray erectors 308–309
tray loaders 309–310

ultraviolet sterilisation
  syrup materials 118–119
  water 34
undercover gassing 240

volumes Bunsen 208
volumes Ostwald 208

water
  analyses, hard and soft areas 17
  dechlorination 23–24
  impurities, effects on product quality 17
  specification for soft drinks 18, 343–344
  sterilisation 22–23
  treatment 16–36
wrap-around cartons 312–313